small arms survey 2011

states of security

THE GRADUATE INSTITUTE | GENEVA

INSTITUT DE HAUTES ÉTUDES
INTERNATIONALES ET DU DÉVELOPPEMENT

GRADUATE INSTITUTE OF INTERNATIONAL
AND DEVELOPMENT STUDIES

CAMBRIDGE
UNIVERSITY PRESS

CAMBRIDGE UNIVERSITY PRESS

Cambridge, New York, Melbourne, Madrid, Cape Town, Singapore, São Paulo, Delhi, Tokyo

Cambridge University Press
The Edinburgh Building, Cambridge CB2 8RU, UK

Published in the United States of America by Cambridge University Press, New York

www.cambridge.org
Information on this title: www.cambridge.org/9780521146869

© Small Arms Survey, Graduate Institute of International
and Development Studies, Geneva 2011

This publication is in copyright. Subject to statutory exception
and to the provisions of relevant collective licensing agreements,
no reproduction of any part may take place without the written
permission of Cambridge University Press.

First published 2011

Printed in the United Kingdom at the University Press, Cambridge

A catalogue record for this publication is available from the British Library

Library of Congress Cataloguing in Publication data

ISBN 978-0-521-19712-0 Hardback
ISBN 978-0-521-14686-9 Paperback

FOREWORD

The provision of effective and accountable public security continues to be a major challenge in the 21st century. Violence manifests itself in many different ways in places as diverse as the streets of Chicago, the north of Mexico, and several regions in Yemen. To meaningfully address this variety is a challenge for well-resourced, stable countries with legitimate leadership. It is all the more difficult for fragile countries with weak capacity, sometimes-contested leadership, or fractured or under-resourced security institutions. Yet it is in these settings that basic security matters most. We know that violence has significant and long-lasting negative effects on development. Even in the absence of violence, the lack of an effective and accountable level of security and justice delivery slows down development. Violence also creates negative global externalities that affect all of us.

While national authorities as well as multilateral and bilateral agencies are increasingly prioritizing security and justice development as part of their peacebuilding and state-building efforts, much more remains to be done. Many populations continue to be under-served. In a range of settings, communities, groups, and individuals may feel compelled to provide for their own security, often with uneven results. Certain states opt to shift some of the security 'burden' from the public to the private sector. While it is not yet clear what the long-term consequences of these decisions may be, we must guard against actions that widen social and economic inequalities. Moreover, global issues, such as the arms trade, have profound effects on violence and fragility, but remain under-addressed.

These issues are complex and political. Informed decisions require good analysis and the sharing of expertise. I therefore welcome the *Small Arms Survey 2011: States of Security*. It builds on the *World Development Report: Conflict, Security and Development,* as well as on recent work of the OECD–DAC International Network on Conflict and Fragility on state-building and armed violence reduction. This edition of the *Small Arms Survey* shows how—in places such as Côte d'Ivoire, Haiti, and Madagascar—the effective and accountable delivery of basic security is often compromised. The volume also provides important new quantitative analysis on the growth of the global private security sector and informal security providers who are filling the gap. The case studies and analysis contained in the *Small Arms Survey 2011* will benefit policy-makers, researchers, and all those concerned with understanding and responding to modern security and development challenges.

—**J. Brian Atwood**
Chair, Development Assistance Committee (DAC)
Organisation for Economic Co-operation and Development (OECD)

CONTENTS

ABOUT THE SMALL ARMS SURVEY

The Small Arms Survey is an independent research project located at the Graduate Institute of International and Development Studies in Geneva, Switzerland. Established in 1999, the project is supported by the Swiss Federal Department of Foreign Affairs and current contributions from the Governments of Australia, Belgium, Canada, Denmark, Finland, Germany, the Netherlands, Norway, Sweden, the United Kingdom, and the United States. The Survey is grateful for past support received from the Governments of France, New Zealand, and Spain. The Survey also wishes to acknowledge the financial assistance it has received over the years from different United Nations agencies, programmes, and institutes.

The objectives of the Small Arms Survey are: to be the principal source of public information on all aspects of small arms and armed violence; to serve as a resource centre for governments, policy-makers, researchers, and activists; to monitor national and international initiatives (governmental and non-governmental) on small arms; to support efforts to address the effects of small arms proliferation and misuse; and to act as a clearinghouse for the sharing of information and the dissemination of best practices. The Survey also sponsors field research and information-gathering efforts, especially in affected states and regions. The project has an international staff with expertise in security studies, political science, law, economics, development studies, sociology, and criminology, and collaborates with a network of researchers, partner institutions, non-governmental organizations, and governments in more than 50 countries.

NOTES TO READERS

Abbreviations: Lists of abbreviations can be found at the end of each chapter.

Chapter cross-referencing: Chapter cross-references are fully capitalized in brackets throughout the book. One example appears in Chapter 5, which reviews the relationship between multinational corporations and their security providers: 'Two major international initiatives, both spearheaded by the Swiss government, have sought to address the lack of regulation of private security companies and prevent improper use of force and human rights violations (PRIVATE SECURITY COMPANIES).'

Exchange rates: All monetary values are expressed in current US dollars (USD). When other currencies are also cited, unless otherwise indicated, they are converted to USD using the 365-day average exchange rate for the period 1 September 2009 to 31 August 2010.

Small Arms Survey: The plain text—Small Arms Survey—is used to indicate the overall project and its activities, while the italicized version—*Small Arms Survey*—refers to the publication. The *Survey,* appearing italicized, relates generally to past and future editions.

Small Arms Survey

Graduate Institute of International and Development Studies

47 Avenue Blanc, 1202 Geneva, Switzerland

t +41 22 908 5777 **f** +41 22 732 2738

e sas@smallarmssurvey.org **w** www.smallarmssurvey.org

ACKNOWLEDGEMENTS

This is the 11th edition of the *Small Arms Survey*. Like previous editions, it is a collective product of the staff of the Small Arms Survey project, based at the Graduate Institute of International and Development Studies in Geneva, Switzerland, with support from partners. Numerous researchers in Geneva and around the world have contributed to this volume, and it has benefited from the input and advice of government officials, advocates, experts, and colleagues from the small arms research community and beyond.

The principal chapter authors were assisted by in-house and external contributors who are acknowledged in the relevant chapters. In addition, chapter reviews were provided by: Rina Alluri, Philip Alpers, Alejandro Alvarez, David Atwood, Stephen Baranyi, Christopher Beese, James Bevan, Brian Concannon, Wendy Cukier, Neil Davison, Rinaldo Depagne, Bill Godnick, Claudio Gramizzi, Lucient Grouse, Scott Horton, Richard D. Jones, Adèle Kirsten, William Kullman, Stuart Maslen, Massilon Miranda, Emma Naughton, Daniël Prins, Désiré Razafindrazaka, Mark Sedra, Allan Stam, Håvard Strand, Djacoba Tehindrazana-rivelo, Stephen Teret, Salil Tripathi, Erwin van Veen, Dominique Wisler, Luc Zandvliet, and David Zounmenou. Anne-Marie Buzatu, Robin Coupland, André du Plessis, and Tamar Rahamimoff-Honig also provided valuable assistance on the chapters.

Small Arms Survey 2011

Editors	Eric G. Berman, Keith Krause, Emile LeBrun, and Glenn McDonald
Coordinators	Emile LeBrun and Glenn McDonald
Publications Manager	Tania Inowlocki
Designer	Richard Jones, Exile: Design & Editorial Services
Cartographer	Jillian Luff, MAP*grafix*
Copy-editor	Tania Inowlocki
Proofreader	Donald Strachan

Principal chapter authors

Introduction	Robert Muggah
Chapter 1	Patrick Herron, Jasna Lazarevic, Nic Marsh, and Matt Schroeder
Chapter 2	Glenn McDonald
Chapter 3	Pierre Gobinet
Chapter 4	Nicolas Florquin
Chapter 5	Elizabeth Umlas
Chapter 6	Oliver Jütersonke and Moncef Kartas
Chapter 7	Savannah de Tessières
Chapter 8	Athena Kolbe and Robert Muggah
Chapter 9	Sarah Parker

Eric G. Berman, Keith Krause, Emile LeBrun, and Glenn McDonald were responsible for the overall planning and organization of this edition. Tania Inowlocki managed the production of the *Survey* and copy-edited the book; Jillian Luff produced the maps; Richard Jones provided the design and the layout; Donald Strachan proofread the *Survey*; and Margaret Binns compiled the index. John Haslam, Daniel Dunlavey, and Josephine Lane of Cambridge University Press provided support throughout the production of the *Survey*. Fabio Dondero, Janis Grzybowski, Samar Hasan, Sarah Hoban, Takhmina Karimova, Matthias Félix Nowak, and Emilia Richard fact-checked the chapters. Olivia Denonville helped with photo research. Cédric Blattner, David Olivier, Benjamin Pougnier, and Carole Touraine provided administrative support.

The project also benefited from the support of the Graduate Institute of International and Development Studies, in particular Philippe Burrin and Monique Nendaz.

We are extremely grateful to the Swiss government—especially the Department for Foreign Affairs and the Swiss Development Cooperation—for its generous financial and overall support of the Small Arms Survey project, in particular Serge Bavaud, Siro Beltrametti, Erwin Bollinger, Jean-François Cuénod, Thomas Greminger, Cristina Hoyos, Jürg Lauber, Armin Rieser, Paul Seger, Julien Thöni, Claude Wild, and Reto Wollenmann. Financial support for the project was also provided by the Governments of Australia, Belgium, Canada, Denmark, Finland, Germany, the Netherlands, Norway, Sweden, the United Kingdom, and the United States.

The project further benefits from the support of international agencies, including the International Committee of the Red Cross, the UN Development Programme, UNICEF, the UN High Commissioner for Refugees, the UN Office for the Coordination of Humanitarian Affairs, the UN Office for Disarmament Affairs, the UN Institute for Disarmament Research, the UN Office on Drugs and Crime, and the World Health Organization.

In Geneva, the project has benefited from the expertise of: Aino Askgaard, Katherine Aguirre Tobón, Daniel Ávila Camacho, Silvia Cattaneo, Paul Eavis, Philip Kimpton, Patrick Mc Carthy, Jennifer Milliken, Suneeta Millington, and Pieter M. van Donkersgoed.

Beyond Geneva, we also received support from a number of colleagues. In addition to those mentioned above, and in specific chapters, we would like to thank: Steve Costner, Paul Holtom, Roman Hunger, Kimberly-Lin Joslin, Tajiro Kimura, Guy Lamb, Martin Langer, Steven Malby, Neelie Pérez Santiago, John Probyn, Mélanie Régimbal, and Jorge Restrepo.

Our sincere thanks go out to many other individuals (who remain unnamed) for their continuing support of the project. Our apologies to anyone we have failed to mention.

**—Keith Krause, *Programme Director*
Eric G. Berman, *Managing Director***

A customer approaches a business entrance as a security guard totes a shotgun in Santo Domingo, Dominican Republic, September 2005.
© Miguel Gomez/AP Photo

Introduction

INTRODUCTION

Since the emergence of the modern state, the provision of security to its population has been regarded as one of its core obligations. In its ideal form, the social contract entails a government's agreement to protect the governed. The perceived authority and legitimacy of states thus largely depends on their ability to secure their boundaries and maintain public order.

In practice, however, prospects for a legitimate social contract are far-off for many people around the world. Instead, individuals and communities are compelled to purchase or provide their own security if they are to have any at all. Not only do governments routinely fail in their obligation to guarantee public order but some perpetrate widespread violence against their own population. In parts of Africa, Asia, and Latin America, public security providers are almost completely absent from the lives of people living in rural areas.

In this context, it is not surprising that scholars and policy-makers are revisiting certain fundamental questions about security provision in the 21st century: Who actually provides security, and under whose authority? What policies, activities, and technologies ensure that security is provided? Who are the beneficiaries of security provision, and who loses out? These are not esoteric, theoretical questions, but practical, empirical ones. The answers often lie in the multiple, overlapping, and hybrid security-promotion efforts adopted by governments and communities around the world.

The *Small Arms Survey: States of Security* draws attention to the changing balance of state security provision and alternative arrangements, be they commercial security provision or informal community-based solutions. Building on new analysis and field research undertaken by the Small Arms Survey and its partners, the 2011 edition of the *Survey* also considers some of the effects of the absence of state security provision on human and national security. In so doing, it shines a spotlight on the rapid rise of the private security industry in both fragile and stable environments. As always, the role of small arms is highlighted.

ROOM FOR IMPROVEMENT: STATE SECURITY PROVISION

The provision of state security—and its absence—are influenced by a range of factors, including the quality and capacity of governance, resource distribution, and cultural and social norms. In many states, public safety takes a back seat to other concerns, including the protection of national assets, the preservation of a political regime, or the private interests of elites. In countries where the state is weak or predatory, public authorities may neglect their security obligations entirely.

A profound influence on a state's willingness and ability to provide security is the experience of armed conflict and chronic fragility. In 2010, about 25 countries were at war, some of them suffering from multiple low- and medium-intensity conflicts (UNDP, forthcoming, p. 11). The vast majority of these armed conflicts were internal, involving one or more non-state armed groups contesting state authority.

Yet the end of armed conflict is no guarantee of a return to widespread security. As previous editions of the *Small Arms Survey* have shown, the post-conflict period can give rise to new social tensions, increased criminality, and the deepening of shadow economies (Small Arms Survey, 2009, ch. 7); indeed, many post-conflict countries are among the most corrupt in the world (TI, 2010, pp. 2–3). Without the effective implementation of political settlements, peace accords, and the consolidation of the rule of law at war's end, disparities and resentment will probably persist, reinforcing mutual mistrust between state forces and civilians.

The international community has deployed significant resources to develop norms and assist states in developing and rebuilding effective, fair security systems. The Organisation for Economic Co-operation and Development estimates that in 2007 and 2008 multilateral and bilateral development agencies alone invested USD 800–900 million per year in restoring or strengthening security sectors (OECD, 2009, p. 16). The United Nations and the World Bank, bilateral agencies, and regional and non-governmental organizations all view the establishment of transparent, accountable, welfare-oriented security sectors as a core task and a precondition for good governance and development. Notwithstanding agreement on such goals, there is still no consensus on the best means of achieving them.

In fact, since early successes in Eastern Europe, Southern Africa, and Latin America, the outcomes of security sector reform (SSR) have not been heartening. This relatively poor performance in SSR reflects political resistance and weak capacity in certain settings. It also highlights the difficulties of securing commitment from a wide range of government actors (van de Goor and van Veen, 2010, p. 98). An increasing number of attempts have been made to implement SSR in conflict-affected areas, in the absence of any commitment to reconciliation or mutual coexistence. As a result, some security sector 'reformers' are suggesting a return to basics: locally led projects that are tied to tangible, community-specific gains (Ball, 2010, p. 41).

Contemporary Haiti is an important case for gauging the effectiveness of current approaches to SSR. The paradigmatic fragile state, Haiti has seen repeated investment in security promotion, particularly over the past decade. Recent security improvements documented in this volume (Chapter 8) contradict claims that SSR is a lost cause in Haiti. At the same time, analysts have begun to advocate alternative, informal security arrangements as a more effective route to providing safety and public order, thus side-stepping the persistent problem of weak governance (Colletta and Muggah, 2009). In the African context, one observer has proclaimed that the future of post-conflict security provision is 'non-state' (Baker, 2010).

FILLING A NEED: NON-STATE SECURITY ARRANGEMENTS

In places where the state's presence is limited, alternative security arrangements proliferate. These range from organized tribal militias to private commercial security providers and informal community security arrangements. Some build on traditional structures that pre-date formal state or colonial administrations. Other, newer systems arise in response to specific failures of state security forces—or the threats they pose to populations.

State and non-state security arrangements are not necessarily separate; they may be authorized by some of the same entities or may involve some of the same security providers. As discussed in Chapter 5, 'hybrid' forces encompassing private as well as current or former state forces are not uncommon. In states affected by and emerging from war, including Afghanistan and Iraq, state functions are routinely supported, and in some cases substituted, by multinational and locally recruited and administered security service providers. As governments around the world cut budgets, including those relating to public security, core state functions—from policing to prison surveillance—are being outsourced, as described in Chapter 4.

The rise of private security companies has been accompanied by the growth of a bewildering array of localized forms of security promotion. Their legitimacy, authority, and capacity rest on complex social relationships that enable different groups to coexist and form alliances. In Côte d'Ivoire, for example, local security arrangements include vigilante groups, the *dozos* (traditional hunters), and patriotic militias—not to mention a burgeoning market in private security services. As discussed in Chapter 7, some of these groups are ethnically aligned and have become vectors of insecurity themselves.

While these local, often ephemeral, informal arrangements can provide substantial security benefits, their service distribution is inherently unequal. They provide security as a private 'club' good——available to members—and not as a public good available to all. Not all populations benefit, nor do those who benefit do so evenly. In some polarized societies, security for some spells insecurity for others, as when non-state forces are enmeshed in political, economic, or ethnic patronage networks. Despite such embroilment, however, they may be no worse than state forces in this respect.

Some analysts urge placing informal and private security forces under legal, transparent, and accountable state control. This approach is undoubtedly promising in settings where the government is able to absorb former competitors or enemies in this way. It is less realistic if the state itself is largely derelict in the provision of public security, as in the case of Madagascar, described in Chapter 6. Further, for some communities, the legitimacy of non-state security providers stems from the very distance they maintain from the state apparatus.

Non-state providers can be inefficient and corrupt, but they are a fact of life in practically every corner of the world. From Colombia and Jamaica to Afghanistan and Sudan, political leaders and elites draw on such networks to bolster their power and reward their followers. Despite the many problems they can generate, such arrangements are often legitimate in the eyes of local populations. Especially where formal security institutions are considered inept or predatory, local residents are unlikely to forfeit existing patronage or identity-based social survival systems for the promise of formal untested ones (de Waal, 2009, p. 2). It is important to recognize the role these systems play since communities are embedded in, and not detachable from, prevailing social structures.

CHAPTER HIGHLIGHTS

Seven of the nine chapters in the *Small Arms Survey 2011* examine various aspects of the theme of security—who provides it, who benefits from it, and how the modalities and consequences of varying forms of security provision differ, particularly if small arms are part of the picture. The first thematic chapter discusses the adoption of emerging weapons technology among Western police forces and implications for use-of-force doctrines (Chapter 3). The next chapter charts the growth of the private security industry worldwide and estimates the number of personnel and their weapons (Chapter 4). The following chapter highlights the use of private security companies by multinational corporations—particularly in the extractive sector—including implications for the misuse of small arms (Chapter 5).

Case study chapters this year focus on three states confronted with particularly difficult security challenges. In Madagascar, the public security apparatus—characterized by institutionalized abuse of power for personal enrichment—

has reneged on any semblance of security provision (Chapter 6). In Côte d'Ivoire—a country divided since 2002 between government-controlled and rebel-controlled areas—the populations in both areas express little faith in their official security providers; meaningful reunification and democratic oversight of the two forces are a long way off (Chapter 7). Meanwhile, in Haiti, public confidence in the police as security providers has improved, despite deep weaknesses in the state's justice and security sectors (Chapter 8). The thematic section ends with a chapter reviewing legislative controls over civilian possession of firearms in 42 jurisdictions around the world (Chapter 9).

This edition also presents the 2011 Small Arms Trade Transparency Barometer and an estimate of the value of the annual authorized trade in light weapons (Chapter 1). In addition, it features an assessment of developments related to small arms control at the UN in 2010 and looks back on ten years of UN action (or inaction) on small arms and light weapons (Chapter 2).

UPDATES SECTION

Chapter 1 (Light weapons transfers): As part of the Small Arms Survey's multi-year project to estimate the annual value of authorized transfers of small arms and light weapons, their parts, accessories, and ammunition, this chapter examines the global trade in light weapons, using sources that include information obtained directly from governments, the Arms Transfers Database of the Stockholm International Peace Research Institute, and the UN Register of Conventional Arms.

The chapter estimates the annual total value of international authorized transfers of light weapons at USD 1.1 billion, including USD 755 million for anti-tank guided weapons, USD 102 million for man-portable air defence systems, and USD 257 million for four other types of non-guided light weapons. This year's findings bring the estimated value of the annual trade in small arms and light weapons (including their ammunition) to nearly USD 7.1 billion. The multi-year study will be completed in 2012 with an examination of the value of the international trade in parts and accessories for these weapons.

Chapter 2 (UN process): A decade after UN member states adopted the Programme of Action (PoA), it is not clear whether the UN small arms process has had a significant impact on national practice. This chapter reviews some recent positive developments, including the start of negotiations on an arms trade treaty, but it also highlights several causes for concern. In 2010, there were few functioning points of contact for the PoA and its offshoot, the International Tracing Instrument (ITI), and little exchange of information on ITI implementation. Preliminary research suggests that only 50 to 60 states are taking their UN small arms commitments seriously. Although it is difficult to draw firm conclusions based on the limited information that is currently available, the UN membership's continuing reluctance to embrace independent scrutiny of PoA (and ITI) implementation suggests it has a case to answer.

STATES OF SECURITY

Chapter 3 (Emerging technology): The firearms used by Western police forces have not undergone significant technological changes in recent years. For innovation, police forces seek inspiration from emerging trends in the development of military weapons. This chapter reviews the procurement and use of new weapon types by police

Definition of small arms and light weapons

The Small Arms Survey uses the term 'small arms and light weapons' to cover both military-style small arms and light weapons as well as commercial firearms (handguns and long guns). It largely follows the definition used in the *Report of the UN Panel of Governmental Experts on Small Arms* (UNGA, 1997):

Small arms: revolvers and self-loading pistols, rifles and carbines, assault rifles, sub-machine guns, and light machine guns.

Light weapons: heavy machine guns, hand-held under-barrel and mounted grenade launchers, portable anti-tank and anti-aircraft guns, recoilless rifles, portable launchers of anti-tank and anti-aircraft missile systems, and mortars of less than 100 mm calibre.

The term 'small arms' is used in this volume to refer to small arms, light weapons, and their ammunition (as in 'the small arms industry') unless the context indicates otherwise, whereas the terms 'light weapons' and 'ammunition' refer specifically to those items.

forces in France, the United Kingdom, and the United States, with particular emphasis on the adoption of so-called 'less-lethal' technologies. These cutting-edge weapons, already widely adopted in some countries, allow police to engage targets that are farther away, providing them more flexibility in the use of force. 'Less-lethal' weapons have their weaknesses, however, and appropriate police doctrine and training for them—sometimes absent or underdeveloped—are essential to maximizing their utility and avoiding their misuse.

Chapter 4 (Private security companies): The 'outsourcing' of security to private companies has increased dramatically in recent years. While this trend has been reported widely, the arms used by private security companies (PSCs) have received little attention. Based on a review of 70 countries, this chapter examines the scale of the private security industry at the global level, assesses the extent to which it is armed, and asks whether PSC equipment contributes to or threatens security. It finds that there are more private security personnel worldwide—between 19.5 and 25.5 million individuals—than police officers. But private security companies hold less than 4 million firearms compared to the 26 million held by law enforcement and the 200 million held by armed forces. Companies tend to be better armed in Latin America and in conflict-affected areas than elsewhere. While some private security companies have been involved in the illegal acquisition and possession of firearms, a lack of systematic record-keeping makes evaluating their management and use of firearms difficult. In general, the growth of the private security sector has outpaced regulation and oversight mechanisms.

Chapter 5 (Multinational corporations): Multinational corporations (MNCs) are important clients of private security companies, which they employ to protect personnel and assets in a wide range of countries. The diversity of security personnel in the employ of MNCs and the sources of the weapons they hold have implications for the safety and security of the communities in which multinational companies operate. Focusing in particular on extractive MNCs operating in high-risk environments, this chapter finds that multinationals sometimes hire individuals with poor human rights records in government security forces. The regulatory and oversight systems are generally weak at all levels, creating conditions for violence, including the excessive use of armed force, by private security contractors working for MNCs. In recent years MNCs have attempted to set standards of practice in the use of private security, but adherence has not been monitored and few penalties exist for failure to abide by them.

Chapter 6 (Madagascar): Since its independence in 1960, the island nation of Madagascar has remained deeply influenced by its colonial experience. After almost a century of resistance to occupation, Malagasy society emerged divided along clan and class lines. This colonial heritage—and the country's strategic location—have played a decisive role in preventing Madagascar from developing effective official security forces for the maintenance of law and order. This chapter analyses the historical roots of a security sector that is today characterized by severely underpaid regular forces and an inflated number of high-ranking officers pursuing their own political and economic agendas. The army, in particular, has been manipulated by successive heads of state and their entourages; it continues to be embroiled in the struggles over political power and access to the country's wealth of resources. Since the ousting of President Marc Ravalomanana in 2009, the precarious situation has been exacerbated by high rates of armed robbery, the presence of international criminal networks, and collusion of members of the security sector in the plundering of the island.

Chapter 7 (Côte d'Ivoire): Since 2002, the Republic of Côte d'Ivoire has been divided into a rebel-held area in the north and the government-run south. It has thus featured two parallel security apparatuses, two treasuries, and two administrations. Based on a national household survey, interviews, and focus groups conducted by the author in 2010, this chapter compares reported levels of insecurity and perceptions of security providers in the government-run and rebel-held zones. It finds that levels and types of insecurity are comparable across the zones, stemming primarily from banditry and resource-based violence. Public trust in official security providers in both zones is low, though slightly higher for the government forces in the south than for the rebel Forces Nouvelles in the north. As a result, civilians in both zones call on private security actors to fill the gap.

Chapter 8 (Haiti): After decades of political instability, dire poverty, and devastating natural disasters, Haiti is often portrayed as the archetypal failed state. But when it comes to security, at least, Haitians themselves are reporting certain improvements. Drawing on a series of household surveys fielded from 2005 through 2010, including three conducted by the authors, this chapter finds that security has steadily improved in Haiti over the past decade, most dramatically following the installation of an elected government in 2007. It also reveals that, despite media reports to the contrary, violence and crime were remarkably low immediately after the January 2010 earthquake. Although Haiti has one of the lowest ratios of police officers to inhabitants in the world and a police force with a sometimes brutal human rights record, following the earthquake more than two-thirds of the population said they would turn first to the police if faced with a threat to their person or property—a surprising level of confidence not previously documented. Haitians also report a lower level of gun ownership than is widely assumed outside the country; in 2010 just 2.3 per cent of Port-au-Prince area households acknowledged owning firearms. While the country undoubtedly faces many long-term development and governance challenges, these new findings offer reasons for optimism regarding the capacity and credibility of the Haitian National Police.

Chapter 9 (Civilian possession): Civilians own an estimated three-quarters of the world's firearms. What are the different approaches states take to regulate civilian access and use of these weapons? This chapter presents a comparative analysis of civilian possession legislation in 42 jurisdictions (28 countries and 14 sub-national entities). While variations abound, most of the jurisdictions adopt many of the same general measures, including prohibiting access to certain weapons they consider ill-suited to civilian use, such as larger-calibre, military-style weapons. Most have some form of owner licensing, as well as firearm registration or record-keeping; most also prevent certain civilians, such as criminals, from owning firearms. The specific approach states take to civilian gun control is influenced by cultural, historical, and constitutional factors, yet all but two of the countries under review regard civilian access to firearms as a privilege rather than a basic right.

CONCLUSION

Individuals and communities around the world experience security provision in a variety of ways. While the state has the primary obligation to protect its population, it has never been the only security provider. The recent, global rise of the private security industry is being met with unease, yet the key question is whether security arrangements—be they public or private, formal or informal—are in fact enhancing rather than impairing security, not only for the principal beneficiaries of such arrangements, but also for those who are left unprotected.

Through its examination of trends in security provision around the world, and the role of small arms and light weapons in that equation, this edition of the *Small Arms Survey* seeks to extend our understanding of patterns and causes of armed violence, and the best means of responding to it. Future editions of the *Survey* will continue this task, in particular by assessing, beginning in the 2012 volume, more than a decade of national and multilateral efforts to address the proliferation and misuse of small arms and light weapons. ■

—Robert Muggah

BIBLIOGRAPHY

Baker, Bob. 2010. 'The Future is Non-State.' In Sedra, pp. 208–28.

Ball, Nicole. 2010. 'The Evolution of the Security Sector Reform Agenda.' In Sedra, pp. 29–45.

Colletta, Nat and Robert Muggah. 2009. 'Context Matters: Interim Stabilisation and Second Generation Approaches to Security Promotion.' *Journal of Conflict, Security & Development,* Vol. 9, Iss. 4, pp. 425–53.

de Waal, Alex. 2009. 'Fixing the Political Marketplace: How Can We Make Peace without Functioning State Institutions?' *Fifteenth Christen Michelsen Lecture.* Bergen: Social Science Research Council.

OECD (Organisation for Economic Co-operation and Development). 2009. *Security System Reform: What Have We Learned?* Paris: OECD.

Sedra, Mark, ed. 2010. *The Future of Security Sector Reform.* Waterloo, Ontario: The Centre for International Governance Innovation.

Small Arms Survey. 2009. *Small Arms Survey 2009: Shadows of War.* Cambridge: Cambridge University Press.

TI (Transparency International). 2010. *Corruption Perceptions Index 2010.* Berlin: TI.
    <http://www.transparency.org/policy_research/surveys_indices/cpi/2010/results>

UNDP (United Nations Development Programme). Forthcoming. *Transitional Governance: A Framework for Consolidating Peace.* New York: UNDP.

UNGA (United Nations General Assembly). 1997. *Report of the Panel of Governmental Experts on Small Arms.* A/52/298 of 27 August.
    <http://www.un.org/Depts/ddar/Firstcom/SGreport52/a52298.html>

van de Goor, Luc and Erwin van Veen. 2010. 'Less Post-Conflict, Less Whole of Government and More Geopolitics?' In Sedra, pp. 88–101.

Members of a US Army mortar team fire on Taliban positions with a 120 mm mortar in Kunar province, north-eastern Afghanistan, in January 2010. © Brennan Linsley/AP Photo

Larger but Less Known
AUTHORIZED LIGHT WEAPONS TRANSFERS

1

INTRODUCTION

While the Kalashnikov-pattern assault rifle has become the symbol of contemporary warfare, light weapons play just as significant a role. Anti-tank missiles can destroy even the most heavily armoured vehicles. Modern man-portable air defence systems can shoot down aircraft from distances of up to eight kilometres. In heavily populated areas, indiscriminate mortar attacks can kill or injure hundreds of civilians. Despite these potential dangers, the international trade in light weapons is significantly less transparent than the trade in small arms. This chapter sheds new light on international transfers of light weapons through an analysis of available data and the strengths and shortcomings of the sources from which this data is drawn.

This study is the third instalment of the Small Arms Survey's multi-year assessment of authorized international transfers of small arms and light weapons, their parts, accessories, and ammunition, previously valued at USD 4 billion per year (Small Arms Survey, 2006, pp. 66–67). This chapter estimates the annual total value of international authorized transfers of light weapons at USD 1.1 billion. Combining this value with the revised estimate for authorized transfers of firearms (USD 1.68 billion[1]) and ammunition for small arms and light weapons (USD 4.3 billion) yields a running (incomplete) total of nearly USD 7.1 billion per year.[2] The *Small Arms Survey 2012* will assess international transfers in parts and accessories for small arms and light weapons. It will also provide an estimate for the entire annual international trade in small arms and light weapons, their parts, accessories, and ammunition. The main findings of this chapter include:

- The annual trade in light weapons is estimated to be USD 1.1 billion. This includes USD 755 million for anti-tank guided weapons (ATGWs), USD 102 million for man-portable air defence systems (MANPADS), and USD 257 million for four types of non-guided light weapons.[3]

- Despite recent increases in the number of countries reporting transfers of small arms and light weapons to the United Nations Register of Conventional Arms (UN Register), the overall quality and amount of information on light weapon transfers remain low.

- The international trade in MANPADS appears notably small. Only 18 of the 74 countries under review imported any MANPADS between 2003 and 2009, and only 12 imported more than 100 units. Given data limitations, however, these figures are probably underestimates.

- The wars in Iraq and Afghanistan have contributed to significant increases in the procurement of anti-tank guided weapons. For example, the UK's imports of Javelin ATGWs from 2005 to 2009 exceeded total imports for the years 2000 to 2004 by 5,331 units—a 4,000 per cent increase.

- The 2011 Small Arms Trade Transparency Barometer identifies Switzerland, the United Kingdom, Germany, Serbia, and Romania as the most transparent of the major small arms and light weapons exporters. The least transparent major exporters are Iran and North Korea, both scoring zero.

- In 2008 the top exporters of small arms and light weapons (those with annual exports of at least USD 100 million), according to available customs data, were (in descending order) the United States, Italy, Germany, Brazil, Switzerland, Israel, Austria, South Korea, Belgium, the Russian Federation, Spain, Turkey, Norway, and Canada (see Box 1.1).
- In 2008 the top importers of small arms and light weapons (those with annual imports of at least USD 100 million), according to available customs data, were (in descending order) the United States, Canada, the United Kingdom, Germany, Australia, France, and Pakistan.

The top three exporters in 2008 were the United States, Italy, and Germany.

This chapter begins by defining key terms and concepts. It then provides an assessment of transparency in the trade in light weapons, along with the annual update of the Small Arms Trade Transparency Barometer. The chapter then outlines the methods used to calculate an estimated annual value for light weapons transfers. The sections that follow present a detailed analysis of the data mined for six light weapons categories: non-guided light weapons—mortars, grenade launchers, recoilless guns, and portable rocket launchers—and portable missile systems (ATGWs and MANPADS).[4] The chapter concludes by reflecting on our current understanding of the global authorized trade in light weapons and the gaps in that understanding.

TERMS AND DEFINITIONS

This chapter is based on a definition of 'light weapons' provided in *Small Arms Survey 2008: Risk and Resilience,* which is derived from the 1997 report by the UN Panel of Government Experts on Small Arms (UNGA, 1997).[5] The *Small Arms Survey 2008* modifies the 1997 definition by identifying specific weight limits for light weapons and their ammunition, by increasing the Panel's calibre threshold for mortars from 100 mm to 120 mm, and by adding man-portable, rail-launched rockets to the Panel's list of light weapons (Small Arms Survey, 2008, pp. 8–11).

The *Small Arms Survey 2010* further refines the definition to distinguish 'light weapons' from 'light weapons ammunition'. Specifically, it defines MANPADS, rockets in single-shot disposable launch tubes, and rockets fired from rails as light weapons rather than ammunition.

In line with these definitions, this chapter uses the term 'light weapons' to refer to the following items:

- mortar systems up to and including 120 mm;
- hand-held (stand-alone), under-barrel, and automatic grenade launchers;
- recoilless guns;
- portable rocket launchers, including rockets in single-shot disposable launch tubes; and
- portable missiles and launchers, namely ATGWs and MANPADS.

This list of light weapons excludes heavy machine guns and anti-materiel rifles, for which transfers data is often aggregated with small arms and is usually impossible to distinguish from that of other firearms. For the purposes of analysing transfers, the Small Arms Survey traditionally has treated these two types of weapons as 'firearms'—while the UN Governmental Panel of Experts lists them as 'light weapons'.[6] Improvised explosive devices are not covered in this chapter since the Small Arms Survey's definition of authorized transfers does not apply to most international transfers of these weapons (see below). To the extent possible, light weapons designed for or used exclusively on platforms larger than light vehicles are also excluded,[7] as are parts and accessories,[8] which will be addressed in the *Small Arms Survey 2012.*

Box 1.1 Trends in the small arms trade

Each year, the United Nations Commodity Trade Statistics Database, or UN Comtrade, receives data on arms transfers from more than 100 countries, making it one of the richest and most consistent sources of data on the small arms trade. Yet since many countries contribute partial data–or none at all–and since UN Comtrade's categorizations and aggregation of different types of equipment do not permit transfers of some types of light weapons and their munitions to be identified, the resulting figures reflect only a portion of the actual trade. Nevertheless, given that these limitations are fairly constant over time, UN Comtrade is still useful for tracking trends in arms transfers.

An analysis of UN Comtrade data reveals that the total value of exports of small arms, light weapons, their ammunition, and associated parts and accessories in 2008 (the latest year for which there is a complete dataset) was USD 4.3 billion (see Annexe 1.1). Fourteen countries recorded exports of USD 100 million or greater, earning them 'top exporter' status–the most ever.[9] Transfers from each of the largest exporters, Italy and the United States, exceeded USD 500 million. Four countries–Israel, Norway, South Korea, and Spain–exported more than USD 100 million for the first time. Five other countries–the Czech Republic, Japan, the United Kingdom, Sweden, and Finland–exported USD 50-99 million. In descending order, the top importers–those importing weapons worth a total of at least USD 100 million–were the United States, Canada, the United Kingdom, Germany, Australia, France, and Pakistan. In addition, 15 countries imported weapons valued at USD 50-99 million (see Annexe 1.2).

Perhaps most remarkable is the continued growth in US imports, which exceeded USD 1 billion in 2007 and rose again in 2008 to 1.27 billion. Furthermore, preliminary data for 2009 shows the United States as having reported more than 1.77 billion in imports, which would represent a startling 39 per cent increase over 2008. Between 2000 and 2009, US imports increased by some 246 per cent. This rise is probably attributable to purchases by both the US military and civilians.[10]

Table 1.1 Exporter rankings for 2008

Ranking	Exports in 2008 (USD million)	Notes comparing 2008 with 2007
Top first tier (> USD 500 million)		
United States	715	
Italy	562	
Top second tier (USD 100-500 million)		
Germany	472	
Brazil	273	
Switzerland	211	
Israel	179	Moved up from major first tier
Austria	173	
South Korea	165	Moved up from major first tier
Belgium	124	
Russian Federation	119	
Spain	116	Moved up from major first tier
Turkey	111	
Norway	101	Moved up from major first tier
Canada	100	
Major first tier (USD 50-99 million)		
Czech Republic	94	
Japan	92	
United Kingdom	75	Moved down from top second tier
Sweden	69	Moved up from major second tier
Finland	65	
Major second tier (USD 10-49 million)		
France	43	
Croatia	43	
Portugal	39	
China	37	Moved down from top second tier
Poland	36	
Taiwan	32	
Serbia	31	
Mexico	28	
Singapore	16	
Cyprus	14	Exported less than 10 million in 2007
Netherlands	13	Exported less than 10 million in 2007
Romania	13	
Argentina	13	
Bosnia and Herzegovina	10	
India	10	Exported less than 10 million in 2007

For the purposes of this chapter, authorized international transfers of light weapons are defined as cross-border movements of light weapons that are authorized by the importing, exporting, or transit states. For a detailed discussion of the definition of international transfers, see Small Arms Survey (2009, p. 9, box 1.1).

TRANSPARENCY IN LIGHT WEAPONS TRANSFERS AND PROCUREMENT

Overall, the amount and quality of information on light weapons transfers remains low, despite a significant increase in the number of countries participating in one or more national or international mechanisms for reporting on arms transfers in recent years. The quality of the information submitted by governments to the main publicly available information sources—public procurement boards, UN Comtrade, and the UN Register—varies significantly in terms of completeness and detail, and few submissions provide a complete accounting of light weapons transfers. Public

A Chinese salesman describes his products to a Russian customer during the China International Defense Electronics Exhibition in Beijing, April 2006.
© Elizabeth Dalziel/AP Photo

procurement boards are rarely used for light weapons,[11] and UN Comtrade's reporting mechanism makes disaggregating light weapons transfers virtually impossible.[12]

Of the reporting mechanisms for transfers of small arms and light weapons, the UN Register contains the richest data on light weapons transfers. Since its establishment in 1991, the Register has become the primary international mechanism for reporting on the transfer of conventional weapons. Small arms and light weapons were largely excluded from the reporting mandate until 2003,[13] when the calibre threshold for mortar systems on which states are to report was reduced to 75 mm, and MANPADS were added.[14] The UN also established a voluntary mechanism for reporting on transfers of small arms and light weapons in 2003. While reporting was initially extremely limited, it has risen steadily in recent years. Only five governments submitted data on small arms transfers in 2005, yet that figure had risen to 47 by 2009[15] (UNODA, 2010a, p. 21). Nonetheless, reporting remains inconsistent, in terms of both frequency and quality. Seventeen of the countries that have submitted small arms reports at some point since 2003 failed to do so in 2009 (UNODA, 2010b, pp. 15–18). Also noteworthy is the absence of data on most light weapons transfers (MANPADS being one notable exception) from the three largest producers of light weapons: China, the Russian Federation, and the United States (Small Arms Survey, 2008, pp. 34–35).

Other important potential sources of information on light weapons transfers are individual government agencies that compile data on national light weapons transfers. In response to queries sent to more than 80 governments over a five-month period in 2010, the governments of Bosnia, Canada, Colombia, Germany, Liechtenstein, Norway, Portugal, Slovakia, Sweden, and the UK provided data on their light weapons imports. Officials from the Netherlands and Thailand completed questionnaires on light weapons procurement practices. At least two other countries—South Africa and the United States—routinely publish detailed procurement data, including data on imports, in publicly available government reports. National data obtained directly from governments is often significantly more detailed and complete than that reported to the UN Register.

Forty-seven states submitted information on transfers of small arms or light weapons that occurred in 2008 to the UN Register. Twenty of these states provided data on transfers of light weapons; 9 of these states provided detailed data, including all of the following: (1) whether the transfer was an import or export; (2) the states of origin and destination; (3) the weapon type and model; and (4) the quantity of weapons transferred to each destination. Among the eleven states that withheld some or all of these details, weapon model and quantity transferred were the most frequently omitted information. Reporting states withhold information by declaring it 'classified' or by simply leaving parts of the reporting form blank. Panama, for example, declared that it imported light weapons in 2008 but did not identify the exporting state or provide any information on the quantities or models.[16]

The other 27 states that submitted information either reported 'nil' with reference to small arms and light weapons transfers (six states) or only provided data on small arms (21 states). It should be noted that, for the purposes of this study, a 'nil' entry is assumed to be an accurate report of an absence of light weapons transfers. It remains unclear why the 21 states that submitted information on small arms did not report on light weapons transfers. Some may not have engaged in any transfers of light weapons. In other cases, data on light weapons transfers may have been withheld because the government considers the information restricted (classified).

As explained above and illustrated by Map 1.1, the light weapons trade remains far from transparent despite the development of international mechanisms for reporting on light weapons. Until the majority of UN member states consistently submit detailed and complete reports on their light weapons transfers to the UN Register, our understanding of the international trade in light weapons will remain limited.

Reporting remains inconsistent, both in terms of frequency and quality.

Map 1.1 **Reporting on light weapons transfers to the UN Register, 2008**

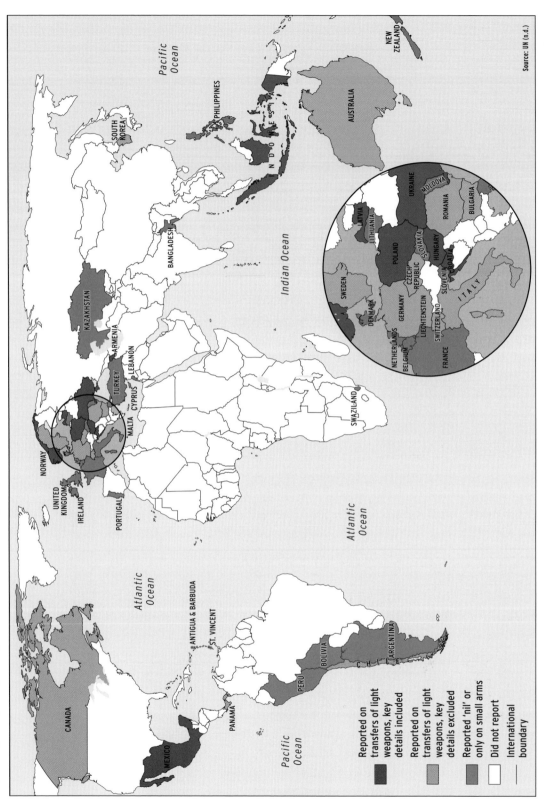

Source: UN (n.d.)

Reported on transfers of light weapons, key details included

Reported on transfers of light weapons, key details excluded

Reported 'nil' or only on small arms

Did not report

International boundary

Notes: This map illustrates the quality of reporting to the UN Register of Conventional Arms on light weapons transfers in 2008. All of the shaded states submitted data on international transfers of small arms and light weapons to the Register. Differences in shading reflect the comprehensiveness and specificity of the information submitted by individual states. States that are shaded red reported on light weapons transfers and included the following details about individual transfers: (1) a clear distinction between imports and exports; (2) the states of origin and destination; (3) the weapon type and model; and (4) the quantity of weapons transferred to each destination. Pink shading is used for states that reported on light weapons transfers but omitted one or more of the key details identified above. States that are shaded grey either reported 'nil' (no transfers) for 2008, or submitted data only on small arms transfers. For the purposes of this chapter, 'nil' entries are considered transparent reports of transfers. Yet by definition, 'nil' reports do not include the key details upon which the categorization scheme is based; consequently, they could not be included in the assessment of reporting quality. States that only submitted data on small arms transfers are included in the same category as 'nil' transfers because it is often unclear whether the state in question did not import or export any light weapons, or failed to report on their light weapons transfers.

THE 2011 TRANSPARENCY BAROMETER

The Small Arms Trade Transparency Barometer was introduced in the *Small Arms Survey 2004* to evaluate countries' transparency in reporting on their small arms and light weapons exports. The Barometer has evolved significantly since its introduction. Points are awarded for timeliness, access and consistency, clarity, comprehensiveness, and inclusion of data on deliveries as well as for reporting on licences granted and refused. The 2011 Barometer examines countries claiming—or believed—to have exported USD 10 million or more of small arms and light weapons, including their parts, accessories, and ammunition, during at least one calendar year between 2001 and 2009.

The 2011 Barometer assesses national transparency in small arms export activities for 2009, based on reporting in 2010.[17] The three main sources are: (1) national arms export reports;[18] (2) the UN Register; and (3) UN Comtrade (see Table 1.2).

As its name indicates, the Barometer is designed to measure—and promote—transparency. It evaluates the quantity, precision, timeliness, and usefulness of the data countries make publicly available and can also be used to highlight trends in national reporting. Although it does not assess the veracity of the data states provide, it can highlight discrepancies between different sources.

This edition assesses the reporting practices of 49 countries: the 48 countries covered in the 2010 Barometer plus the Philippines—believed to have exported roughly USD 14 million worth of relevant materiel in 2009. Additional countries may feature in future Barometers, if and when more information about their international small arms transfers becomes available.

This year's Transparency Barometer identifies Switzerland, the United Kingdom, and Germany as the three most transparent countries. They have held the same top spots for three consecutive years (Lazarevic, 2010, pp. 193–97).[19] Belgium and the United States broke into the top ten this year, replacing Sweden and Denmark (which now rank 11th and 14th, respectively). The least transparent countries are Iran and North Korea, both scoring zero for five successive years (Lazarevic, 2010, pp. 187–99). Although the latter is conspicuously opaque about its international arms transfers, there is clear evidence of its involvement in illicit small arms exports (see Box 1.4). The average score fell slightly since last year (from 11.50 to 11.40), although the average score of the top 10 countries rose from 17.45 to 18.00 points. However, about two-thirds of the countries reviewed received fewer than half the maximum number of points available (that is, less than 12.50 out of 25.00), suggesting that, despite progress among some states, there remains much scope for improved reporting.

The biggest country-specific declines were experienced by Bosnia and Herzegovina and Saudi Arabia. Both countries' scores were reduced by four points. Bosnia and Herzegovina did not report to UN Comtrade on 2009 activities and its national arms export report was less detailed than in previous years, hence its scoring fell by 31 per cent to 9.00 points. Saudi Arabia's scoring was assessed through its reporting to UN Comtrade. For 2009 activities, Saudi Arabia provided information on re-exports of a single UN Comtrade category.[20] As a result, its total score decreased by 55 per cent to 3.25 points. Moreover, several member states of the European Union (EU) experienced a decrease in their points for the categories *clarity* and *licences refused,* as a result of changes in the presentation of *EU Report* data (see Box 1.2).

There were several positive trends in reporting in 2010. For the first time, Serbia and Spain provided the UN Register with 'voluntary background information' on their international small arms and light weapons transfers. Thailand reported to the UN Register for the first time since 2004. Romania has started including information on transit/transshipment in its national arms export report. Belgium experienced the greatest point increase with its score rising by

Switzerland, the United Kingdom, and Germany are the most transparent countries.

Table 1.2   Small Arms Trade Transparency Barometer 2011, covering major exporters*

	Total (25 max)	Export report (year covered)** / EU Annual Report***	UN Comtrade	UN Register	Timeliness (1.5 max)	Access and consistency (2 max)	Clarity (5 max)	Comprehensiveness (6.5 max)	Deliveries (4 max)	Licences granted (4 max)	Licences refused (2 max)
Switzerland	21.00	x (09)	x	x	1.50	1.50	4.00	5.25	3.00	4.00	1.75
United Kingdom	20.00	x (09) / EU Report	x	x	1.50	2.00	4.00	5.00	3.50	2.50	1.50
Germany	18.75	x (09) / EU Report	x	x	1.50	1.50	3.75	3.50	3.50	3.50	1.50
Serbia[1]	18.50	x (08)	x	x	1.50	1.00	3.25	4.75	3.50	2.50	2.00
Romania	18.00	x (09) / EU Report	x	x	1.50	2.00	2.50	4.00	3.00	3.00	2.00
Belgium	17.75	x (09) / EU Report	x	x	1.50	2.00	2.75	3.50	3.50	3.00	1.50
Netherlands	17.00	x (09) / EU Report	x	x	1.50	2.00	3.50	4.00	2.50	2.50	1.00
Spain	16.50	x (09) / EU Report	x	x	1.50	1.50	2.50	3.25	4.00	2.50	1.25
Norway	16.25	x (09)	x	x	1.50	1.50	4.00	3.75	3.50	2.00	0.00
United States[2]	16.25	x (08)	x	x	1.50	1.50	2.75	4.50	3.00	3.00	0.00
Sweden	15.50	x (09) / EU Report	x	x	1.50	2.00	3.25	4.25	3.00	1.50	0.00
Italy	15.25	x (09) / EU Report	x	x	1.50	1.50	3.25	4.50	3.00	1.50	0.00
Czech Republic[3]	15.00	x (09) / EU Report	x	x (08)	1.50	1.50	2.75	3.50	3.50	1.50	0.75
Denmark	14.75	x (09) / EU Report	x	x	1.50	1.50	3.50	3.25	3.00	2.00	0.00
Finland[4]	14.00	x (08) / EU Report	x	x	1.50	1.50	3.00	3.00	3.00	2.00	0.00
Poland[5]	14.00	EU Report	x	x	1.50	1.00	1.75	3.75	4.00	2.00	0.00
France[6]	13.50	x (09) / EU Report	x	x (08)	1.50	1.50	3.25	2.75	3.00	1.50	0.00
Slovakia	13.50	x (09) / EU Report	x	x	1.50	1.50	2.25	3.25	3.00	2.00	0.00
Montenegro[7]	12.50	x (09)	x	-	1.50	0.50	2.75	4.25	1.50	2.00	0.00
Austria[8]	12.25	x (08) / EU Report	x	x	1.50	1.50	1.75	2.50	3.50	1.50	0.00
Portugal[9]	12.25	x (08) / EU Report	x	x	1.50	1.50	2.00	2.75	3.00	1.50	0.00
Canada[10]	11.75	x (06)	x	x (08)	1.50	1.00	2.25	4.00	3.00	0.00	0.00
Bulgaria	11.50	x (09) / EU Report	x	x	1.50	1.50	2.00	2.00	3.00	1.50	0.00
Israel	10.75	-	x	x	1.50	1.00	1.75	3.50	3.00	0.00	0.00

Country										
Croatia[11]	10.50	-	x	1.50	0.50	1.75	3.75	3.00	0.00	0.00
Australia	10.25	-	x	1.50	1.00	1.50	3.25	3.00	0.00	0.00
Hungary[12]	10.25	EU Report	x(08)	1.50	0.50	1.25	2.50	3.00	1.50	0.00
South Korea	10.25	-	x	1.50	1.00	1.50	3.25	3.00	0.00	0.00
Pakistan[13]	9.75	-	x(08)	1.50	0.50	1.50	3.25	3.00	0.00	0.00
Thailand	9.75	-	x	1.50	0.50	1.50	3.25	3.00	0.00	0.00
India	9.25	-	x	1.50	1.00	1.50	2.25	3.00	0.00	0.00
Mexico	9.25	-	x	1.50	1.00	1.50	2.25	3.00	0.00	0.00
Turkey[14]	9.25	-	x(08)	1.50	0.50	1.50	2.75	3.00	0.00	0.00
Argentina	9.00	-	x	1.50	1.00	1.50	2.00	3.00	0.00	0.00
Bosnia and Herzegovina	9.00	x(09)	x	1.50	1.00	1.50	1.25	1.50	1.50	0.75
Japan	9.00	-	x	1.50	1.00	1.25	2.25	3.00	0.00	0.00
Philippines[15]	9.00	-	x(08)	1.50	0.50	1.50	2.50	3.00	0.00	0.00
Brazil	8.50	-	x	1.50	1.00	1.00	2.00	3.00	0.00	0.00
Cyprus[16]	8.50	EU Report	x(08)	1.50	0.50	1.50	2.00	3.00	0.00	0.00
United Arab Emirates[17]	8.50	-	x(08)	1.00	0.00	1.50	3.00	3.00	0.00	0.00
China	8.00	-	x	1.50	1.00	1.00	1.50	3.00	0.00	0.00
Ukraine	8.00	x(09)	x	1.50	1.50	1.00	2.00	2.00	0.00	0.00
Taiwan	7.75	-	-	1.50	0.50	1.50	1.75	2.50	0.00	0.00
Singapore	7.00	-	x	1.50	1.00	1.00	1.50	2.00	0.00	0.00
Russian Federation	6.00	-	x	1.50	1.00	0.50	1.00	2.00	0.00	0.00
Saudi Arabia	3.25	-	x	1.50	0.00	0.50	0.75	0.50	0.00	0.00
South Africa[18]	2.00	x(09)	x	1.50	0.50	0.00	0.00	0.00	0.00	0.00
Iran	0.00	-	x	0.00	0.00	0.00	0.00	0.00	0.00	0.00
North Korea	0.00	-	x	0.00	0.00	0.00	0.00	0.00	0.00	0.00

* Major exporters are countries that export–or are believed to export–at least USD 10 million worth of small arms, light weapons, their parts, accessories, and ammunition in a given year. The 2011 Barometer includes all countries that qualified as a major exporter at least once during the 2001–09 period.

** x indicates that a report was issued.

*** The Barometer assesses information provided in the EU's *Twelfth Annual Report* (CoEU, 2011), reflecting military exports by EU member states in 2009.

Scoring system

The scoring system for the 2011 Barometer is identical to that used in 2010, providing comprehensive, nuanced, and consistent thresholds for the various categories. The Barometer's seven categories assess: timeliness as well as access and consistency in reporting (categories i-ii), clarity and comprehensiveness (iii-iv), and the level of detail provided on actual deliveries, licences granted, and licences refused (v-vii). For more detailed information on the scoring guidelines, see the Small Arms Survey (n.d.b).

Explanatory notes

Note A: The Barometer is based on each country's most recent arms export report, made publicly available between 1 January 2009 and 31 December 2010.

Note B: The Barometer takes into account national reporting to the UN Register from 1 January 2009 to 12 January 2011 as well as information states have submitted to UN Comtrade for their 2009 exports up to 31 December 2010.

Note C: The fact that the Barometer is based on three sources–national arms export reports, reporting to the UN Register, and UN customs data–works to the advantage of states that publish data in all three outlets. Information provided to each of the three sources is reflected in the scoring. The same information is not credited twice, however.

Country-specific notes

1) Serbia published a national arms export report in 2010 that was limited to data from 2008. For the first time, the country is evaluated on a 25-point scale as it can now fulfil the criteria of reporting during three consecutive years and can therefore earn all the points potentially available under 'access and consistency'.

2) The US report is divided into several documents. For the purposes of the Barometer, the 'US annual report' refers to the State Department report pursuant to Section 655 on direct commercial sales, as well as the report on foreign military sales that is prepared by the Department of Defense. The reports are assessed together to provide a composite picture of US government transparency. The State Department did not publish the direct commercial sales report for 2009 by the cut-off date. The United States is therefore evaluated on the basis of its most recent report, covering activities in 2008.

3) The Czech Republic did not submit data to the UN Register for its 2009 activities by the cut-off date. It is therefore evaluated on the basis of its most recent submission, covering export activities in 2008.

4) Finland did not issue a national report for its arms export activities in 2009 by the cut-off date. It is therefore evaluated on the basis of its most recent national report, covering activities in 2008.

5) The authors discovered Poland's (first) national arms export report in February 2011, after the cut-off date for review. Although dated 2010, the report was posted on the website of Poland's Ministry of Foreign Affairs only in February 2011. The report will be assessed as part of the 2012 Barometer if no other national report is published in 2011. For the 2011 edition, Poland is evaluated based on its contribution to the *EU Annual Report*.

6) France did not submit data to the UN Register for its 2009 activities by the cut-off date. It is therefore evaluated on the basis of its most recent submission, covering activities in 2008.

7) For the first time, Montenegro is evaluated on a 25-point scale as it can now fulfil the criteria of reporting during three consecutive years and can therefore earn all the points potentially available under 'access and consistency'.

8) Austria did not issue a national report for its arms export activities in 2009 by the cut-off date. It is therefore evaluated on the basis of its most recent national report, covering activities in 2008.

9) Portugal published a national arms export report in 2010 that was limited to data from 2008. Until last year, Portugal's *Statistical Yearbook of National Defence* was evaluated. Recently, Portugal put online its *Annual Reports on Exports of Military Equipment*. The latter report has been evaluated for the 2011 edition of the Transparency Barometer and will be used for future editions.

10) Canada did not issue a national report on export activities in 2009 by the cut-off date; it is therefore evaluated according to its most recent national report, covering export activities in 2006. Canada did not submit data on 2009 activities to the UN Register by the cut-off date; it is therefore evaluated on the basis of its most recent submission, covering activities in 2008.

11) The authors discovered Croatia's (first) national arms export report in January 2011, after the cut-off date for review. Although dated October 2010, the report was not posted on the website of the Croatian Ministry of Economy, Labour, and Entrepreneurship, but rather uploaded on the website of the South Eastern and Eastern Europe Clearinghouse for the Control of Small Arms and Light Weapons (SEESAC). The report will be assessed as part of the 2012 Barometer if no other national report is published in 2011. Croatia did not submit data to the UN Register for its 2009 activities by the cut-off date. It is therefore evaluated on the basis of its most recent submission, covering activities in 2008.

12) Hungary is one of two EU member states under review that do not publish a national report; however, it does contribute information to the *EU Annual Report*. Hungary did not submit data to UN Comtrade for its 2009 activities by the cut-off date. It is therefore evaluated on the basis of its most recent submission, covering activities in 2008.

13) Pakistan did not submit data to the UN Register for its 2009 activities by the cut-off date. It is therefore evaluated on the basis of its most recent submission, covering activities in 2008.

14) Turkey did not submit data to the UN Register for its 2009 activities by the cut-off date. It is therefore evaluated on the basis of its most recent submission, covering activities in 2008.

15) The Philippines did not submit data to the UN Register for its 2009 activities by the cut-off date. It is therefore evaluated on the basis of its most recent submission, covering activities in 2008.

16) Cyprus is one of two EU member states under review that do not publish a national report; however, it does contribute information to the *EU Annual Report*. Cyprus did not submit data to the UN Register for its 2009 activities by the cut-off date. It is therefore evaluated on the basis of its most recent submission, covering activities in 2008.

17) The United Arab Emirates did not submit data to UN Comtrade for its 2009 activities. It is therefore evaluated on the basis of its most recent submission, covering activities in 2008.

18) South Africa did not submit data to UN Comtrade for its 2009 activities. It is therefore evaluated on the basis of its most recent submission, covering activities in 2008.

Source

Lazarevic (2011)

Box 1.2   **Licence refusals in the EU Annual Report**

EU member states have been exchanging data on their export licence approvals and refusals since 1999.[21]

Every year, the EU publishes a report on these exchanges of information. The *First Annual Report* was published in 1999. Since then, the data presented and illustrated in the reports has improved. Initially, states provided data on the total value of actual exports, the number of export licences granted, and the number of notified denials without specifying the country of destination or the EU Common Military List (ML) categories (CoEU, 1999; 2000). The *Third Annual Report* provides the same data according to sub-regions (CoEU, 2001). The fourth and fifth *Annual Reports* disaggregate the data by destination country and provide reasons for licence refusals, but data is still not disaggregated by ML categories (CoEU, 2002; 2003). The sixth to the eleventh *Annual Reports* finally provide information on ML categories. These details can be used to identify how many licences each EU member state refused and for what reasons (CoEU, 2004-07; 2008a; 2009).

With the *Twelfth Annual Report,* however, the EU adopted a new approach. The breakdown of refusals is no longer national, but instead aggregated at the EU level (CoEU, 2011); as a result, it is no longer possible to determine how many licences each EU member state has refused, for what ML category, or for what reasons.

Many EU member states thus lose at least 0.5 points in the Transparency Barometer as reflected in the scoring of Denmark, France, Italy, Poland, Portugal, Slovakia, and Sweden. If the information on licence refusals had not been buried in an EU total, but rather disaggregated by member state, then Italy, for example, would have had the same score as in the 2010 Transparency Barometer (Herron et al., 2010, p. 15).

three points (from 14.75 to 17.75) due to better reporting in the Belgian regions[22] on temporary exports, on end users, and on licences granted and refused. The greatest increase in percentage terms was Taiwan, whose score rose 24 per cent since last year thanks to better reporting to UN Comtrade.[23]

The promotion of national, regional, and global transparency mechanisms enhances confidence, which in turn can improve global standards in reporting on small arms transfers. The South Eastern and Eastern Europe Clearinghouse for the Control of Small Arms and Light Weapons (SEESAC) recently launched an initiative serving the interests of transparency, namely a regional report on arms exports for South-eastern Europe, whose content is also available in an online database (Bromley, 2010; SEESAC, n.d.).[24] It documents how countries in the region have sought to increase transparency in their arms export activities. Similarly, Saferworld has developed a database and released a report; both compare information supplied by EU member states on their arms transfers and shed light on missing information and shortcomings in reporting (Isbister and Okechukwu, 2010; Saferworld, n.d.).

ESTIMATING INTERNATIONAL LIGHT WEAPONS TRANSFERS

The opacity of much of the international trade in light weapons means that generating an estimate of the annual value of international transfers requires extrapolation from the documented trade. This section gives a brief overview of how the study uses data provided by the most transparent states as a basis for estimating values of light weapons imports by less transparent states. Annexe 1.3 presents a more detailed account of this process.

The fundamental assumption of the study is that if the values of light weapons transfers of a representative sample of states are known, it is possible to use this data as a basis from which to make reasonable estimates of the values of transfers of other states. From this methodological starting point, developing a global estimate proceeded in three stages: (a) generating a representative sample of the documented trade; (b) identifying the factors that best account for variations in spending on light weapons among states; and (c) deriving estimated import values for states outside the sample.

The documented trade: generating a sample

The first stage was to gather information on documented transfers of light weapons for as large a sample of countries as possible. Data obtained directly from governments, the Arms Transfers Database of the Stockholm International Peace Research Institute (SIPRI),[25] and the UN Register serves to build a picture of documented light weapons imports over multiple years.[26] This study makes use of data relating to imports between 1998 and 2010, although only very few states provide records for the entire period; the majority of the data relates to the years between 2003 and 2009.[27]

Countries are excluded from the sample if the authors determined that available import data is overly aggregated, unclear, or incomplete, for example if specific types of light weapons are explicitly excluded. The study also sets thresholds for inclusion based on the number of years of available data.[28] These thresholds strike a balance between, on the one hand, maximizing the size of the sample and, on the other, ensuring the data for a specific country is as representative as possible of typical annual light weapons imports by that country. Because of differences in the availability of data on different types of light weapons, varying thresholds are used for MANPADS and their missiles (a minimum of four years of data within the period[29]), ATGWs and their missiles (a minimum of three years of data), and non-guided light weapons (a minimum of two years of data).

The authors identified 73 countries[30] that meet these criteria for MANPADS and their missiles, 25 countries for ATGWs and their missiles, and 26 countries for non-guided light weapons. The sum of the average annual light weapons imports to these countries is USD 242 million, a figure that represents the 'documented trade'. The country sample and the documented value together serve as the basis for generating the global estimate of the authorized trade in these weapons, as described further below.

The average annual documented trade in light weapons totals USD 242 million.

Explaining variation

The second stage consisted of identifying the factors that best account for variations in spending on light weapons among states. Through an analysis of the documented trade, the authors identified the following four factors:

1. *The size of its armed forces.* As the size of an armed force increases, so does the number of light weapons required to equip it.

2. *The value of a state's military expenditure per member of its armed forces.* The more a state spends on its soldiers generally, the more likely its soldiers are to be equipped with greater quantities of light weapons and with higher-value types and models.

3. *The extent to which a state is involved in armed conflict.* Troops in active combat are more likely to be equipped with more (and higher-value) light weapons than troops that are engaged solely in peacetime actions. Moreover, weapons are likely to be used more frequently in armed conflict settings and will therefore have to be replaced more quickly.

4. *The availability of domestically produced light weapons.* The capacity to produce light weapons domestically reduces the need for, and acquisition of, imported weapons.

Having identified these four variables—armed force size, military expenditure per member of the armed forces, conflict status, and production capacity—the authors gathered data related to each category for almost every country in the world.[31]

Estimating imports

The third stage consisted of deriving estimated import values for non-sample states. The quantitative methods used were similar to those used to estimate the value of the annual authorized trade in ammunition for light weapons in Small Arms Survey (Herron et al., 2010). This process consisted of four main steps.

The first was to divide countries into nine categories based on armed force size (large, medium, and small) and military expenditure per member of the armed forces (high, medium, and low) (see Table 1.3). The second step was to use light weapons import values for sample countries to generate 'typical' annual import values for each of the nine groups.[32] The third was to generate provisional estimates of annual imports by non-sample countries by multiplying the relevant 'typical import value' by the size of the armed force of the non-sample country. The fourth was to modify, where applicable, provisional import estimates for non-sample countries upwards or downwards to take into account a state's conflict status and capacity to produce light weapons domestically.

These steps were carried out separately, and with slight variations, for MANPADS and their missiles; ATGWs and their missiles; and non-guided light weapons. Detailed information on this process is available in Annexe 1.3.

For each of the three light weapons groupings, the sum of the estimates for all non-sample countries yields the US dollar value estimates of the undocumented trade shown in Table 1.4. When added together, the resulting group sub-totals generate an estimated annual value of undocumented light weapons transfers of USD 872 million. Combining this figure with the documented trade of USD 242 million yields a combined estimate of USD 1.1 billion for the annual authorized global trade in light weapons.

Limitations of the estimation model

The methods described above have limitations. One set of caveats relates to underlying assumptions. For example, the methodology assumes that the sample of countries used as the basis of extrapolation is representative of the global trade as a whole. Yet since the sample is not random but rather determined by transparency, a bias in the sample, which would either inflate or deflate the total value, cannot be ruled out.

Table 1.3 Parameters for light weapons import country categories

Military expenditure (USD) per active service person per year	Armed force size		
	> 1,000,000	27,000–1,000,000	< 27,000
> 100,000	High-large	High-medium	High-small
20,000–100,000	Medium-large	Medium-medium	Medium-small
< 20,000	Low-large	Low-medium	Low-small

Table 1.4 Estimated annual values of international transfers of light weapons

	MANPADS (USD million)	ATGWs (USD million)	Other light weapons (USD million)	Total (USD million)
Documented	66	129	47	242
Undocumented	36	626	210	872
Total	102	755	257	1,114

Another set of caveats relates to inherent limitations of the data. One such issue concerns US dollar values for light weapons. Since much of the documented trade is reported in quantities only, it was necessary to convert all such transfers into values using generic unit prices for particular light weapons types, adding an unknown level of imprecision to the value of the import. Perhaps most importantly, determining the veracity of existing data is often difficult; the authors could not always verify that a sample country's reporting accurately reflected its light weapons imports.

Because of these and other limitations, the figure of USD 1.1 billion should be read as a best estimate rather than a definitive accounting. It is worth noting, however, that steps have been taken to prevent an artificial inflation of the figure. For example, wherever possible, the authors excluded any imports of items purchased for use on large vehicles, even if the items fell within this chapter's definition of a light weapon.

A more detailed discussion of limitations of the methods used in this study and the steps taken to mitigate the problems associated with them is presented in Annexe 1.3. Box 1.3 assesses the utility of the Internet as a supplemental research tool.

THE LIGHT WEAPONS TRADE

This section provides an assessment of the major producers, exporters, and importers of light weapons; noteworthy developments in the technology and use of light weapons; and patterns of light weapons transfers in recent years. It is divided into two parts. The

---

Box 1.3   Using new media to research the arms trade

The Internet has revolutionized research on the arms trade. Information on the production, trade, and holdings of weapons that was previously inaccessible to the public is now available in online databases and publications that are accessible to anyone with an Internet connection. While the full potential of the Internet as an arms trade research tool has yet to be tapped, new media–such as blogs, file-sharing sites, and Web forums–contain a wealth of information that is beginning to shape public understanding of the arms trade.

Among the millions of videos uploaded to video-sharing sites such as YouTube, many contain footage of weapons. The video clips, whether taken by amateurs or professional broadcasters, often provide important information about arms transfers. For example, it had been rumoured that the Russian Federation had exported sophisticated SA-24 MANPADS to Venezuela, but it had not been possible to confirm the transaction. Then, in April 2009, footage of a Venezuelan military parade in which the missiles were on display was uploaded to YouTube. The footage, which includes several close-up shots of the missiles from different angles, not only allows for a positive identification of the weapons as SA-24 MANPADS, but also includes contextual information about their intended use and deployment provided by Venezuelan President Hugo Chávez (Schroeder and Buongiorno, 2009).[33]

Blogs are another useful source of data and information on the arms trade. Some blogs (weblogs containing news and commentary uploaded by an individual or a small group of people) are written by individuals with unique or in-depth knowledge of particular weapon systems, regional procurement patterns, or national military holdings, such as current or former military personnel. Other rich sources of information are online discussion forums frequented by individuals with substantive knowledge of the armed forces and military procurement practices of a particular country or region.[34] Forums and blogs not only contain valuable information about arms transfers and military procurement, but also offer a chance to communicate directly with well-informed people who would otherwise be difficult to find and contact.

Like all sources, new media have limitations. The video of Venezuela's SA-24 missiles is an unusual case; the authenticity of most videos posted on sites such as YouTube is difficult or impossible to verify, and few are as clear and high-quality as the video of Venezuela's missiles. Some videos are of such poor quality that it is impossible to identify the make or model of the featured weapon with any certainty. Moreover, claims made about the contents of videos may be inaccurate, as a result of either poor analysis or propaganda, and it is often necessary to use other sources to verify video content. Similar caveats apply to blogs and discussion forums. The anonymity of the Internet, including the widespread use of pseudonyms, makes it difficult to verify a source of information.

Nonetheless, new media form an increasingly important supplemental source of information on the international arms trade and particularly on transfers between countries that release little or no public data on their military procurement or arms transfers.

first part looks at the international trade in non-guided light weapons—mortar systems, grenade launchers, recoilless guns, and portable rocket launchers. The second part assesses the trade in portable missiles, namely ATGWs[35] and MANPADS. Both sections draw heavily, but not exclusively, on the datasets generated for this study.

It should be noted that recent technological developments in light weapons and their ammunition are blurring the lines between guided and non-guided light weapons. Several countries are producing Global Positioning System (GPS) and laser-guided rounds for 120 mm mortar systems and are developing similar rounds for smaller-calibre mortar systems. Similarly, advancements in ballistics calculators, range finders, and fusing for ammunition are dramatically increasing the accuracy of grenade launchers and other light weapons. The production and use of guided and other 'smart' rounds for light weapons remains extremely limited, however, and therefore the distinction between guided and non-guided light weapons is still relevant, at least for the time being.

Non-guided light weapons

For the purposes of this chapter, non-guided light weapons include all light weapons except portable missiles and their launchers (which are covered in the following section) and anti-materiel rifles and heavy machine guns, transfers of which are examined in the 2009 edition of the *Small Arms Survey*.[36] Items covered in this section include the following:

- mortar systems up to and including a calibre of 120 mm;[37]
- hand-held, under-barrel, and automatic grenade launchers;
- recoilless guns; and
- portable rocket launchers (including single-shot, disposable units).[38]

This is clearly a diverse range of weapons, yet all share some common characteristics. Infantry forces use all of them and their primary targets are often similar: other infantry (or members of armed groups); lightly armoured vehicles; and bunkers or other hard cover. These weapons are produced in several dozen countries and are used by most armed forces.[39]

Weapons thought to be obsolete have had renewed use in counterinsurgency roles.

All of these weapons are mature technology; mortars and rocket launchers date back centuries, for example. The main technological developments over recent decades have occurred in the production of new forms of ammunition and accessories, such as optics and other devices used in aiming and target acquisition. This divergence between relatively unsophisticated launchers and the availability of more advanced ammunition has led a wide variety of militaries to field light weapons. Well-funded militaries are able to equip their mortars, grenade launchers, rocket launchers, and recoilless guns with advanced ammunition—such as rocket-assisted mortar bombs or proximity-fused grenades—while armed forces with low budgets use similar weapons with much more basic and inexpensive ammunition. To sum up, these light weapons are extensively deployed by armed forces across the globe; the varying sophistication of ammunition means that they have a niche in all armed forces.[40]

In general, the non-guided light weapons used by contemporary armed forces are designed to be rugged, reliable, easily portable, comparatively affordable, and easy to use and maintain. Infantry engaged in counterinsurgency operations can deploy them in harsh environments. Some models have enjoyed a resurgence following the experience of US and associated armed forces fighting in Iraq and Afghanistan. Anti-tank rocket systems previously viewed as obsolete, for example, have been widely used by forces in Iraq and Afghanistan against insurgents ensconced in buildings, bunkers, or other fortifications and caves (see the discussion on the M72 rocket system, below). This deployment and use of light weapons by well-resourced armies fighting in counterinsurgency campaigns has had a marked effect on the international trade. Since 2005, the largest purchasers of light weapons have been countries heavily involved in Iraq and Afghanistan. This procurement reflects the much greater use of weapons by forces engaged in high-intensity conflict compared to peacetime training.[41]

The trade in these four types of light weapons is discussed in greater detail in the following sections.

Mortars

Mortars fire projectiles indirectly, their ammunition travelling at high ballistic arcs and at slower speeds than other artillery of similar calibres. Since the First World War, the main innovations in mortar technology have been the mounting of mortars on vehicles and the development of automatic loading mechanisms (both of which usually add so much weight that the systems are no longer considered 'light weapons'). More significant are technological improvements to mortar ammunition, including GPS and laser guidance, and rocket assistance.[42] These developments provide greater accuracy and, in the case of rocket assistance, greater range.

From 2006 to 2009, seven of the 26 countries studied reported a total of 388 imports of mortars up to a calibre of 120 mm. These weapons can be disaggregated by calibre into three categories (see Table 1.5).

Table 1.6 reports the five most significant importers of mortars from 2006 to 2009, accounting for 384 of the 388 mortars imported over the period. The highest total was 173 by Bangladesh, which imported 60 mm and 82 mm

US Army soldiers fire a 120 mm mortar during a fire mission at the combat outpost Zerok in East Paktika province, Afghanistan, September 2009.
© Dima Gavrysh/AP Photo

Table 1.5 Total identified imports of mortar systems, 2006-09

Calibre	Quantity imported
Up to 60 mm	168
61-82 mm	173
83-120 mm	47
Total imports	**388**

Table 1.6 Top five importers of mortars, 2006-09

Importing country	Quantity imported
Bangladesh	173
Georgia	105
Mexico	62
Portugal	22
Lithuania	22
Total	**384**

Note: These figures combine all calibres of mortar examined in this sample.

mortars from China. The second-largest importer of mortars was Georgia, which acquired 60 mm and 82 mm mortars from Bosnia and Herzegovina and Bulgaria. Other countries imported fewer than 100 mortars over the four-year period. The low level of procurement suggests that the imported units were not intended to replace a large proportion of existing national stocks—a process known as 'peak procurement'.[43] It is consistent with statements by military officials interviewed for this study who reported that mortar systems are replaced very infrequently; they have an estimated service life of 25 years, and possibly much longer in practice.[44] In some cases, officials reported that equipment obtained during the cold war, some of which is more than 40 years old, is still in service.[45]

The identity of the importers, along with interviews with military officials, suggest that the imported mortars were procured for specific military units (such as peacekeepers), were replacements for systems nearing the end of their shelf life, or were supplements for armed forces engaged in, or preparing for, armed conflict.[46]

Grenade launchers

Grenade launchers fire a small projectile (a grenade) that usually contains high explosives, gas or other irritants, smoke, or incendiary materials. This section covers three main types of grenade launchers, all of which use cartridge-based ammunition fired from a conventional barrel with an enclosed breach. *Under-barrel grenade launchers* are mounted to rifle barrels and fire a single projectile. *Hand-held grenade launchers* are self-contained weapons that can be fitted with sights, grips, and a butt. They often have a small magazine and a semi-automatic firing mechanism. *Automatic grenade launchers* are tripod-mounted, crew-served weapons capable of firing hundreds of rounds per minute. The standard US automatic grenade launcher is the Mark 19, which was first fielded in Vietnam during the mid-1960s. Since 1984, updated versions have been exported to some 30 countries (Foss, Gourley, and Tigner, 2008; Kemp, 2007).

Modern projected grenades are used against dispersed troops, personnel, soft-skinned vehicles, and some structures. Like mortars, the most significant recent technological developments in this area concern the ammunition rather than the launchers themselves. Advanced ammunition with airburst capabilities is produced by several companies, including Singapore Technologies Kinetics (STK), Nammo of Norway, and Arsenal of Bulgaria. Airburst grenades usually feature proximity fuses and are designed to detonate above or near troops that are partially hidden. Other innovations in ammunition include 'kicker' charges, which bounce the grenade off the ground so that it detonates in the air.[47] Grenade launchers were imported by 12 of the 26 countries in the sample. These imports are

Table 1.7 **Total identified imports of grenade launchers, 2006–09**	
Type	**Quantity imported**
Under-barrel	1,912
Hand-held	13
Automatic	342
Unspecified	290
Total imports	**2,557**

Table 1.8 **Top importers of grenade launchers by type, 2006–09**		
Type	**Importing country**	**Quantity imported**
Under-barrel	Mexico	1,429
	Latvia	250
	Poland	123
	Armenia	70
	Croatia	37
Hand-held	Slovakia	13
Automatic	Armenia	100
	Latvia	100
	Georgia	98
	Poland	43
Unspecified	Mexico	108
	Latvia	80
	Poland	53
	Lithuania	46

summarized in Table 1.7, which lists the number imported by type over the period 2006–09. Table 1.8 presents the top importers for each category.

Most transfers consisted of some hundred or fewer grenade launchers, which were probably procured for specific military units. Again, such imports are not indicative of 'peak procurement' in which a large proportion of national stocks are procured or replaced.

Recoilless guns

A recoilless gun resembles a conventional gun—its cartridge is loaded into a barrel—except that the rear of the barrel is open and the blast from the explosive propellant is allowed to escape. The projectile is launched from the barrel at a much lower velocity than from a conventional gun (recoilless guns therefore require much more propellant per round). While most recoilless guns have a rifled barrel (and are known as recoilless rifles), some are smoothbore. The advantage of a recoilless gun, as its name suggests, is that the recoil is minimized, meaning the gun does not need a heavy carriage or recuperator. Unlike conventional artillery, recoilless guns are often light enough to be towed by light vehicles or carried by hand. Their projectiles are much heavier than those fired by the grenade or rocket launchers described in this section. Recoilless guns offer direct fire in comparison to the indirect fire provided by mortars. The drawbacks of recoilless rifles include the comparatively slow velocity of the projectile (which limits its utility against modern tank armour), the need for large quantities of propellant, and the heavy back blast, which can be hazardous and expose the location of the operator (Weir, 2005, pp. 201–04).

One of the most widely exported recoilless rifles is the Swedish Carl Gustaf, currently produced by Saab Bofors Dynamics. Its first prototype was made in 1946, and modern variants are still widely used in some 40 countries (Felstead, 2010). It was originally designed to be an anti-tank weapon and, weighing around 10 kg, it is light enough to be carried and fired by one person or a crew of two (Weir, 2005, p. 204). Its utility against tanks diminished as armour improved, but it continues to be used against light vehicles and buildings, bunkers, and other hard cover.

The 84 mm Carl Gustaf rounds are also far more affordable than guided missiles, and various types of ammunition are available for use against a variety of targets.

Over the period 2006–09, three of the 26 countries in the sample imported a total of 187 recoilless guns. Of these, 182 were identified as Carl Gustaf recoilless rifles, while the remainder were unspecified models. Given

Table 1.9 **Importers of recoilless guns, 2006–09**	
Country	**Quantity imported**
Canada	150
Poland	35
Slovakia	2
Total	**187**

the widespread deployment of the weapon, the small number of importing countries reflects the long service life of these guns, which last 25 years or longer.[48] Table 1.9 summarizes the procurement data.

Portable rocket launchers, rockets, and single-shot, disposable units

Rocket launchers are composed of a tube through which a self-propelled projectile is launched and other components, such as sights, grips, and a firing mechanism. Like recoilless guns, the rear of the tube is open, allowing for a strong back blast, and the firer does not experience strong recoil forces. Rocket launchers are sold as reusable units with separate reloadable ammunition or as single-shot, disposable weapons. Single-shot, disposable units contain a launcher and unguided projectile in a single, sealed unit, which is discarded after use. This section includes recoilless guns in single-shot, disposable units because their operational use and procurement are more similar to those of single-shot, disposable rockets than to those of recoilless guns. In practice, there is no clear dividing line between the two as some systems (such as the RPG-29) fire for such a short time that the rocket has finished burning before it leaves the launch tube. Their launching method is therefore almost indistinguishable from that of recoilless guns.

Recent procurement of the M72 light anti-tank weapon, or LAW, is a good example of how the demand for light weapons has been stimulated by the wars in Iraq and Afghanistan. The M72 first entered service with the US military in 1963 and was used extensively during the Vietnam War (Ohmen, 2005). It is a single-shot, disposable unit; the launcher is supplied with a rocket ready to be fired. By the 1980s, it had been replaced by heavier systems—the AT-4 and the Shoulder-launched Multipurpose Assault Weapon, or SMAW—both of which were deemed to be more effective against contemporary tanks equipped with modern armour. The M72 was brought back into service following US experience in Iraq; the Marine Corps took old weapons out of storage, and new units have been procured. The advantages of this supposedly obsolete weapon are its small size, light weight, low cost, and relatively small back blast. For these reasons, it is much better suited to urban warfare than weapons designed to destroy heavily armoured vehicles. It is also less expensive than other portable rockets and missiles, costing a reported USD 2,500 per unit—a fraction of the cost of a Javelin anti-tank guided missile (*Defense Industry Daily,* 2005). M72 variants have also been deployed by other armed forces fighting in Iraq and Afghanistan, such as Australia, Canada, and the UK. Israel reportedly ordered 28,000 units (*Defense Industry Daily,* 2008); in the period 2006–09, the UK imported 3,280 M72 variants from Norway and the United States. The producer Nammo Talley has developed variants of the M72 specifically for counterinsurgency tasks (*Jane's International Defence Review,* 2006).

The term 'rocket-propelled grenade', or RPG, usually refers to a family of anti-tank systems first developed by the Soviet Union shortly after the Second World War. Variants are now in production in many countries around the world.[49] RPGs consist of a launcher and rocket-propelled explosive projectiles. An RPG launcher differs from grenade launchers in that it has an open rear and exhaust gases from the rocket are ejected from the back when it is launched. The ammunition for RPGs has been continuously updated, and modern tandem warheads are capable of penetrating even the reactive armour of contemporary tanks (Richardson, 2008).

Over the period 2006–09, five of the 26 countries studied reported imports of single-shot, disposable systems, totalling 20,818 units (see Table 1.10).

Table 1.10 **Importers of single-shot, disposable systems, 2006–09**	
Importing country	**Quantity imported**
Canada	12,000
United Kingdom	5,810
Slovenia	2,300
Lithuania	381
Mexico	327
Total	**20,818**

The most frequently imported single-shot, disposable systems were variants of the M72. Between 2006 and 2009 Canada imported 12,000 and the UK imported 4,810. These were most likely intended for use in Iraq or Afghanistan.

Box 1.4 North Korean arms transfers: insight from the UN Panel of Experts

On 11 December 2009, a Soviet-era cargo aircraft flying from Pyongyang, North Korea, was detained by Thai authorities during a stopover in Bangkok. The crew reportedly told Thai authorities that their aircraft carried 'oil-drilling equipment' and that they intended to refuel and head farther south (Barrowclough, 2009). While inspecting the plane, however, Thai authorities discovered 35 tons of arms and related materiel with an estimated value of USD 18 million. The shipment reportedly contained large quantities of light weapons, including rocket-propelled grenade launchers (RPG-7s), thermobaric RPG rounds (TGB-7Vs), and MANPADS (UNSC, 2010, para. 64).

The seizure in Thailand was one of several arms shipments from North Korea interdicted since 2009, when a UN arms embargo was expanded to include small arms and light weapons exports (UNSC, 2010, para. 18; 2006, para. 8; 2009, para. 9). The Security Council resolution expanding the embargo called on states to 'inspect all cargo' to and from North Korea 'in their territory, including seaports and airports' (UNSC, 2009, para. 11). The subsequent surveillance and interdiction of North Korean arms shipments has shed important light on illicit arms transfers from one of the most opaque and secretive countries in the world.

Documents seized in Bangkok reveal much about the techniques used by the North Korean regime to conceal shipments of illicit weapons. The various parties to the transfer were located in several different countries. The owner of the cargo plane was based in the United Arab Emirates, while the aircraft was registered in Georgia, leased by a shell company in New Zealand, and chartered by another shell company in Hong Kong (UNSC, 2010, para. 64). The Stockholm International Peace Research Institute determined that the cargo plane was previously registered by other companies linked to various well-known arms dealers, including Tomislav Dmanjanovic and Viktor

The only country that reported specific information on imports of reusable rocket launchers was Bangladesh, which imported 200 Type-69 RPGs (a variant of the RPG-7) from China. The lack of reusable RPG imports may be explained, in part, by the sample of countries for which import data was available. Most of these countries use NATO-standard weaponry, which does not include reusable RPGs.

Portable missile systems

This section provides a brief overview of the international trade in portable missiles, specifically ATGWs and MANPADS. It begins with a brief description of these weapons, their main producers, and their roles on the battlefield. Key characteristics of the international trade in portable missiles are then identified through an analysis of the multi-country dataset compiled for this study.

Background

For the purposes of this chapter, the term 'portable missiles' refers to two types of weapons: anti-tank guided weapons and man-portable air defence systems. ATGWs are missile systems originally designed for use against tanks and

Bout (Barrowclough, 2009). The airway bill was falsified, stating that the cargo consisted of '145 crates of "mechanical parts"' and providing misinformation about the route and destination of the flight, all of which are well-known masking techniques (UNSC, 2010, para. 64). The shipper was identified as Korea Mechanical Industry Co., Ltd., located in North Korea, and the consignee as Top Energy Institute in Iran, but the ultimate destination of the weapons was obscured by multiple flight plans (UNSC, 2010, para. 64).

UN investigators have identified three additional interdicted shipments, descriptions of which shed further light on North Korea's arms trafficking infrastructure. According to a report by the investigators, several government agencies are involved in the organization of arms transfers, including the Worker's Party of Korea and the Second Economic Committee, the latter believed to play 'the largest and most prominent role' (para. 55). While the mode of transport varied from shipment to shipment, similar methods were employed to disguise the cargo and the delivery routes. Based on the information collected to date, the UN Panel of Experts concludes that North Korea 'has established a highly sophisticated international network for the acquisition, marketing and sale of arms and military equipment' (para. 55).

The UN Panel report is a valuable supplement to existing public sources of data on North Korean arms transfers.[50] While the shipments documented by the Panel provide important insight into the nature of North Korea's exports of small arms and light weapons, the full extent of this trade remains unknown. Over time, however, the Panel of Experts may provide a more complete picture of North Korea's secretive arms trade.

Thai police officers and soldiers remove boxes of weaponry from a foreign-registered cargo plane at Don Muang airport in Bangkok, December 2009. The aircraft was reportedly loaded with North Korean weapons and flying to Iran when it was intercepted. © AP Photo

other armoured vehicles, though they have been employed against a wide array of other targets. Some ATGWs are lightweight, shoulder-fired weapons designed for use by dismounted infantry (foot soldiers) against armoured vehicles at close range. Other systems fire larger missiles that are capable of destroying tanks and other heavily armoured vehicles from distances exceeding 8 km. Many of the latter are fired from several different platforms,[51] including aircraft, naval vessels, tracked and wheeled vehicles, and pedestal mounts.

ATGWs employ a variety of guidance systems. Wire-guided missiles are guided by signals delivered to the missile by a thin wire that unravels as the missile travels to the target. The Russian Malyutka and the US TOW[52] are common wire-guided ATGWs. Infrared-seeking ATGWs such as the US Javelin guide themselves to the target after locking onto its infrared signature. Beam-riding missiles 'ride' a laser beam directed at the target by the operator. Examples include the Russian Kornet, the South African Ingwe, and the Swedish BILL.[53]

MANPADS are lightweight, portable surface-to-air missiles that are fired either from the operator's shoulder or from a pedestal mount. MANPADS are often categorized by their guidance systems: passive infrared seeking, semi-autonomous command-line-of-sight, and laser beam-riding.[54] Engagement ranges vary significantly from system to system. A first-generation Soviet SA-7 has a maximum effective range of about 3,400 metres whereas some newer MANPADS can hit targets 8 km away (O'Halloran and Foss, 2008). Like ATGWs, the missiles fired from MANPADS are also fired from aircraft, ships, and land vehicles.[55]

Soldiers of the US Army's 101st Airborne Division fire a TOW missile at the building housing Saddam Hussein's sons Odai and Ousai in Mosul, Iraq, on 22 July 2003.
© US Army/Sgt. Curtis G. Hargrave/AP Photo

Major producers of portable missiles include China, France, the Russian Federation, Sweden, and the United States. The US TOW missile is among the most numerous and widely deployed portable missile in the world; since 1970, militaries in more than 40 countries have procured more than 650,000 TOW missiles (Foss, 2004). The MILAN, which is produced by the European conglomerate MBDA, has been deployed almost as widely, albeit in smaller numbers.[56] Far fewer MANPADS have been produced than ATGWs, but they have proliferated as widely;[57] the US military estimates that more than 150 countries have deployed MANPADS since the 1960s.[58] The Russian Federation and China are also prominent international suppliers of ATGWs and MANPADS, having exported various systems to several dozen countries (O'Halloran and Foss, 2008; Jones and Ness, 2007). The *Small Arms Survey 2008* estimates that, as of 2007, at least 21 countries were producing ATGWs and 21 countries were producing MANPADS (Small Arms Survey, 2008, pp. 34–35).

Roles for ATGWs have changed significantly since the first systems were produced in the 1950s. The 'classic target set' for ATGWs, as identified by Jane's Information Group, included heavily armoured vehicles such as main battle tanks and lightly armoured and unarmoured vehicles (Gibson and Pengelley, 2004). Exposed infantry was also included, but the main focus was on destroying armoured vehicles. This target set has expanded to reflect the increasing focus on counterinsurgency, urban operations, and other non-traditional military operations over the past 30 years. In 2005, defence analyst Doug Richardson identified 28 major battles dating back to the early 1980s in which ATGWs were used. In only seven of these battles were ATGWs employed against tanks and other heavily armoured vehicles. In every other case, the targets were 'unarmoured vehicles, trucks, buildings, mud huts, bunkers, caves, small boats, and even individual snipers' (Richardson, 2005). This expanded list of targets is not likely to shrink anytime soon, as evidenced by the prolific and varied use of ATGWs in Iraq and Afghanistan.

> The top four recipients of ATGWs accounted for nearly 90 percent of these transfers.

MANPADS are notably less versatile than ATGWs and therefore their role on the battlefield has changed little over the past 40 years. Low-flying military aircraft remain the primary targets of most MANPADS, with some newer systems also reportedly capable of engaging unmanned aerial vehicles and cruise missiles.

Transfers of portable missiles, 2000–09

This section provides an overview of international transfers of portable missiles. Most of the analysis is based on data collected for estimating US dollar values for annual international transfers of ATWGs and MANPADS (as described above). This data includes records of imports of ATGWs by 25 countries and imports of MANPADS by 74 countries.[59] Transfers to and from nearly every region of the world are reflected in this data, though not all regions are represented evenly; nearly half of the countries included in the MANPADS dataset are European, for example. Similarly, data on some countries' imports stretches back more than a decade, while data on other countries only captures imports over a few years.

Anti-tank guided weapons. The 25-country dataset on imports of ATGWs highlights several key characteristics of the international trade in these weapons. Notable is the disproportionate significance of just a few states: nearly half of the countries studied reported zero imports of ATGWs, and transfers to the 13 importing countries were highly concentrated among the largest recipients. The largest importer, Slovakia, accounted for 44 per cent of all documented transfers, and the top four recipients—Slovakia, the UK, Turkey, and Norway—accounted for nearly 90 per cent of these transfers. The same is true among exporters, the top three of which accounted for approximately 90 per cent of all imports by the 25 countries studied.

The picture of the trade painted by these transfers may be partly distorted by limitations in the data sample used, in terms of both the number of countries and the length of time for which data was available. Nonetheless, the dataset does provide some insight into the global trade. First, the data suggests that the international trade in ATGWs may be somewhat idiosyncratic. A good example is the transfer of 10,498 Malyutka missiles from Hungary to Slovakia in 2009.[60] Nearly everything about this transfer is unusual. Slovakia has one of the smaller militaries in Europe and, while it is active in Afghanistan and Cyprus, only about 500 Slovak troops are participating in these operations. Similarly, Hungary is not a known producer of the Malyutka, and while it imported several thousand missiles in the 1960s and 1970s, the quantity transferred to Slovakia appears to outnumber total (documented) imports (SIPRI, n.d.b).

Also notable are the types and quantities of imported ATGWs, which are roughly consistent with prevalent assumptions about leading suppliers and their respective markets. Soviet/Russian, US, and French-designed ATGWs accounted for nearly 97 per cent of all imported missiles (see Table 1.11). These countries are established producers and exporters with large client bases and it is not surprising that their systems topped the list of imported ATGWs. The remaining imports consisted of 432 Israeli Spike missiles and launchers and 344 Chinese Red Arrow ATGWs. The comparatively low number of imported Spike missiles is not reflective of its growing market share,[61] and the quantity of these missiles is likely to increase as deliveries are made against new contracts.[62]

Finally, the dataset underscores the essential role of ATGWs in modern counterinsurgency operations and the recent impact of these operations on the international trade in ATGWs. Despite the high cost per unit, which can exceed USD 100,000 (*Jane's Defense Weekly*, 2010), ATGWs have been used extensively in recent counterinsurgency campaigns, including in Iraq and Afghanistan. In Afghanistan alone, an average of nearly 100 Javelin missiles are fired in combat operations each month (*Jane's Defense Weekly*, 2010). British troops account for the majority of this usage (70 missiles per month), whereas the much larger US force only fires an estimated 25 missiles per month. US troops rely more heavily on the larger and more powerful TOW anti-tank guided missile, roughly 2,000 of which were fired in Afghanistan in the first half of 2010 alone.[63] French troops in Afghanistan have used the Eryx and MILAN missiles (*Jane's International Defence Review*, 2010).

Data on British imports of Javelin missiles reflects the high rate of ATGW usage in Afghanistan and its implications for the international trade in these weapons. The UK was the second-largest importer of ATGWs in the 25-country dataset, importing more than 5,600 US Javelin missiles and launchers. Javelin imports have increased significantly during the UK's deployment in Afghanistan, jumping from 135 launch units from 2000 to 2004 to more than 5,466 missiles and launch units in the five-year period ending in 2009 (UKMoD, 2010). Similar patterns are apparent in US budget data, which indicates that funding for the procurement of TOW missiles has

Table 1.11 Imported ATGWs by weapon type, quantity, and percentage of total imports, 2000–09[64]		
Producer (type)	**Quantity imported**	**Percentage of total**
Russian Federation/ Warsaw Pact (Malyutka, Kornet, Spiral, Spandrel)	11,549	48%
France (Eryx, MILAN)	5,957	25%
United States (Javelin, TOW)	5,731	24%
Israel (Spike)	432	2%
China (Red Arrow)	344	1%
Unspecified/unclear	15	<1%
Total	**24,028**	**100%**

increased dramatically since 2005. The number of TOWs procured by the US Army jumped from 200 missiles in US fiscal years 2000 to 2004 to 17,160 missiles in the second half of the decade. While some of these missiles would have been purchased as part of the normal procurement cycle, the total number procured would probably have been much lower had the TOW not become so important a part of US operations in Afghanistan (US Army, 2006–2010).

The popularity of ATGWs among troops serving in Iraq and Afghanistan—and the consequent rise in procurement, including imports—is explained by several characteristics of the weapons that make them well suited for fighting in such environments. Many ATGWs are accurate well beyond the range of most enemy weapons,[65] which is useful in Afghanistan given the long lines of sight in the terrain where much of the fighting occurs and the great distances separating the combatants during engagements. Many modern ATGWs also require significantly less skill than sniper rifles and are less likely to cause collateral damage than artillery and close air support (*Jane's International Defence Review,* 2010). These attributes help to explain why the international trade in ATGWs is increasingly significant.

Man-portable air defence systems. The wars in Iraq and Afghanistan have sparked little additional demand for MANPADS by coalition forces, which is not surprising given that none of the opposing armed groups in either country have military aircraft. Some analysts believe that the absence of an enemy air threat in these theatres has resulted in reductions in funding for research, development, and procurement of MANPADS and other land-based air defences (*Jane's International Defence Review,* 2009). Data on international transfers of MANPADS is not sufficiently detailed to identify specific cases in which procurement or development was foregone or scaled back, let alone the reasons behind such decisions. What the data does show, however, is a particularly low level of MANPADS imports among most countries for which data is available.

> The data shows a remarkably low level of MANPADS imports.

Of the 74 countries studied, only 18 imported MANPADS or their primary components (missiles or launchers). Also notable is the small number of imported MANPADS, the combined total of which was just 4,935 units[66] (vs. imports of 24,028 ATGWs by only 12 importing countries).[67] Of this total, imports by a single country, Venezuela, account for 1,800 units, or 36 per cent, of total documented imports. When Venezuela's imports are excluded, total transfers drop to just 3,135 units for the remaining 73 countries.[68] This figure is remarkably low, especially since the dataset includes several countries with large and well-equipped armed forces.

What explains this low level of import activity among most countries? Is the international market for MANPADS in long-term decline? While spending on research, development, and procurement of MANPADS may have declined in some countries in recent years, there is little evidence of a significant, permanent contraction in the global MANPADS market. Foreign demand for certain systems, including the Swedish RBS-70 and the Stinger missile, remains strong, according to industry representatives.[69] Similarly, the UK government's procurement of more than 7,000 Starstreak missiles—approximately 20 per cent of which are to be used with shoulder-fired launchers—since 2000 is a clear signal that MANPADS remain a staple in British arsenals (UKMoD, 2010).

The development of new MANPADS by China and the Russian Federation is also indicative of continued interest in MANPADS. China has introduced several new MANPADS in recent years, the latest of which—the QW-19—was unveiled in November 2010 (Hewson, 2010). Similarly, the Russian Federation is reportedly developing an entirely new system called the Verba, or 'Willow'. Unlike other Russian MANPADS developed since the end of the cold war, the Verba 'will not be a further upgrade of [the] Igla family [. . .] it will have [a] new missile and launcher', according to an industry representative familiar with the programme (Pyadushkin, 2010, p. 2). Poland has also developed a new MANPADS, called the Piorun (Holdanowicz, 2009). It is highly unlikely that government and industry officials in these countries would invest the resources necessary to develop sophisticated new missile systems if domestic and international markets for MANPADS were disappearing.

The modest import figures for the majority of the countries studied are probably the result of data gaps and long procurement cycles. The trade in MANPADS is more transparent than the trade in many other weapons, but data on MANPADS transfers is incomplete. One of the most significant shortcomings is the exclusion of missile-only transfers from the reporting requirements for the UN Register. Member states are only required to submit data on transfers of complete MANPADS or launchers, not shipments consisting solely of missiles. Consequently, an unknown, but perhaps significant, number of imported missiles are not captured in the data submitted to the Register. Since the Register is the only public source of data on many countries' MANPADS transfers, the exclusion of data on missile-only transfers may help to explain the low import totals.

Also missing from the dataset is information on transfers to states that do not consistently report on their imports. As explained above, only data on countries that reported on their imports of MANPADS for a minimum number of years is included in the study.[70] While this data includes many of the world's largest arms producers and importers, there are some notable exceptions. One is Jordan, which appears to be one of the larger importers of MANPADS in recent years. According to Jordanian import data, the country imported 182 Igla launchers in 2007 alone. But Jordan did not meet the criteria for inclusion and thus is not part of the dataset.

For the abovementioned reasons, the international trade in MANPADS is almost certainly larger than import data used in this study would suggest, but it is unclear how much larger. The number of additional transfers could be

A mannequin holds a Chinese personal anti-aircraft missile at the fifth China Air Show in Zhuhai, China, November 2004.
© Eugene Hoshiko/AP Photo

minimal, as many states probably have not imported any MANPADS or their missiles in recent years. Six of the seven states for which comprehensive import data was obtained directly from governments (and whose data is therefore considered most complete and reliable) did not report any imports of MANPADS, and imports by the seventh state, Germany, consisted of just two launchers. This is not surprising given the long shelf lives of many MANPADS, low usage rates by many importers, and active or latent domestic production in three of the seven states. Without more complete data on MANPADS transfers, however, it is impossible to determine how closely current estimates of international transfers correspond with the actual global trade.

CONCLUSION

This chapter seeks to determine the annual value of international authorized transfers of light weapons and to gain a better understanding of this trade. It draws on data from dozens of sample countries to derive, through extrapolation, an estimated annual value of USD 1.1 billion for the international trade in light weapons. Transfers of portable missiles—MANPADS and ATGWs—represent the bulk of this total. Imports of ATGWs have dramatically increased among key countries in this sample over the past five years, partially as a result of high demand for ATGWs to equip troops in Iraq and Afghanistan. No comparable spike in MANPADS imports is apparent, in part because the target set against which MANPADS are employed is far more limited.

The trade in light weapons is extremely opaque. Information regarding the procurement practices of many countries remains difficult to obtain. Of the three reporting mechanisms on procurement and international transfers of small arms and light weapons reviewed for this study, only the UN Register contains national data that is sufficiently detailed and complete to be used in generating the estimated annual global value of transfers. The number of states that routinely report on imports of light weapons to the UN Register is still fairly limited, however. Data was thus collected through direct outreach to more than 80 governments, SIPRI's database on arms transfers, and field research. These additional sources yielded several hundred additional records, but much of the international trade in light weapons remains undocumented and poorly understood.

Despite these shortcomings, the light weapons trade is significantly more transparent today than it was a decade ago. The addition of small arms and light weapons transfers as an optional reporting category to the UN Register in 2003 has captured hundreds of transfers, many of which would have otherwise gone unreported. This expanding dataset sheds new light on the international trade in light weapons, particularly transfers to and from Europe. Contributions to the Register by more countries—and more specific and consistent reporting by countries that already contribute—would dramatically improve public data on light weapons transfers and, consequently, public understanding of this critically important trade. ▰

LIST OF ABBREVIATIONS

ATGW	Anti-tank guided weapon
EU	European Union
GPS	Global Positioning System

MANPADS	Man-portable air defence system
MILAN	*Missile d'infantrie léger antichar* (light infantry anti-tank missile)
ML	EU Common Military List (categories)
RPG	Rocket-propelled grenade (shoulder-launched anti-tank weapon)
SIPRI	Stockholm International Peace Research Institute
TOW	Tube-launched, optically tracked wire-guided weapon system
UN Comtrade	United Nations Commodity Trade Statistics Database
UN Register	United Nations Register of Conventional Arms

ANNEXES

Online annexes at <http://www.smallarmssurvey.org/de/publications/by-type/yearbook/small-arms-survey-2011.html>

Annexe 1.1 Annual authorized small and light weapons exports for major exporters (yearly exports of more than USD 10 million), 2008

This annexe provides UN Comtrade data on transfers of small arms and light weapons from major exporters in 2008.

Annexe 1.2 Annual authorized small and light weapons imports for major importers (yearly imports of more than USD 10 million), 2008

This annexe provides UN Comtrade data on transfers of small arms and light weapons from major importers in 2008.

Annexe 1.3 Methodology

This annexe provides a detailed summary of the methodology used in this chapter.

ENDNOTES

1   This value includes an estimated USD 100 million in undocumented firearms transfers.

2   See Small Arms Survey (2009, pp. 28–31; 2010, pp. 17–20) for the methods used to arrive at estimates for the authorized trade in firearms and ammunition for small arms and light weapons.

3   These figures do not include the trade in heavy machine guns, covered in the *Small Arms Survey 2009* estimate of the value of the international trade in firearms (Small Arms Survey, 2009, p. 29).

4   This chapter does not cover heavy machine guns or anti-materiel rifles, which are frequently categorized as light weapons. These items were included in the Small Arms Survey's assessment of the international trade in firearms, which is summarized in Small Arms Survey (2009).

5   The Panel's report defines light weapons as 'designed for use by several persons serving as a crew' (UNGA, 1997, para. 25). It includes the following categories of weapons: 'heavy machine-guns; hand-held under-barrel and mounted grenade launchers; portable anti-aircraft guns; portable anti-tank guns, recoilless rifles; portable launchers of anti-tank missile and rocket systems; portable launchers of anti-aircraft missile systems; [and] mortars of calibers of less than 100 mm' (para. 26). The Panel also specifies that light weapons are transportable 'by two or more people, a pack animal or a light vehicle' (para. 27(a)).

6   The term 'firearms' also covers the various 'small arms' that the Panel enumerates. See, for example, Small Arms Survey (2009, pp. 8–11).

7   Data on arms transfers often does not indicate whether imported guided missiles are intended for use with a man- or crew-portable launcher, a light vehicle, or larger platforms. If data does specify that the item in question is to be fired from a naval or aerial platform, or from tracked vehicles, it has been excluded. If the context provides no indication as to the platform from which the missile is to be fired, the data is included. In many cases, ambiguity regarding the intended use of the weapon means that some items configured for use on large platforms (that is, items that are not considered light weapons) may be included in data used for this study. This approach differs from that of the *Small Arms Survey 2008*, which covers all transfers of guided missiles that *could* be fired from a light vehicle or crew-portable launcher as a light weapon, regardless of the intended use (Small Arms Survey, 2008, pp. 8–11).

8   For the purposes of this chapter, 'accessories' are items that are not integral to the operation of portable versions of the system. An example would be the Giraffe radar system for the RBS-70 pedestal-mounted air defence system. The radar system enhances the performance of the RBS-70 but is not required for the basic operation of the system.

9   China is also assumed to have significant undocumented exports that would bring its total to more than USD 100 million. This study has revealed that Sweden's exports of light weapons hovered around USD 32 million in 2008, which may be sufficient to put its total exports over the USD 100 million threshold.

10   The overall figure for the increase in US exports between 2000 to 2009 has been adjusted for inflation. The other year-on-year comparisons have not.

11   For example, on 15 August 2010 a search of the European Defence Agency's Defence Contracts Opportunities and the European Union's Tenders Electronic Daily sub-categories for light weapons yielded only six completed contract award notices (all for 60 mm illuminating mortar bombs).

12   UN Comtrade was never intended to be a reporting mechanism for conventional weapons transactions. In addition to aggregating light weapons and other types of items in the same categories, the database fails to identify the weapon model or the type of transfer, such as permanent export or intra-military transfers to troops stationed abroad.

13   The reporting format for small arms and light weapons divides light weapons into the following categories: (1) heavy machine guns; (2) hand-held, under-barrel, and mounted grenade launchers; (3) portable anti-tank guns; (4) recoilless rifles; (5) portable anti-tank missile launchers and rocket systems; (6) mortars of calibres of less than 75 mm; and (7) other weapon types.

14   MANPADS were added as a sub-category (VIIb) of Category VII, 'Missiles and missile launchers'.

15   National reports on arms transfers submitted in 2009 cover deliveries that occurred in 2008.

16   Panama did not provide information on transfers of items in Categories 2 (hand-held, under-barrel, and mounted grenade launchers) and 3 (mortars of calibres of less than 75 mm).

17   There are exceptions to these yearly timeframes. For example, the Barometer takes account of reports submitted in 2010 that cover export activities in 2008 or earlier. In addition, states that failed to provide any information in 2010 are evaluated on their most recent submissions and reports, provided they were issued no earlier than 1 January 2009 (see the notes to Table 1.2). See Lazarevic (2010) for full details of the scoring methodology and a description of the changes to the Transparency Barometer scoring system since its introduction in 2004.

18   This includes information EU states have contributed to the *EU Annual Report* on military exports (CoEU, 2011).

19   For comparisons of 2010 rankings and scores, consult the online version of the 2010 Transparency Barometer (Small Arms Survey, n.d.a).

20   The UN Comtrade category in question is 930120, a mixed category, which includes conventional materiel such as torpedo tubes as well as some light weapons (such as rocket and grenade launchers).

21   In December 2008, the Council Common Position 2008/944CFSP replaced the European Union Code of Conduct on Arms Exports, in force since June 1998 (CoEU, 1998; 2008b, paras. 4, 8; 2011, p. 1).

22   Each Belgian region (Brussels, Flanders, and Wallonia) reports separately on its arms exports. The reports of all three regional parliaments are taken into account in determining the national score for Belgium.

23   Taiwan's score has been generated using the data it submits to UN Comtrade, as published by the International Trade Centre in its Trade Map database (ITC, n.d.).

24   The five states contributing to the regional report are: Albania, Bosnia and Herzegovina, the former Yugoslav Republic of Macedonia, Montenegro, and Serbia.

25   The SIPRI database contains data on transfers of MANPADS and ATGWs from a variety of sources, including official national and intergovernmental reports, press articles, and outputs of non-governmental organizations. For more information, see SIPRI (n.d.a).

26   Following a decision that documented imports would capture the greatest proportion of the global trade, import data (rather than export data) is used to build a global value of transfers. This decision was based on a lack of transfer data for a number of states believed to be significant exporters but insignificant importers (such as China and the Russian Federation).

27   Data is clustered around these years because of the expansion of reporting to the UN Register in 2003 (see above). While transfer records are available across a longer time period for some light weapon types in the SIPRI database, the completeness of transfers recorded in the database for a particular state in a particular year is not always easy to judge. As a result, this study relies heavily on data submitted to the UN Register. For more on the selected data sources and their implications for the conclusions, see Annexe 1.3.

28   'Years of available data' includes years for which the authors were able to ascertain that a particular state imported no light weapons, for example, if it submitted an explicit 'nil' report to the UN Register.

29   The threshold for inclusion in the MANPADS model is slightly more complicated than for the other models. Countries are included if there is a minimum of four consecutive years of data—or a total of five or more years of data—on their imports between the years 2003 and 2009.

30   Data on Venezuela's imports in 2009 is also included in the documented trade figure, but has not been used for extrapolation.

31   Data on armed force size and military spending is drawn from IISS (2009), SIPRI (2009), and CIA (n.d.). Data on conflict status was generated from UCDP/PRIO (n.d.) and data on production capacity from Jones and Ness (1997), Leff (2008), and O'Halloran and Foss (2008; 2009).

32   'Typical' light weapons import values of a particular group consist of averages of import values of sample countries belonging to that group.

33   Footage of the parade, including the SA-24 MANPADS, can be viewed at YouTube (2009).

34   For example, forum members of the Bangladesh Military Forces website have discussed military procurement by Bangladesh (BMF, n.d.).

35   The term 'anti-tank guided weapon' is a slight misnomer as it fails to convey the broad array of targets against which these weapons are increasingly used and the proliferation of new warheads for existing ATGWs that are optimized for targets other than heavily armoured vehicles (see 'Warhead/Target Matrix' in Gibson and Pengelley (2004)). Nonetheless, this chapter uses the term for two reasons. First, 'ATGW' and 'anti-tank guided missile' are still widely used and therefore the introduction of a new term could be confusing for some readers. Secondly, the term is not inaccurate, since many of the weapons in this category have retained their effectiveness against armoured vehicles even as their roles have expanded. See Gibson and Pengelley (2004); Foss (2009).

36   Descriptions of the weapons described in this chapter are drawn, in part, from Small Arms Survey (2008, pp. 20–27).

37   Mortar systems up to 120 mm are included. An assessment of their weight shows that the majority of identified systems can be transported by a light vehicle (the threshold for being defined as light weapons). The weight parameters used are set out in Small Arms Survey (2008, pp. 8–11).

38   Here, the term 'rockets' refers to unguided rocket-propelled projectiles.

39   Confidential author communication with industry personnel.

40   Confidential author communication with industry personnel.

41   Confidential author communication with industry personnel.

42   See Herron et al. (2010, p. 34).

43   See Small Arms Survey (2006, p. 9).

44   Service life estimate from UN (2008).

45   Confidential author interviews with government officials and industry personnel.

46   For more information about this type of low-level procurement, see Small Arms Survey (2006, p. 9).

47   See Foss, Gourley, and Tigner (2008); Kemp (2007); Williams (2008).

48   Service life estimate from UN (2008).

49   Single-shot, disposable rocket launchers also fire rocket-propelled grenades. Nevertheless, the terms 'RPG' and 'rocket-propelled grenade' are used here to describe projectiles fired from a reusable launcher which was developed in the Soviet Union and allied countries, as this follows the common use of the term.

50   According to the UN Panel of Experts, customs data submitted to UN Comtrade recorded a combined total of USD 22.9 million in arms transfers from North Korea between 2000 and 2009. North Korea does not submit national reports on any of its exports of weapons or military goods to the UN Register or to UN Comtrade (Lazarevic, 2010, pp. 101–02); see also the transparency section, above. The figure of USD 22.9 million is based on UN Comtrade information from importing countries only and cannot be regarded as comprehensive. The UN Panel of Experts estimates the total international trade in North Korean arms to be worth at least USD 100 million per year (UNSC, 2010, para. 65).

51   Use on certain platforms (including naval, air, or tracked vehicles) places these items outside this chapter's definition of light weapons.

52   TOW stands for tube-launched, optically tracked wire-guided weapon system.

53   For more information on ATGWs, see Small Arms Survey (2008, pp. 18–20).

54   China also reportedly employs a fourth type of guidance system, semi-active laser guidance, in a version of its QW-3 missile (O'Halloran and Foss, 2008, p. 11).

55   For more information on MANPADS, see Small Arms Survey (2008, pp. 16–18).

56   MILAN stands for *missile d'infantrie léger antichar* (light infantry anti-tank missile).

57   The US government estimates that more than one million MANPADS have been produced worldwide since the 1960s (GAO, 2004, p. 10).

58   Figure derived from MSIC (n.d.).

59   As explained above, three different models were used to calculate the global US dollar value estimate for transfers of MANPADS, ATGWs, and other light weapons. The ATGW model includes data on imports by Antigua and Barbuda, Argentina, Armenia, Bangladesh, Canada, Cyprus, the Czech Republic, Denmark, Germany, Indonesia, Liechtenstein, Lithuania, Mexico, Moldova, the Netherlands, Norway, Poland, Portugal, Romania, Slovakia, Slovenia, South Africa, Turkey, the United Kingdom, and the United States. The MANPADS model includes data on all of these countries plus Albania, Australia, Austria, Azerbaijan, Belarus, Belgium, Belize, Bolivia, Bosnia and Herzegovina, Brazil, Brunei Darussalam, Bulgaria, Burkina Faso, Chile, Costa Rica, Croatia, Cuba, France, Georgia, Guatemala, Hungary, Iceland, Israel, Japan, Kazakhstan, Kyrgyzstan, Latvia, Lebanon, the former Yugoslav Republic of Macedonia, Malaysia, the Maldives, Malta, Mauritius, Mongolia, Namibia, New Zealand, Pakistan, Paraguay, Senegal, Serbia, Singapore, South Korea, Suriname, Sweden, Switzerland, Tajikistan, Trinidad and Tobago, and Ukraine. Imports by Venezuela were included but were not used for purposes of extrapolation. For a detailed explanation of how these countries were selected, see Annexe 1.3.

60   Hungary's 2009 submission to the UN Register contained no data on the transfer of ATGWs to Slovakia.

61   Jane's Information Group recently observed that '[b]eneath TOW's class of heavy ATGW, two designs have really got the Western-leaning market covered: [Raytheon's] Javelin and Rafael's Spike' (*Jane's International Defence Review*, 2010).

62   Because of licensed production arrangements, deliveries of Spike missiles to many countries will only represent a fraction of the total number of missiles procured, even at their peak. A 2003 deal with Poland for 2,675 missiles and 264 launchers included licensed production of 70 per cent of the components for—and final assembly of—the missiles in Poland (Small Arms Survey, 2008, p. 20; Holdanowicz, 2007). Similarly, 60 per cent of the 2,600 Spike missiles procured by Spain in 2007 were to be produced locally, as were 70 per cent of the missiles ordered by Germany in 2009 (Ben-David, 2007; Wagstaff-Smith, 2009).

63   According to industry and government officials, most TOW missiles are fired from vehicle-mounted launchers. Since some are fired from pedestal-mounted launchers, they are still considered portable missiles and, as mentioned above, are included in the dataset when no information on intended platforms is provided.

64   Data on national ATGW imports by the 25 countries studied often does not cover the entire ten-year period.

65   The Javelin and TOW missiles have engagement ranges of 2,500 and 3,750 metres, respectively—well beyond the effective ranges of the AK series assault rifles, machine guns, sniper rifles, and other weapons commonly used by insurgents in Afghanistan (Jones and Ness, 2007, pp. 501, 509).

66   This total excludes transfers of missiles in which it is clear that they are intended for platforms outside the scope of this study, including aircraft, heavy land vehicles, and naval vessels.

67   Only countries that provided data on quantities of imported ATGWs are counted.

68   Furthermore, nearly 19 per cent of this total consists of MANPADS imported by the United States for purposes other than short-range air defence: 549 launchers from Bulgaria were 'demilitarized/destroyed' and 34 launchers from Ukraine were imported for research on countermeasures. When these transfers are excluded, total imports by the 73 countries drops to just 2,552 units.

69   Confidential author interviews with industry officials.

70   As noted above, the threshold for inclusion is four consecutive years of reporting, or reporting for five or more years between 2003 and 2009.

BIBLIOGRAPHY

Barrowclough, Anne. 2009. 'North Korean Arms Plane Linked to East European Arms Traffickers.' Times Online. 16 December.
    <http://www.timesonline.co.uk/tol/news/world/asia/article6956963.ece>

Ben-David, Alon. 2007. 'Spain Signs for Spike-LR System.' *Jane's Defence Weekly*. 17 January.

BMF (Bangladesh Military Forces). n.d. 'Forum Discussions.' Website. <bdmilitary.com>

Bromley, Mark. 2010. *Regional Report on Arms Exports in 2008*. Belgrade: South Eastern and Eastern Europe Clearinghouse for the Control of Small Arms and Light Weapons.

CIA (Central Intelligence Agency). n.d. *The World Factbook*. Accessed 30 August 2009. <https://www.cia.gov/library/publications/the-world-factbook/>

CoEU (Council of the European Union). 1998. European Union Code of Conduct on Arms Exports. 5 June.
    <http://www.consilium.europa.eu/uedocs/cmsUpload/08675r2en8.pdf>

——. 1999. *Annual Report in Conformity with Operative Provision 8 of the European Union Code of Conduct Arms Exports*. C 315/1. 3 November.
    <http://eur-lex.europa.eu/LexUriServ/LexUriServ.do?uri=OJ:C:1999:315:0001:0004:EN:PDF>

——. 2000. *Second Annual Report According to Operative Provision 8 of the European Union Code of Conduct on Arms Exports*. C 379/1. 29 December.
    <http://eur-lex.europa.eu/LexUriServ/LexUriServ.do?uri=OJ:C:2000:379:0001:0006:EN:PDF>

——. 2001. *Third Annual Report According to Operative Provision 8 of the European Union Code of Conduct on Arms Exports*. C 351/1. 11 December.
    <http://eur-lex.europa.eu/LexUriServ/LexUriServ.do?uri=OJ:C:2001:351:0001:0009:EN:PDF>

——. 2002. *Fourth Annual Report According to Operative Provision 8 of the European Union Code of Conduct on Arms Exports*. C 319/1. 19 December.
    <http://eur-lex.europa.eu/LexUriServ/LexUriServ.do?uri=OJ:C:2002:319:0001:0045:EN:PDF>

——. 2003. *Fifth Annual Report According to Operative Provision 8 of the European Union Code of Conduct on Arms Exports*. C 320/1. 31 December.
    <http://eur-lex.europa.eu/LexUriServ/LexUriServ.do?uri=OJ:C:2003:320:0001:0042:EN:PDF>

——. 2004. *Sixth Annual Report According to Operative Provision 8 of the European Union Code of Conduct on Arms Exports*. C 316/1. 21 December.
    <http://eur-lex.europa.eu/LexUriServ/LexUriServ.do?uri=OJ:C:2004:316:0001:0215:EN:PDF>

——. 2005. *Seventh Annual Report According to Operative Provision 8 of the European Union Code of Conduct on Arms Exports*. C 328/1. 23 December.
    <http://eur-lex.europa.eu/LexUriServ/LexUriServ.do?uri=OJ:C:2005:328:0001:0288:EN:PDF>

——. 2006. *Eighth Annual Report According to Operative Provision 8 of the European Union Code of Conduct on Arms Export*. C 250. 16 October.
    <http://eur-lex.europa.eu/LexUriServ/LexUriServ.do?uri=OJ:C:2003:320:0001:0042:EN:PDF>

——. 2007. *Ninth Annual Report According to Operative Provision 8 of the European Union Code of Conduct on Arms Exports*. C 253/1. 26 October.
    <http://eur-lex.europa.eu/LexUriServ/LexUriServ.do?uri=OJ:C:2003:320:0001:0042:EN:PDF>

——. 2008a. *Tenth Annual Report According to Operative Provision 8 of the European Union Code of Conduct on Arms Exports*. C 300/1. 22 October.
    <http://eur-lex.europa.eu/LexUriServ/LexUriServ.do?uri=OJ:C:2003:320:0001:0042:EN:PDF>

——. 2008b. Council Common Position 2008/944/CFSP of 8 December 2008 Defining Common Rules Governing Control of Exports of Military Technology and Equipment ('EU Common Position'). *Official Journal*, L 335/99. 13 December.
    <http://eur-lex.europa.eu/LexUriServ/LexUriServ.do?uri=OJ:L:2008:335:0099:0099:EN:PDF>

——. 2009. *Eleventh Annual Report According to Article 8(2) of Council Common Position 2008/944/CFSP Defining Common Rules Governing Control of Exports of Military Technology and Equipment*. 2009/C 265/01.
    <http://eur-lex.europa.eu/LexUriServ/LexUriServ.do?uri=OJ:C:2009:265:FULL:EN:PDF>

——. 2011. *Twelfth Annual Report According to Article 8(2) of Council Common Position 2008/944/CFSP Defining Common Rules Governing Control of Exports of Military Technology and Equipment.* 2011/C 9/01. 13 January.
    <http://eur-lex.europa.eu/LexUriServ/LexUriServ.do?uri=OJ:C:2011:009:FULL:EN:PDF>

Defense Industry Daily. 2005. 'Marines Fought the LAW, and the LAW Won.' 10 March.
    <http://www.defenseindustrydaily.com/marines-fought-the-law-and-the-law-won-0151/>

——. 2008. 'Israel: LAW on Order.' 14 September. <http://www.defenseindustrydaily.com/Israel-LAW-on-Order-05069/>

Felstead, Peter. 2010. 'Operational Use Buoys Carl Gustaf Development.' *International Defence Review.* 4 October.

Foss, Christopher. 2004. 'Anti-Armour Weapons: Making an Impact.' *Jane's Defence Weekly.* 9 June.

——. 2009. 'ATGWs Still Hit the Spot: Anti-tank Guided Weapons.' *Jane's Defence Weekly.* 7 September.

——, Scott Gourley, and Brooks Tigner. 2008. 'Firepower on the Move: AGLs Find Favour on the Battlefield.' *International Defence Review.* 21 June.

GAO (United States Government Accountability Office). 2004. *Nonproliferation: Further Improvements Needed in U.S. Efforts to Counter Threats from Man-Portable Air Defense Systems.* May.

Gibson, Neil and Rupert Pengelley. 2004. 'Warheads Widen Infantry Weapon Effects for Urban Warfare.' *Jane's International Defence Review.* 1 December.

Herron, Patrick, et al. 2010. 'Emerging from Obscurity: The Global Ammunition Trade.' In Small Arms Survey. *Small Arms Survey 2010: Gangs, Groups, and Guns.* Cambridge: Cambridge University Press, pp. 6–39.

Hewson, Robert. 2010. 'China's Air Defences Get Mobile and Multiple.' *Jane's Defence Weekly.* 3 December.

Holdanowicz, Grzegorz. 2007. 'ZM Mesko Completes Spike-LR Missile Tests.' *Jane's Missiles & Rockets.* 1 November.

——. 2009. 'Poland Displays New Grom Variant.' *Jane's Missiles & Rockets.* 6 October.

IISS (International Institute for Strategic Studies). 2009. *The Military Balance 2009.* London: Routledge.

Isbister, Roy and Nneka Okechukwu. 2010. *More than Box-ticking: Arms Transfer Reporting in the EU.* London: Saferworld. November.

ITC (International Trade Centre). n.d. 'Trade Map.' <http://www.trademap.org/stTermsConditions.aspx>

Jane's Defence Weekly. 2010. 'The Perfect Shot: Infantry Support Weapons.' 9 June.

Jane's International Defence Review. 2006. 'Nammo Introduces M72 LAW Projectiles.' 1 July.

——. 2009. 'Beneath the Radar: Land-based Air Defence Is Back in Business.' 10 November.

——. 2010. 'Changing Their Spots: Anti-tank Missiles Tackle Wider Target Sets.' 21 June.

Jones, Richard and Leland Ness. 2007. *Jane's Infantry Weapons 2007–2008.* Coulsdon, Surrey: Jane's Information Group.

Kemp, Ian. 2007. 'Automatic Grenade Launchers.' *Armada International.* 1 October.
    <http://www.thefreelibrary.com/_/print/PrintArticle.aspx?id=171018007>

Lazarevic, Jasna. 2010. *Transparency Counts: Assessing State Reporting on Small Arms Transfers.* Occasional Paper No. 25. Geneva: Small Arms Survey.

——. 2011. *Small Arms Trade Transparency Barometer 2011.* Unpublished background paper. Geneva: Small Arms Survey.

Leff, Jonah. 2008. *Global Production of Light Weapons, 1957–2007.* Unpublished background paper. Geneva: Small Arms Survey.

MSIC (Missile and Space Intelligence Center). n.d. 'MAN-Portable Air Defense Systems: Worldwide Threat.' Redstone Arsenal, AL: MSIC, Defense Intelligence Agency, United States Department of Defense.

O'Halloran, James and Christopher Foss. 2008. *Jane's Land-Based Air Defence, 2008–2009.* Coulsdon, Surrey: Jane's Information Group.

——. 2009. *Jane's Land-Based Air Defence, 2009–2010.* Coulsdon, Surrey: Jane's Information Group.

Ohmen, Christopher. 2005. 'Marine Corps Brings Back Old Weapon.' *Marine Corps News.* 9 February.
    <http://www.military.com/NewsContent/0,13319,usmc3_020905.00.html>

Pyadushkin, Maxim. 2010. *Report on the International Defence Exhibition of Land Forces 2010.* Unpublished background paper. Geneva: Small Arms Survey.

Richardson, Doug. 2005. 'MBDA Plans Its Next-generation Close-combat Weapons.' *Jane's Missiles & Rockets.* 1 July.

——. 2008. 'RPG Attack Halts Israeli Merkava.' *Jane's Missiles & Rockets.* 1 January.

Saferworld. n.d. 'Arms Transfer Reporting Database.' Accessed 15 December. <http://www.saferworld.org.uk/eureporting/>

Schroeder, Matt and Matt Buongiorno. 2009. *Missile Watch*, Vol. 2, No. 2.
    <http://www.fas.org/blog/ssp/2009/11/missile-watch-global-update-april-october-2009.php>

SEESAC (South East and Eastern Europe Clearinghouse for the Control of Small Arms and Light Weapons). n.d. 'Regional Reports: Online Database.' Accessed 20 January. <http://www.seesac.org/arms-exports-reports/regional-reports/1/>

SIPRI (Stockholm International Peace Research Institute). 2009. *SIPRI Yearbook 2009: Armaments, Disarmament, and International Security.* Oxford: Oxford University Press.

——. n.d.a. 'Arms Transfers Project Sources.' Accessed 1 September 2010. <http://www.sipri.org/databases/armstransfers/background/sources>

——. n.d.b. Arms Transfers Database. Accessed September 2010. <http://www.sipri.org/research/armaments/transfers/databases/armstransfers>

Small Arms Survey. 2006. *Small Arms Survey 2006: Unfinished Business.* Oxford: Oxford University Press.

——. 2008. *Small Arms Survey 2008: Risk and Resilience.* Cambridge: Cambridge University Press.

——. 2009. *Small Arms Survey 2009: Shadows of War.* Cambridge: Cambridge University Press, ch. 1.

——. n.d.a. 'Small Arms Trade Transparency Barometer 2010.' Geneva: Small Arms Survey.
    <http://www.smallarmssurvey.org/fileadmin/docs/Weapons_and_Markets/Tools/Transparency_barometer/SAS-Transparency-Barometer-2010.pdf>

——. n.d.b. 'The Transparency Barometer.' <http://www.smallarmssurvey.org/weapons-and-markets/tools/the-transparency-barometer.html>

UCDP/PRIO (Uppsala Conflict Data Program/Peace Research Institute, Oslo). n.d. 'UCDP/PRIO Armed Conflict Dataset V4-2009.'
<http://www.prio.no/CSCW/Datasets/Armed-Conflict/UCDP-PRIO/>

UKMoD (United Kingdom Ministry of Defence). 2010. 'Letter to Patrick Herron Regarding Release of Information under the Freedom of Information Act 2000.' 23 July.

UN (United Nations). 2008. *Manual on Policies and Procedures Concerning the Reimbursement and Control of Contingent-Owned Equipment of Troop/Police Contributors Participating in Peacekeeping Missions.* A/C.5/63/18. New York: UN.

—. n.d. *Register of Conventional Arms.* Accessed August 2010. <http://disarmament.un.org/un_register.nsf>

UN Comtrade (United Nations Commodity Trade Statistics Database). n.d. UN Comtrade Website. Accessed 19 January 2011.
<http://comtrade.un.org/db/default.aspx>

UNGA (United Nations General Assembly). 1997. *Report of the Panel of Governmental Experts on Small Arms.* A/52/298 of 27 August.
<http://www.un.org/Depts/ddar/Firstcom/SGreport52/a52298.html>

UNODA (United Nations Office of Disarmament Affairs). 2010a (undated). 'Objective Information on Military Matters and Transparency in Armaments: Fact Sheet.' <http://www.un.org/disarmament/convarms/Register/DOCS/2010-04-27MILEX&RegisterFactsheetFINAL.pdf>

—. 2010b. *Transparency in Armaments: Reporting to the United Nations Register on Conventional Weapons—Fact Sheet.* New York: UNODA.
<http://www.un.org/disarmament/convarms/Register/DOCS/2010-11-01_RegisterFactSheet.pdf>

UNSC (United Nations Security Council). 2006. Resolution 1718. S/RES/1718 of 14 October.
<http://daccess-dds-ny.un.org/doc/UNDOC/GEN/N06/572/07/PDF/N0657207.pdf?OpenElement>

—. 2009. Resolution 1874, adopted 12 June. S/RES/1874 of 12 June.
<http://daccess-dds-ny.un.org/doc/UNDOC/GEN/N09/368/49/PDF/N0936849.pdf?OpenElement>

—. 2010. *Report of the Panel of Experts Established Pursuant to Resolution 1874 (2009).* S/2010/571 of 5 November.
<http://www.un.org/ga/search/view_doc.asp?symbol=S/2010/571>

US Army (United States Department of the Army). 2006. *Missile Procurement, Army—Committee Staff Procurement Backup Book: Fiscal Year (FY) 2007 President's Budget.* February. <http://asafm.army.mil/Documents/OfficeDocuments/Budget/BudgetMaterials/fy07/pforms//missiles.pdf>

—. 2007. *Missile Procurement, Army—Committee Staff Procurement Backup Book: Fiscal Year (FY) 2008/2009 Budget Estimates.* February.
<http://asafm.army.mil/Documents/OfficeDocuments/Budget/BudgetMaterials/fy08-09/pforms//missiles.pdf>

—. 2008. *Missile Procurement, Army—Committee Staff Procurement Backup Book: Fiscal Year (FY) 2009 Budget Estimates.* February.
<http://asafm.army.mil/Documents/OfficeDocuments/Budget/BudgetMaterials/fy09/pforms//missiles.pdf>

—. 2009. *Missile Procurement, Army—Committee Staff Procurement Backup Book: Fiscal Year (FY) 2010 Budget Estimates.* May.
<http://asafm.army.mil/Documents/OfficeDocuments/Budget/BudgetMaterials/FY10/pforms//missiles.pdf>

—. 2010. *Missile Procurement, Army—Committee Staff Procurement Backup Book: Fiscal Year (FY) 2011 Budget Estimates.* February.
<http://asafm.army.mil/Documents/OfficeDocuments/Budget/BudgetMaterials/FY11/pforms//missiles.pdf>

Wagstaff-Smith, Keri. 2009. 'Eurospike Wins German Guided Missile Contract.' *Jane's Defense Industry.* 29 June.

Weir, William. 2005. *50 Weapons That Changed Warfare.* Franklin Lakes, NJ: Career Press.

Williams, Anthony. 2008. 'Coming out of Their Shell.' *Defence Management Journal,* No. 41.
<http://www.defencemanagement.com/article.asp?id=342&content_name=Land&article=10169>

YouTube. 2009. 'IGLA-S/SA-24 Grinch Manpads en Venezuela.' Uploaded 19 April.
<http://www.youtube.com/watch?v=u_XT0nzvIGQ>

ACKNOWLEDGEMENTS

Principal authors

Authorized transfers: Patrick Herron (Small Arms Survey), Nicholas Marsh (PRIO), Matt Schroeder (FAS)

Small Arms Trade Transparency Barometer: Jasna Lazarevic (Small Arms Survey)

Contributors

Authorized transfers: Janis Grzybowski, Benjamin King

Small Arms Trade Transparency Barometer: Thomas Jackson

This picture, brought back from a Mount Everest expedition, allegedly shows the footprints of the Abominable Snowman or Yeti, circa 1961. © Popperfoto/Getty Images

Fact or Fiction?

THE UN SMALL ARMS PROCESS

2

INTRODUCTION

In a continuing saga, 2010 saw the UN small arms process struggling with its future. On the positive side, the Fourth Biennial Meeting of States produced a substantive outcome document on the implementation of the UN Small Arms Programme of Action (PoA) (UNGA, 2001), including detailed text on process issues. The year also witnessed a successful start to UN negotiations on an Arms Trade Treaty (ATT).

Yet some of the failures were striking. Exceptionally low rates of reporting and information exchange in 2010 suggest that UN member states were largely indifferent to the International Tracing Instrument (ITI), five years after its adoption by the UN General Assembly (UNGA, 2005). A decade after the finalization of the PoA, UN member states continued to balk at any form of independent assessment of implementation. They may have something to hide; the Small Arms Survey's examination of the national points of contact (NPCs), which states are supposed to establish under the PoA, indicates that just over one-quarter of the UN membership has functioning NPCs in place.

The UN small arms calendar for 2011 features a new type of meeting for the PoA, an Open-ended Meeting of Governmental Experts (MGE). The UN General Assembly's broad ('omnibus') resolution on small arms now runs to 31 operative paragraphs. But do all these words—spoken and written—really amount to anything? This chapter reviews the latest developments in the UN small arms process and situates them against the broader canvas of UN discussions on this issue over the past decade. It identifies some of the achievements of that process, while also highlighting several causes for concern. Its principal conclusions include the following:

- UN member states have begun to translate the relatively vague language of the PoA into more specific prescriptions for action.
- The UN small arms process has become increasingly structured, with biennial meetings, expert meetings, and review conferences now scheduled for the PoA.
- A lack of commitment to the PoA and ITI on the part of many states is clear, underlined, in particular, by the continued inability to agree on any type of formal, independent evaluation of implementation.
- To their credit, UN member states have finally agreed to begin negotiations on legally binding principles designed to underpin the international transfer of conventional arms.
- The ATT negotiations may be facing the same resistance to effective international arms control that the PoA has encountered to date.

The first part of the chapter ('2010 Update') reviews the key developments in the UN small arms process for 2010. These concern both the PoA and the ITI, and also the first phase of the ATT negotiations. Although the future ATT is supposed to cover the full range of conventional arms, not just small arms and light weapons, an effective ATT would give more detailed expression to the PoA's general norms on transfer controls and, in this sense, related negotiations form part of the UN small arms process. The chapter's second part ('A bird's-eye view') steps back from the developments of 2010 to consider what progress has been made—on paper and on the ground—in translating the PoA's broad principles into concrete action. It highlights, in particular, the need for independent scrutiny of PoA implementation. The chapter concludes by taking stock of the UN small arms process a decade after the July 2001 adoption of the PoA.

2010 UPDATE

This section reviews three key developments in the UN small arms process that took shape in 2010: the outcome of the Fourth Biennial Meeting of States; the General Assembly's omnibus resolution on small arms; and the start of ATT negotiations. The chapter's evaluation of these results, both positive and negative, lays the groundwork for the more analytical section that follows ('A bird's-eye view').

The Fourth Biennial Meeting of States

The BMS4 discussion of border controls proved more controversial than expected.

The Fourth Biennial Meeting of States to Consider the Implementation of the UN Programme of Action (BMS4)[1] followed the most successful in the series, BMS3,[2] held in July 2008, and preceded important PoA meetings in 2011 and 2012 (see Box 2.1). Key elements of the preparatory process that had contributed to the BMS3 success were retained for BMS4. They included the early designation of the chair (Ambassador Pablo Macedo of Mexico), focusing the meeting on a limited number of themes, and the use of facilitators (called 'Friends of the Chair' in this case) to allow for more in-depth work on these topics, both during the preparatory phase and at the meeting itself.

BMS4 took up three themes from BMS3, namely international cooperation and assistance, consideration of the implementation of the International Tracing Instrument, and an 'other issues' session that, as at BMS3, allowed states to call attention to any topic they felt 'important for the implementation of the Programme of Action' (UNGA, 2010c, para. 51). The other meeting themes drawn from the PoA were border controls and 'follow-up', meaning the meetings and mechanisms designed to give practical effect to the PoA.

Based on consultations with delegations in the margins of the plenary sessions, the Friends of the Chair[3] transformed earlier discussion papers[4] into the raw components of the BMS4 outcome document. The final day of the meeting, 18 June, saw protracted discussions about whether to open the draft outcome that had been distributed the previous evening to line-by-line amendments. In the event, after a few changes were squeezed in,[5] the outcome document was adopted by consensus late in the afternoon, but with a note that reflected the displeasure of some delegations with the procedure (UNGA, 2010b, para. 23).[6]

Border controls

The discussion of border controls at BMS4 proved more controversial than many had expected given the clear interest all states have in ensuring effective control over their air, sea, and land borders. This was partly the result of the issue's incomplete incorporation into the PoA,[7] partly the consequence of discussions focusing on the question of responsibility for border control between neighbouring states. As is the case elsewhere in the BMS4 outcome, the border controls section is divided into a narrative part and a more prescriptive 'way forward' part. Overall, the section covers the key elements of effective border control:

Box 2.1 The UN small arms process: a thumbnail sketch

The UN Programme of Action does not ignore follow-up. Neither does it accord it pride of place. The instrument's 'follow-up' section envisages biennial meetings of states 'to consider the national, regional and global implementation of the Programme of Action' (UNGA, 2001, para. IV.1.b), as well as a conference, held in 2006 'to review progress made in [its] implementation' (para. IV.1.a). There is no formal mechanism for the monitoring and evaluation of implementation; the PoA simply provides for voluntary national reporting, with the UN Office for Disarmament Affairs disseminating whatever information states make available (para. II.33). Although not mentioned in the PoA, the UN General Assembly's annual 'omnibus' resolution on small arms, drafted by its First Committee (Disarmament and International Security), is another core element of Programme follow-up–helping to tie the various meetings together and set the general direction on substance and process.[8]

While they produced no substantive outcome, the first two biennial meetings, held in July 2003 and July 2005, helped keep international attention on the small arms issue in the period following the September 2001 terrorist attacks on the United States. Many states and other stakeholders hoped the 2006 Review Conference would do significantly better. Yet four weeks of meeting time[9] yielded nothing but a procedural report devoid of substance. Conflicting interpretations of the Review Conference mandate pitted states that sought a focus on implementation against those that aimed to bring new issues into the PoA framework.[10] Among the Review Conference casualties was agreement, of any kind, on further PoA meetings. Subsequent sessions of the General Assembly First Committee filled this gap, however, with decisions to convene a third biennial meeting by 2008 and a second review conference by 2012 (UNGA, 2006a, para. 4; 2008d, para. 14).

Preparations for BMS3 got under way early, with the designation of the chair in December 2007 and the identification of three topics for focused consideration in early 2008. The selection of specific meeting themes contrasted with the wide-ranging–and ultimately unproductive–discussion format favoured by BMS1 and BMS2.[11] BMS3 also distinguished itself from its predecessors by reaching agreement on a substantive outcome that fleshes out skeletal PoA text on international cooperation and assistance, brokering, and stockpile management and surplus disposal (UNGA, 2008b).[12] In the more divisive climate prevailing after the 2006 Review Conference, the outcome was adopted by vote.[13]

Consensus-based decision-making, which had characterized the initial phase of UN small arms deliberations, made its return to the process in 2009 with the General Assembly's unanimous adoption of that year's omnibus resolution on small arms. It included agreement on the convening of BMS4 in June 2010, an MGE by 2011, and a Second Review Conference by 2012 (UNGA, 2009e, paras. 6, 15-16).

- cooperation *between* states, including at the regional level and comprising such activities as the exchange of information and experiences; the harmonization of legislation, practices, and tools; and joint action (UNGA, 2010c, paras. 4, 7, 11–12, 15);

- cooperation and coordination among different authorities responsible for border control *within* a state (paras. 7, 11, 13, 16);

- capacity building (paras. 8, 13–14); and

- the need to take into account related activities, including trafficking in drugs and precious minerals, organized crime, and terrorism (paras. 5, 8).

At the Mariposa border crossing in Nogales, Arizona, US customs and border protection agents search for cash and weapons in the back of a vehicle headed for Mexico, May 2009. © Matt York/AP Photo

The 'way forward' section also encourages states to ensure that the prevention of small arms trafficking figures among their priorities for national border management (para. 13). As customs authorities around the world tend to place arms smuggling relatively low on their list of concerns (Wurche, 2010), implementation of this recommendation could make a considerable difference to small arms control efforts worldwide. Yet such advances also presuppose stronger linkages between the UN small arms process and other relevant actors, such as the World Customs Organization.

Although the border control section covers the bases, its language, even in the 'way forward' part, is entirely non-prescriptive. The section presents a series of options and best practices for border control; it does not require—or even push—states to do anything about their borders. Other 'way forward' sections of the BMS4 outcome, such as those in the BMS3 document, are also largely cast as recommendations ('States are encouraged to . . .'), rather than firm commitments. Yet, in contrast to this other text, the lack of specificity, even clarity,[14] in much of the border controls language leaves the impression that UN member states are entirely 'off the hook' in this area.

International cooperation and assistance

International cooperation and assistance was a theme at the Third Biennial Meeting of States, with language in the BMS3 outcome covering, in some detail, the matching of needs and resources, needs assessment by recipient states, and national reporting.[15] The challenge at BMS4 was to build upon—rather than simply repeat—this text.

The BMS4 section on international cooperation and assistance recaps many of the BMS3 priorities, including those mentioned above,[16] but in each case there are one or more new elements. In relation to the matching of needs and resources, the BMS4 outcome makes a link to regional efforts, citing a potential role for the UN regional disarmament centres (UNGA, 2010c, paras. 30j–k). It also references recent initiatives undertaken by the UN Institute for Disarmament Research (UNIDIR) and the UN Office for Disarmament Affairs (UNODA), including the new template for PoA and ITI reporting (paras. 22, 30i).[17]

Yet the key value added in the BMS4 international cooperation and assistance section lies elsewhere. First, the BMS4 section puts greater emphasis on cooperation, 'including joint or coordinated action', than does its BMS3 counterpart (UNGA, 2010c, para. 30a).[18] This could help rebalance discussions of PoA Section III,[19] since the issue of cooperation, despite its critical importance to Programme implementation,[20] has often been eclipsed at UN small arms meetings by its more appealing twin ('assistance'). Second, the BMS4 outcome, building on text agreed at BMS3 (UNGA, 2008b, outcome, para. 7c), highlights the need to assess the effectiveness of cooperation and assistance (UNGA, 2010c, para. 30e).[21] This emphasis on effectiveness is the logical follow-up to BMS3 discussions that focused on improved identification, communication, and matching of needs and resources. Strengthening the delivery of assistance is one challenge; ensuring that states measurably benefit from such assistance is another. Finally, the adoption of the BMS4 outcome by consensus, including its affirmation of the BMS3 text on international cooperation and assistance (UNGA, 2010c, para. 29), means that states that had balked at the latter[22] are now bound by both BMS3 and BMS4 outcomes.

Follow-up

While the BMS3 outcome addresses the question of PoA follow-up, much of the relevant text is tucked away in the document's 'other issues' section and thus lacks the normative strength found elsewhere, in particular in its 'way forward' sections.[23] The UN General Assembly's 2008 omnibus resolution incorporated several elements of the BMS3 outcome's 'forward-looking implementation agenda for the Programme of Action' (UNGA, 2008b, outcome, para. 29), especially in the area of national reporting,[24] but it is the BMS4 outcome that does most to elaborate upon the PoA's basic provisions for follow-up (see Box 2.1). This was the first PoA meeting with a dedicated session on follow-up.[25]

The BMS4 outcome elaborates upon the PoA's basic provisions for follow-up.

The BMS4 outcome sets out the following parameters for PoA follow-up:

- a six-year cycle for biennial meetings of states and review conferences (UNGA, 2010c, para. 44);[26]
- no agreement on whether MGEs should be part of the six-year cycle, following the first in May 2011, but an acknowledgement that they 'had a potential role to play in [the PoA] implementation architecture' if adequately prepared and 'action-oriented' (para. 32);
- early designation of the chair of a PoA meeting—if possible, one year in advance (paras. 34, 45);
- '[i]n order to ensure continuity among meetings,' collaboration between the chair of a PoA meeting and the chair and chair-designate of the previous and following meetings (para. 45);
- early development of PoA meeting agendas (paras. 34, 46);
- agreement to clearly define and distinguish the mandates of the different kinds of PoA meeting, but no indication as to what this means, in concrete terms, for any BMS, MGE, or review conference (paras. 34, 48);[27]
- agreement to link, and ensure the complementarity of, different PoA meetings and, to that end, a recommendation to include in national reports information on progress made in the implementation of measures set out in preceding PoA meeting outcomes (paras. 34, 39);
- use of a new reporting template, developed by UNODA, to increase comparability among national reports, facilitate the matching of needs and resources, and simplify the provision of updated information on implementation (paras. 35, 41);
- shifting the reporting schedule to a biennial basis, timed to coincide with BMSs and review conferences, with a view to increasing the number and quality of reports (paras. 35, 38);

- analysis of national reporting, including a comprehensive, ten-year assessment of progress made in the implementation of the Programme of Action as an input for the 2012 Review Conference (paras. 36, 40); and
- possible establishment of a voluntary sponsorship fund benefitting states that would otherwise be unable to participate in PoA meetings (paras. 37, 43).

The BMS4 discussion of follow-up included consideration of preparations for the May 2011 MGE and the 2012 Review Conference. As noted above, there was some scepticism regarding the formal incorporation of MGEs into the PoA meeting cycle. States therefore recommended that the 2012 Review Conference address the question of additional MGEs (UNGA, 2010c, para. 44). In relation to the May 2011 meeting, previously scheduled by the General Assembly, states emphasized the need to limit the number of issues under discussion, presumably in order to foster a 'pragmatic, action-oriented' exchange (paras. 32, 47). BMS4 did not have much to say about the 2012 Review Conference, although states did recommend that it 'assess and, as necessary, strengthen the follow-up mechanism of the Programme of Action' (para. 49). BMS4 was not to be the last word on PoA follow-up.

The International Tracing Instrument

The International Tracing Instrument, adopted by the UN General Assembly in December 2005, and since then applicable to all UN member states, provides for biennial meetings and reports on the implementation of the

A Colombian police officer holds a weapon whose markings indicate it belongs to the armed forces of Venezuela, September 2008. The gun was seized from a criminal gang that allegedly works for drug traffickers in Medellín, Colombia. © Luis Benavides/AP Photo

Instrument (UNGA, 2005, paras. 36–37). The first such meeting was held within the framework of BMS3 in July 2008. BMS4 was therefore the second time UN member states could take stock of progress made in implementing the ITI's weapons marking, record-keeping, and tracing provisions.

The outcome on the ITI, annexed to the BMS4 report, is not a carbon copy of the BMS3 text, but neither does it offer much value added. Most importantly, it encourages states to use the UNODA reporting template—redesigned in 2010 with both the PoA and ITI in mind—emphasizing its utility to the comparison of implementation reports and the evaluation of ITI effectiveness (UNGA, 2010d, para. 10d).[28] In line with evolving practice for PoA reporting, the 2010 outcome also encourages states 'to submit their reports well in advance of biennial meetings and review conferences' (para. 10d). In addition, the outcome recognizes, in veiled terms, the value of converting paper-based records into electronic form (para. 4b); encourages the development of legislation providing for the mutual exchange of information and intelligence, which is useful for tracing (para. 10g); and underlines, somewhat more forcefully than its BMS3 predecessor, 'the important role that civil society plays in promoting the full implementation of the International Instrument' (para. 10i).[29]

Fundamentally, however, the 2010 ITI outcome is enfeebled by omission. Four-and-a-half years after the Instrument's adoption by—and simultaneous application to—UN member states, the draft outcome's end-of-year deadline for the submission of point of contact information (UNGA, 2010a, para. 10c) proved too much for some states and was removed from the final version. Nor is there any commitment to exchange information on national marking practices.[30] All this despite the firm commitment states have made under the ITI to exchange both types of information 'as soon as possible after the adoption' of the Instrument (UNGA, 2005, para. 31). In this light, text in the 2010 ITI outcome encouraging 'States that had not yet done so [. . .] to exert every effort to designate national points of contact' (UNGA, 2010d, para. 10c) falls rather flat.

> Many UN member states appeared indifferent to the ITI in 2010.

Also noteworthy by its absence in the 2010 ITI outcome is the 2008 outcome's mention of the importance of import marking to weapons tracing (UNGA, 2008b, ITI outcome, para. 3a).[31] Last, but not least, in contrast to the 2008 ITI outcome and other BMS3 and BMS4 sections, the 2010 text replaces the word 'measures' with the softer term 'understandings' in the paragraph introducing the 'way forward' text (UNGA, 2010d, para. 10 chapeau).[32] The reference in paragraph 5 of the 2010 outcome to an analysis of ITI reporting, conducted by the Small Arms Survey in 2010, hints at the fact that there may, in fact, be even less to the 2010 text than meets the eye.[33]

In contrast to the PoA's voluntary reporting, states have undertaken, without qualification, to report on their implementation of the ITI every two years (UNGA, 2001, para. II.33; 2005, para. 36). In 2008, the first year of ITI reporting, 62 of 192 UN member states reported on their implementation of the Instrument (Cattaneo and Parker, 2008, p. 97). Not a huge number, yet one month before BMS4, during the second round of ITI reporting, this number had dropped further, to 43 (Parker, 2010, p. 52).[34] The low ITI reporting rates possibly stem from the use of an older reporting template, developed before the adoption of the ITI in late 2005.[35] If so, the template finalized by UNODA in 2010, integrating ITI and PoA provisions on marking, record-keeping, and tracing (UNODA, 2010), should boost ITI reporting in future. Nonetheless, the fact that many of the states that submit PoA reports have not yet reported on the ITI[36] may reflect a deeper problem, namely a lack of familiarity with, or even knowledge of, the ITI— or perhaps a simple lack of interest in the Instrument.

Whatever the specific reasons, it appears that many UN member states were indifferent to the International Tracing Instrument in 2010. There are no comprehensive studies of ITI implementation,[37] but there are other indicators of the extent to which states are taking their ITI commitments seriously. Reporting is one, communication of

A police officer holds a seized gun in Tijuana, Mexico, after a presentation to the media, August 2009. © Guillermo Arias/AP Photo

national point of contact information another. Here, too, the commitment is firm; the ITI *requires* states to 'designate one or more national points of contact' and, as noted above, communicate this information to the UN (UNGA, 2005, paras. 25, 31a). The weak 2010 outcome on the ITI is mirrored in the failure of many UN members to take such basic steps for implementation. As of mid-January 2011, the UNODA website listed ITI-specific point of contact information for only 18 of 192 UN member states—just under ten per cent of the UN membership (UNODA, 2011).[38]

The 2010 omnibus resolution

The 2010 session of the UN General Assembly's First Committee (Disarmament and International Security) provided an opportunity to consolidate the progress made at BMS4 and prepare for important PoA meetings in 2011 (the MGE) and 2012 (the Second Review Conference). As it turned out, the Committee's general ('omnibus') resolution on small arms accomplished something in each area, although it also yielded a setback in the area of PoA follow-up.

Elements from BMS4

The small arms omnibus resolution duly endorses the BMS4 report, including the outcome, and 'encourages all States to implement, as appropriate, the measures highlighted' in the 'way forward' sections of the report. It also contains, in a footnote, a reference to paragraph 23 of the main BMS4 report,[39] which, as noted earlier, reflects the dissatisfaction of some delegations with the lack of line-by-line discussion on the draft outcome (UNGA, 2010g, para. 4).

The omnibus resolution includes several paragraphs on international cooperation and assistance, some of which emphasize, in line with the BMS4 outcome, the need to ensure their effectiveness (UNGA, 2010g, paras. 15, 26). Other First Committee resolutions focus to a greater extent on the question of assistance.[40] Following up one of the key themes of the BMS3 outcome, the 2010 resolution on 'practical disarmament measures' emphasizes the contributions of the Group of Interested States[41] and UNODA (its Programme of Action Implementation Support System) to the matching of needs and resources for PoA implementation (UNGA, 2010h, paras. 5–6).

The omnibus resolution includes a vague reference to the border controls question (UNGA, 2010g, para. 16). The resolution also mentions the International Tracing Instrument, but without borrowing from the 2010 ITI outcome;[42] it makes greater use of the BMS4 outcome on follow-up. National reporting on the PoA, including use of the ODA reporting template, is the subject of three paragraphs (paras. 11–13), while another cites the BMS4 proposal to create a voluntary sponsorship fund for enhanced PoA meeting participation (para. 21). There are several omissions from the BMS4 follow-up text,[43] although, as with the lack of specific reference to many other parts of the BMS4 text, they are not important given the resolution's endorsement of the BMS4 report (and outcome) as a whole (para. 4).

The real problem with the omnibus resolution lies in its handling of the BMS4 text on the assessment of progress in PoA implementation. The issue is raised, in general terms, in paragraph 36 of the BMS4 outcome and given more concrete expression in paragraph 40, with its recognition of:

> *the need for a comprehensive assessment of progress in the implementation of the Programme of Action, 10 years following its adoption, as an input for the 2012 Review Conference* (UNGA, 2010c).

The omnibus resolution transforms this provision into a somewhat confusing exercise in self-assessment. The General Assembly:

> Invites *Member States to communicate to the Secretary-General their views on the progress made on the implementation of the Programme of Action, ten years following its adoption, and requests the Secretary-General to present a report containing that information as an input to the 2012 review conference* (UNGA, 2010g, para. 29).

Strictly speaking, states are invited not merely to report on the progress they have made in implementing the PoA, but on progress made in general ('the progress') since the instrument's adoption in July 2001. It is highly unlikely that many (any?) states will want to report on a lack of progress by other countries. In practice, it appears that states are being asked to indicate the progress they have made in implementing the PoA over the full period of its existence in the national reports they submit in advance of the 2012 Review Conference. Critical, independent analysis of PoA implementation is not yet on the UN agenda, it seems.

Future meetings

The 2010 omnibus resolution also devotes significant space to the post-BMS4 schedule of PoA meetings. It sets the dates for the MGE (9–13 May 2011) and, echoing the BMS4 outcome, underlines the importance of 'pragmatic, action-oriented [. . .] agendas for the meeting' (UNGA, 2010g, para. 8).[44] In essence, the MGE should involve an exchange of information and experience among small arms experts, not a political debate of the kind that has featured in many BMSs, and that dominated the 2006 Review Conference. To this end, the omnibus resolution '*[f]urther encourages* States to contribute relevant national expertise' to the MGE (para. 9).

Like the BMS4 outcome, the resolution leaves open the question of additional MGEs, beyond that scheduled for May 2011, although its reference to 'a further [. . .] meeting', in place of the BMS4 outcome's use of the plural ('meetings'), suggests additional scepticism regarding future MGEs (UNGA, 2010g, para. 20; 2010c, para. 44).[45] The omnibus resolution also advances planning for the 2012 Review Conference, scheduling a one-week preparatory committee session for early 2012 (UNGA, 2010g, para. 18) and calling for the 'designation of *one* Chair for both the preparatory committee and the review conference' by May 2011 (the time of the MGE) (para. 19; emphasis added).[46]

The ATT strand

If agreed and reasonably strong, an arms trade treaty would complement the PoA. The PoA addresses small arms transfer control in three paragraphs,[47] but the language is open-ended. While the commitments UN member states make on transfer controls in the PoA tend to be firm,[48] the nature of the required action is not usually specified in any detail.[49] Assuming it is minimally effective, an ATT would provide much clearer guidance on arms export licensing and post-shipment follow-up. It is still too early to know if such an instrument will see the light of day, but in July 2010, after years of hesitation, the UN membership finally began negotiations on an ATT.

The ATT project has its roots in civil society efforts, dating back to the mid-1990s, to promote more responsible arms export practices. The concept found traction with a number of governments in the early 2000s and with the bulk of the UN membership, in December 2006, when the General Assembly adopted its first ATT resolution (UNGA, 2006b).[50] A subsequent report by a UN Group of Governmental Experts (GGE), although short on substance, paved the way for the creation of an Open-ended Working Group (OEWG) (UNGA, 2008c).[51]

Unlike the GGE, the OEWG was open to participation by all UN member states. According to the mandate it received from the General Assembly, the OEWG, in 2009, was to:

> *further consider those elements in the report of the Group of Governmental Experts where consensus could be developed for their inclusion in an eventual legally binding treaty on the import, export and transfer of conventional arms* (UNGA, 2008e, para. 5).

ATT PrepCom Chair Roberto García Moritán of Argentina briefs the media on the coming ATT meeting, New York, July 2010. © Devra Berkowitz/UN Photo

The OEWG held two substantive sessions in 2009[52] and produced a report in which it indicated that it had fulfilled its mandate, stressing, in particular, the 'inclusive' nature of its discussions (UNGA, 2009c, para. 24).[53]

In many ways, the 2009 OEWG sessions were a repeat of the 2008 GGE exercise, this time involving the entire UN membership. Grouped under four substantive headings,[54] the discussions mostly involved a general exchange of views on the principal issues relating to an ATT; only exceptionally did they approach the level of detail needed for any future treaty.[55] At the OEWG's second session, in July 2009, several delegations reiterated previously voiced doubts about the 'feasibility' of an ATT and the need for a legally binding instrument. Only one state was openly sceptical at the end of the session, however. Many of the states that gave closing statements called for an accelerated pace of work—in essence, a shift towards a negotiating mandate or something close to it.[56]

The 2010 PrepCom successfully highlighted many critical issues for the ATT negotiations.

In the event, the General Assembly opted for formal negotiations in its December 2009 ATT resolution. It converted the OEWG sessions that had been planned for 2010 and 2011 into a Preparatory Committee (PrepCom), designed to pave the way for a 'United Nations Conference on the Arms Trade Treaty' that it scheduled for 'four consecutive weeks in 2012' (UNGA, 2009d, paras. 4, 6). Reflecting a debate among UN member states on the rules of Conference decision-making, the General Assembly further specified that the Conference was to 'be undertaken [. . .] on the basis of consensus' (para. 5). What this means, exactly, may only become clear at the time of the Conference, in 2012.[57]

In 2010, there was a need to shift the discussions, however modestly, towards focused consideration of the elements of a treaty text. The PrepCom held its first session on 12–23 July 2010 at UN headquarters in New York. By the middle of the first week, the meeting chair, Ambassador Roberto García Moritán of Argentina,[58] offered an initial draft of the future treaty's structure and preamble (Argentina, 2010a; 2010b). The following meetings, convened by several Friends of the Chair,[59] were devoted to informal (that is, closed)[60] consideration of possible treaty elements in three substantive areas:

- scope;
- common standards or criteria for the import, export, and transfer of conventional arms; and
- implementation and application.

On 22 July, the second-to-last day of the 2010 session, the chair issued an expanded, consolidated version of his earlier draft texts. This *Draft Paper* contains a detailed list of the topics that states had proposed for inclusion in the ATT ('Elements' section), along with draft text for the treaty preamble ('Principles' section) and for a possible 'Goals and Objectives' section (Argentina, 2010c). At the same time, the three Friends issued papers summarizing the discussions in their sessions (Australia, 2010; Egypt, 2010; Trinidad and Tobago, 2010).

The first PrepCom session was generally welcomed as a successful start to the ATT negotiations.[61] After two years of unspecific, often repetitive discussions in the GGE and OEWG, the PrepCom's focus on the nuts and bolts of treaty-making was heralded, in some quarters, as an indication that the ATT had finally gained universal acceptance.[62] Yet while the 2010 PrepCom was certainly successful in highlighting many of the critical issues for the negotiations, there was no convergence of views—let alone specific agreement—on any of these issues.[63]

From statements made throughout the first PrepCom session, it is clear that several countries remain opposed to any ATT that would constrain the decision-making of exporting states or the ability of importing states to secure continued supplies of conventional arms. The tactics have changed—there was minimal outright opposition to an ATT at the 2010 session—but the objective has not. Despite the fact that the papers produced by the Friends are simply compilations of the national positions and proposals made at the session, several countries took issue with them during their closing statements. In some cases, this appeared to signal an intention to hinder progress at the next stage of the negotiations.[64]

The disagreements go beyond questions of principle (such as constraints on national decision-making)[65] to many matters of critical substance (such as the UN Register of Conventional Arms as a basis for treaty scope). Even points that seemed to have been generally accepted at earlier stages of the process—such as the inclusion of small arms and light weapons in any ATT—were called into question during the 2010 session. The incorporation of ammunition in the ATT remains hotly contested, despite its centrality to arms control.[66]

In short, the first PrepCom session was successful in nudging the process, however gently, towards formal treaty negotiations; yet the real work is still to come. As of January 2011, despite attempts to reconcile differences following the 2010 session,[67] there were no signs of convergence on key aspects of the treaty, including the desirability of an effective ATT. The negotiations promise to be exceptionally complex, but there is relatively little time. The July 2010 session accounted for half of the PrepCom's allotted four weeks.[68] The four-week 2012 Conference is unlikely to produce anything of value unless existing gaps between countries are substantially narrowed beforehand.

NGO activists hold placards during a peace march in support of ATT negotiations, New Delhi, India, February 2010.
© Manpreet Romana/AFP Photo

A BIRD'S-EYE VIEW

In July 2011, the UN Programme of Action celebrates its tenth anniversary. What is the significance of the developments described in the last section to the longer-term UN small arms process? This section seeks to place the events of 2010 in the broader context of PoA-related activity over the past decade. It first examines the process of translating the PoA's general norms into more specific prescriptions for action, then considers the extent to which the PoA (and ITI) are in fact spurring any 'action'.

Unpacking the PoA

While the PoA provides a general framework for the regulation of small arms and light weapons, it does not substitute for detailed regulation. More specialized instruments, such as the ITI or a future ATT, fulfil this function. Many PoA commitments are vague. For example, its use of the words 'adequate' or 'effective' to condition prescribed tasks

The PoA does not substitute for detailed regulation.

leaves open the question of what—exactly—states must do in order to meet the relevant standards. There is, in short, a need to 'unpack' the PoA, to provide states with the operational guidance that will help them translate the instrument into concrete action.

As of January 2011, the ATT remained more aspiration than reality; yet several recent initiatives, led by UN member states, have yielded detailed guidance for PoA implementation in a range of areas. In chronological order, they are:

- the International Tracing Instrument (UNGA, 2005);
- the report of the 2007 GGE on brokering (UNGA, 2007);
- the report of the 2008 GGE on surplus ammunition (UNGA, 2008a);
- the BMS3 outcome (UNGA, 2008b); and
- the BMS4 outcome (UNGA, 2010c).

These documents cover the following areas:

- marking, record-keeping, and tracing (ITI, BMS3 and BMS4 outcomes);
- brokering controls (2007 brokering report, BMS3 outcome);
- stockpile management and surplus disposal for weapons and ammunition (BMS3 outcome, 2008 ammunition report);
- border controls (BMS4 outcome); and
- international cooperation, assistance, and national capacity building (BMS3 and BMS4 outcomes).

The chapter briefly examines each of these areas.

Marking, record-keeping, and tracing

The issue of tracing was fast-tracked in the PoA[69] as most states felt the subject required more detailed treatment than what the Programme offered.[70] A GGE, convened in 2002–03,[71] paved the way for an OEWG, which finalized the text of the International Tracing Instrument in June 2005 (UNGA, 2005). Since its adoption by the UN General Assembly, in December 2005, the ITI, a politically binding instrument, applies to all UN member states.

The ITI consolidates and further develops key standards on weapons marking, specifying, with varying levels of detail, the content, placement, and characteristics of marks at different points in the small arms life cycle (UNGA, 2005, sec. III). The section on record-keeping commits states to 'accurate and comprehensive' record-keeping—framed in general terms to take account of constitutional differences between states. It also prescribes minimum time limits for the conservation of weapons records (paras. 11–12). In its section V, the ITI establishes detailed modalities for tracing cooperation that have no parallel in other international agreements. In contrast to the PoA, it also includes a definition of small arms and light weapons (sec. II)—but not ammunition, which is excluded from the Instrument. The BMS3 outcome usefully emphasizes some of the initial steps states need to take in implementing the ITI, but it does not really elaborate upon its provisions. The BMS4 outcome, as noted above, is even more timid in its treatment of the ITI and ITI implementation.

Brokering controls

Brokering was the second issue, along with tracing, which was singled out for priority attention in the PoA.[72] The main PoA paragraph on brokering mentions several of the elements needed for effective regulation—such as registration of brokers, licensing of brokering transactions, and penalties—but in relatively vague, non-prescriptive language (UNGA, 2001, para. II.14). The August 2007 report of the brokering GGE goes further. It provides a definition of brokering, presents the 'recurring elements' of existing national regulatory systems, and offers a series of recommendations designed to strengthen national, regional, and global efforts to tackle illicit brokering (UNGA, 2007, secs. I.B, III, V).

Suspected Russian arms dealer Viktor Bout is escorted by US Drug Enforcement Administration officers after arriving at a New York State airport, November 2010. After months of legal wrangling, he was extradited from Thailand to face US terrorism charges.
© US Department of Justice/Reuters

The BMS3 outcome reiterated several key points from the 2007 brokering report, such as the need for 'a comprehensive approach' and 'the crucial importance of international cooperation' to these efforts (UNGA, 2008b, outcome, paras. 11, 16c). Above all, it 'acknowledged the importance' of the report's recommendations and other findings (paras. 11, 16b), a conclusion reinforced by the General Assembly's repeated call for states to implement the GGE recommendations.[73]

Stockpile management and surplus disposal

While the PoA's provisions on stockpile management and surplus disposal are more detailed than most, articulating key principles and elements for regulation in this area (UNGA, 2001, paras. II.17–19), they nevertheless leave many questions unanswered. When are stockpile management standards and procedures 'adequate' (UNGA, 2001, para. II.17)?

How do states go about clearly identifying stocks that are surplus to national requirements (para. II.18)? What do they need to consider when responsibly disposing of their surpluses (para. II.18)? What are the resource implications of such measures? The BMS3 outcome document provides useful answers to each of these questions,[74] while highlighting the close relationship between the different sectors, especially surplus identification and effective stockpile management.[75]

Ammunition, although essential to effective regulation, has received scant attention in UN small arms agreements. Except when referring to other documents, the PoA makes no mention of ammunition,[76] and this category was deliberately excluded from the ITI—a development that led directly to the establishment of a GGE on surplus ammunition stockpiles.[77] The Group's July 2008 report situates the problem of surplus ammunition stockpiles—for all conventional weapons, not just small arms—within the broader framework of stockpile management, addressing such issues as marking, accounting, public safety, stockpile security, and disposal and destruction (UNGA, 2008a).[78] Many of the report's recommendations apply to the management of arms as well as ammunition. As of January 2011, a set of 'technical guidelines', designed to complement the 2008 report, were being developed by UNODA.[79]

Border controls

Despite its weaknesses, noted earlier in the chapter, the BMS4 outcome on border controls usefully develops the limited text found in the PoA. Among the basic elements of border control it lists are cooperation and coordination *between* and *within* states, capacity building, and—not least—the need to integrate the prevention of small arms trafficking into national border management strategies. For the time being, we are left with these general principles. A proposal by some UN member states to discuss the issue of border controls in greater detail, at the 2011 MGE, was abandoned in the face of opposition by other states.

International cooperation, assistance, and national capacity building

Early consideration of PoA implementation, specifically at the First and Second BMSs, underlined the importance of cooperation, assistance, and national capacity building for this purpose, but in fairly general, non-specific terms; no attempt was made to build upon the provisions of the PoA.[80] At BMS3, and again at BMS4, this discussion became more focused—as reflected in the outcome documents of the two meetings. At BMS3, states considered practical means of improving the identification and communication of needs, along with the matching of needs with resources (UNGA, 2008b, outcome, sec. I).[81] As noted above, the BMS4 outcome, while echoing these points, stresses the

---

Box 2.2 International Small Arms Control Standards

In parallel with the recent efforts at normative development involving UN member states, the UN system's Coordinating Action on Small Arms—or CASA—mechanism has undertaken its own standard-setting initiative. The creation of a comprehensive set of International Small Arms Control Standards (ISACS) is modelled on the UN system's prior development of technical standards for mine action and the disarmament, demobilization, and reintegration of ex-combatants. The ISACS project aims:

> to develop a set of internationally accepted and validated technical standards that provide comprehensive guidance to practitioners and policymakers on legal, policy and operational issues surrounding small arms control (UNGA, 2010e, para. 35).

The ISACS modules are drafted by technical consultants, with inputs from a wide range of experts and practitioners. As of January 2011, 26 modules were in development, with the adoption and launch of final versions planned for late 2011.

Sources: UNCASA (2010); UNGA (2010e, para. 35)[82]

importance of cooperation—not just assistance—in implementing the PoA. It also highlights the need to ensure the effectiveness of cooperation and assistance (UNGA, 2010c, sec. II).

The view at ground level

The development of detailed guidance for PoA implementation should, in theory, help ensure that the commitments states have made in the PoA find practical expression 'at ground level'. But that raises the question of what is known about implementation. A 2010 UN report presents a wide range of activity on small arms, specifically in Africa, the Americas, and Europe (UNGA, 2010e).[83] In some cases this 'activity' involves meetings and discussions.[84] In others, it is more concrete—such as strengthening controls or sharing information for operational purposes.[85] Not all of the news is good, however.[86] Moreover, the report describes many activities that are proposed or planned, not under way or completed.

Most of the information on the implementation of the PoA and ITI comes from national reporting, which is rarely self-critical. Despite encouragement over recent years to share information on 'implementation challenges and opportunities',[87] states are providing relatively little information on the difficulties they encounter in giving practical effect to the two instruments—except to note, usually in general terms, a lack of capacity or need for assistance.[88]

Independent evaluations of PoA implementation have consistently pointed to serious weaknesses.[89] Most recently, the Small Arms Survey sought to determine whether information states had provided to UNODA on their national points of contact for the PoA was accurate and, further, whether the NPCs were operational (Parker, 2010, pp. 26–33).[90] The results are not encouraging. The existence and identity of the NPC could be confirmed in only 52 cases—just over one-quarter of the UN membership (p. 32).[91] The establishment of an NPC is a relatively simple task[92]—the designation of a government official to serve as a liaison on PoA-related matters and the communication of their contact information to UNODA. It serves, in other words, as an indication of some minimal willingness on the part of the country to take its PoA commitments seriously.

> Baseline indicators of political commitment to the PoA and ITI are flashing red.

The current picture of PoA and ITI implementation is quite sketchy; visibility 'at ground level' is very limited. The information states offer in their reports does not, as a whole, include the level of detail that would permit a clear determination of whether specific commitments are being fulfilled—even allowing for the imprecision of many PoA provisions (Parker, 2011). What is clearer, thanks to independent research, is not encouraging. Baseline indicators of political commitment to the PoA (NPCs) and the ITI (reporting rates, information exchange) are flashing red. These admittedly limited assessments give the distinct impression that the UN small arms process is nothing more than a 'paper tiger', limited to declarations of good intent.

The impression that there is little more to the process than paper is reinforced by the UN membership's continuing aversion to any formal assessment of implementation. The guidelines for PoA implementation, described above, are a start in developing a set of 'benchmarks' that could be used for the systematic assessment of implementation. Much more could be done to develop the measurability of the PoA.[93] The critical obstacle is perhaps not the development of such benchmarks but, more simply, the acceptance of independent measurement. Self-assessment of the kind promised in the 2010 omnibus resolution can be meaningful when the assessor is serious about fulfilling its commitments under the relevant instrument. It is wholly insufficient when the real intention is to get on with 'business as usual' and ignore the instrument.

The UN, or civil society, or some combination of both could fulfil the monitoring and evaluation role, preferably with a mandate from UN member states. The greater the trust that monitored actors have in such an exercise, the

An arms monitor of the United Nations Mission in Nepal inspects a weapon surrendered by Nepalese Maoists under the November 2006 Comprehensive Peace Agreement, April 2008. © Agnieszka Mikulska/UN Photo

greater the chance they will act on the results of any evaluation.[94] With or without a UN mandate, however, independent scrutiny of PoA (and ITI) implementation appears long overdue.[95] The BMS4 outcome recommends that the issue of strengthened PoA follow-up be put on the 2012 Review Conference agenda (UNGA, 2010c, para. 49). Such a move would indicate that the UN membership is poised to take the issue more seriously in the coming years. Yet member states' failure, in the 2010 omnibus resolution, to authorize an independent assessment of implementation in advance of the 2012 Review Conference suggests the opposite.

CONCLUSION

A decade after the adoption of the PoA, it is not clear that the UN small arms process has changed much at 'ground level' in terms of concrete implementation. There were, to be sure, some modest successes in 2010 at the diplomatic level. The BMS4 outcome document contributed to the operational guidance for PoA implementation that has been developed in recent years. It also sketched out a more elaborate, and potentially effective, follow-up process for the PoA, extending beyond mere reporting to the focused consideration—and assessment—of reporting. Nevertheless, 2010 also saw persistent indications that most UN member states are not following through on their

PoA and ITI commitments—that these instruments and their associated meetings are, in essence, elaborate fictions that conceal the determination of many countries to carry on with 'business as usual'.

A key setback in 2010 was the General Assembly's use of the general small arms resolution to translate BMS4 text favouring 'a comprehensive assessment' of implementation (UNGA, 2010c, para. 40) into an exercise in self-reporting. Equally important were the exceptionally poor rates of compliance for key markers of PoA and ITI implementation. In 2010, there were few functioning points of contact for the PoA and ITI and little exchange of information on ITI implementation. Based on this evidence, it appears that only 50 to 60 states are taking their UN small arms commitments seriously.[96] It is admittedly difficult to draw firm conclusions based on the limited information that is currently available, but the UN membership's reluctance to embrace independent scrutiny of PoA (and ITI) implementation suggests it has a case to answer.

It is too soon to write off the UN small arms process. It is possible that the 2011 MGE and 2012 Review Conference will provide clear evidence that the UN membership, as a whole, is committed to the concrete, practical work of strengthening small arms control—and submitting such work to independent evaluation. Yet we may also be approaching the point at which the UN small arms process is widely seen as inadequate, paving the way for non-UN initiatives, whether global or regional, that are more ambitious in their design and effective in their implementation.[97]

There are many good reasons to keep the UN at the centre of global activity on small arms. Most importantly, numerous aspects of the problem are transnational in nature. Deficient export controls or weak stockpile management, for example, have global implications. For this reason alone, states *around the world* need to tackle the small arms issue with some minimal level of determination—in particular, by translating their PoA and ITI undertakings into concrete action. Yet phantom NPCs, patchy reporting, and the continued aversion to formal monitoring leave few grounds for optimism. ◾

LIST OF ABBREVIATIONS

ATT	Arms Trade Treaty
BMS	Biennial Meeting of States to Consider the Implementation of the Programme of Action to Prevent, Combat and Eradicate the Illicit Trade in Small Arms and Light Weapons in All Its Aspects ('Biennial Meeting of States')
GGE	Group of Governmental Experts
ISACS	International Small Arms Control Standards
ITI	International Instrument to Enable States to Identify and Trace, in a Timely and Reliable Manner, Illicit Small Arms and Light Weapons ('International Tracing Instrument')
MGE	Open-ended Meeting of Governmental Experts
NPC	National point of contact
OEWG	Open-ended Working Group
PoA	Programme of Action to Prevent, Combat and Eradicate the Illicit Trade in Small Arms and Light Weapons in All Its Aspects
PrepCom	Preparatory Committee
UNIDIR	United Nations Institute for Disarmament Research
UNODA	United Nations Office for Disarmament Affairs

ENDNOTES

1   The full name of BMS4 is the Fourth Biennial Meeting of States to Consider the Implementation of the Programme of Action to Prevent, Combat and Eradicate the Illicit Trade in Small Arms and Light Weapons in All Its Aspects. For meeting documents, see UN (n.d.a). The First, Second, and Third Biennial Meetings of States are referred to as BMS1, BMS2, and BMS3, respectively.

2   For meeting documents, see UN (n.d.c).

3   The Friends of the Chair were Australia (international cooperation and assistance), Colombia (follow-up), Nigeria (other issues), the United States (International Tracing Instrument), and Uruguay (border controls).

4   See UN (n.d.b).

5   See the discussion of the follow-up section of the BMS4 outcome, below.

6   This provision was also referenced in the small arms omnibus resolution. See UNGA (2010g, para. 4, n. 5).

7   The PoA encourages trans-border customs cooperation and information-sharing at the subregional and regional levels (UNGA, 2001, para. II.27), as well as the exchange of experience and training among customs, police, intelligence, and arms control officials for purposes of combating the illicit small arms trade (para. III.7). The BMS3 outcome also mentions the issue of border controls (UNGA, 2008b, outcome, para. 7b).

8   The various elements of PoA follow-up are mentioned in UNGA (2010c, para. 32).

9   The two-week Review Conference was preceded by a two-week Preparatory Committee session.

10   For an analysis of the 2006 Review Conference, see McDonald, Hasan, and Stevenson (2007, pp. 118–26).

11   As mandated in an earlier General Assembly resolution, implementation of the ITI was the fourth BMS3 theme. For more on the BMS3 preparatory process, see Čekuolis (2008).

12   For an analysis of BMS3, see Bevan, McDonald, and Parker (2009, pp. 136–43).

13   The vote saw 134 states in favour and none against, with Iran and Zimbabwe abstaining. See UNGA (2008b, para. 23).

14   See, for example, UNGA (2010c, para. 11).

15   See Bevan, McDonald, and Parker (2009, p. 141).

16   As reflected in the BMS4 outcome, the priorities concern the matching of needs and resources (UNGA, 2010c, paras. 30g–h); the identification and communication of needs by recipient states (paras. 23, 25, 30g); and national reporting (para. 23).

17   See also UNGA (2010c, para. 41).

18   See also UNGA (2010c, paras. 25, 30f). Compare with UNGA (2008b, outcome, para. 7b).

19   Section III is entitled 'Implementation, international cooperation and assistance'.

20   This point is acknowledged in the PoA itself. See UNGA (2001, paras. III.1–2).

21   See also UNGA (2010c, paras. 21, 30a).

22   Iran and Zimbabwe abstained when the adoption of the BMS3 outcome was put to a vote. The United States, which had earlier voted in the General Assembly against convening BMS3, did not attend the meeting.

23   Note that the BMS3 outcome's international cooperation and assistance section also includes provisions relating to follow-up. See Bevan, McDonald, and Parker (2009, p. 142).

24   See Bevan, McDonald, and Parker (2009, pp. 142–43).

25   BMS4 agenda item 6c reads: 'Strengthening of the follow-up mechanism of the Programme of Action, and preparations for the 2011 Experts Group meeting and the 2012 Review Conference' (UNGA, 2010b, para. 9).

26   Two BMSs, two years apart, are to be followed, after two years, by a review conference.

27   The BMS4 draft outcome, issued by the chair in advance of the last day of discussions, 'noted that the [Review Conference] had a mandate to consider whether the Programme of Action was meeting the objectives States had set for it in 2001' (Mexico, 2010, para. 46). At the request of India, however, this language was replaced, just before the end of the meeting, with PoA language stating that the Review Conference is 'to review progress made in the implementation of the Programme of Action' (UNGA, 2001, para. IV.1.a; 2010c, para. 48). A small number of other amendments were made to the draft outcome on the same day, but none had a significant impact on substance.

28   The 2010 ITI outcome also reprises language from 2008, noting that national reports may include 'quantitative data that would enable States to assess the effectiveness of the Instrument in enhancing cooperation in tracing' (UNGA, 2010d, para. 10d; 2008b, ITI outcome, para. 9d).

29   Compare with UNGA (2008b, ITI outcome, para. 9g).

30   Paragraph 7 addresses this issue but is exceptionally weak (UNGA, 2010d).

31   Compare with UNGA (2010d, paras. 6, 10g). Import marks, when present, greatly enhance the chances of a successful trace as they enable investigators to access relatively recent records relating to the weapon. See Bevan (2009, pp. 118–20).

32   Compare with UNGA (2008b, ITI outcome, para. 9 chapeau).

33 'However, States also noted the latest analysis by the Small Arms Survey, which suggested that more work needs to be done to foster wider and deeper cooperation' (UNGA, 2010d, para. 5). See also Parker (2010, pp. 52–68).

34 Figure as of 6 May 2010. This number includes states that simply mentioned the ITI in their national reports. Information countries provided on marking, record-keeping, and tracing that referred only to the Programme of Action was, however, excluded from the total. See Parker (2010, pp. 52–54). Revised figures available in mid-January 2011, as the May 2010 study was being updated, put the 2010 ITI reporting total at approximately 60 states—still lower than for 2008.

35 See UNDP (2003).

36 In 2008, 109 states reported on their implementation of the PoA—as compared with 62 for the ITI (Cattaneo and Parker, 2008, pp. 4, 97).

37 For a review of national preparations for implementation in the months preceding the ITI's adoption by the UN General Assembly, see McDonald (2006, pp. 112–14).

38 The 18 countries were: Algeria, Bahrain, Burundi, Canada, Chile, Colombia, Croatia, Czech Republic, Italy, Jamaica, Mauritius, Namibia, Poland, Portugal, Romania, Senegal, Spain, and Thailand. In two cases, no phone number, fax number, or email address was listed (Bahrain and Thailand). There were multiple entries for two other states (Namibia and Portugal). The multiple entries may have reflected a decision to separate the operational (weapons tracing) and information exchange (overall implementation) functions of the point of contact—a possibility the ITI envisages (UNGA, 2005, para. 25)—but this was not specified.

39 See UNGA (2010b).

40 See UNGA (2010f; 2010h).

41 See UNGA (2010e, paras. 41–42).

42 The omnibus resolution focuses on the ITI's reporting requirements. See UNGA (2010g, paras. 11, 14).

43 The omnibus resolution fails to include such points as the explicit mention of a biennial PoA reporting schedule; collaboration between PoA meeting chairs; and the need for clear definitions of, and distinctions between, PoA meeting mandates. For more on these elements, see the discussion of the BMS4 follow-up section, above.

44 Derived from UNGA (2010c, para. 32).

45 See also UNGA (2010c, para. 32).

46 In 2006, there were two different chairs for the Preparatory Committee and the Review Conference.

47 See UNGA (2001, paras. II.11–13). The PoA also covers brokering (para. II.14) and addresses the implementation of UN Security Council arms embargoes (para. II.15).

48 'States [. . .] undertake [. . .] [t]o assess' (UNGA, 2001, paras. II.1, II.11). 'States [. . .] undertake [. . .] [t]o put in place and implement' (paras. II.1, II.12). Much weaker is: 'To make every effort' (para. II.13).

49 The first part of paragraph II.11 reads: 'To assess applications for export authorizations according to strict national regulations and procedures that cover all small arms and light weapons and are consistent with the existing responsibilities of States under relevant international law, taking into account in particular the risk of diversion of these weapons into the illegal trade' (UNGA, 2001).

50 For information on the early stages of the process, see McDonald, Hasan, and Stevenson (2007, pp. 128–30).

51 See UNGA (2008e); Bevan, McDonald, and Parker (2009, pp. 147–52).

52 The dates were 2–6 March and 13–17 July 2009.

53 See also UNGA (2009c, para. 21).

54 The four substantive agenda items were: '[g]oals and objectives', 'scope', 'principles and draft parameters', and '[o]ther aspects' of an ATT (UNGA, 2009a; 2009b).

55 Author's observations at the second substantive session, 13–17 July 2009. The more concrete, detailed discussions tended to involve the question of ATT 'scope', meaning the range of items the future treaty is to cover.

56 Author's observations at the second substantive session, 13–17 July 2009. For more information on national positions on the ATT, see Bevan, McDonald, and Parker (2009, pp. 148–49). Note that since its change of government in 2009, the United States has ceased to oppose the ATT process. After voting against the General Assembly's 2006 and 2008 ATT resolutions, it voted in favour of the 2009 resolution. Voting records of the UN General Assembly (First Committee resolutions) are available at WILPF (2010).

57 See Caughley (2009).

58 Ambassador Moritán also chaired the earlier ATT meetings (2008 GGE and 2009 OEWG).

59 The Friends of the Chair were Australia (common standards or criteria); Egypt (implementation and application); and Trinidad and Tobago (scope).

60 To the dismay of many non-governmental organizations, these sessions were closed to civil society participants. See Epps (2010).

61 See, for example, Epps (2010).

62 'All U.N. countries had now accepted the principle of a treaty, delegates said' (Worsnip, 2010).

63   Author's observations during the second week of the 2010 PrepCom.

64   Author's observations during the second week of the 2010 PrepCom.

65   Debate continues on the question of whether the standards or criteria for export licensing that would be agreed in an ATT would merely be 'taken into account' by states, perhaps in a superficial manner, or would, instead, pose greater constraints on national decision-making. See Wood and Estévez (2010).

66   On the importance of controls over ammunition, see Greene (2006).

67   See, in particular, the materials from the Boston Symposium on the Arms Trade Treaty (28–30 September 2010), available at UMass Boston (n.d.).

68   Two one-week sessions of the PrepCom have been scheduled for 2011: 28 February to 4 March and 11 to 15 July.

69   See UNGA (2001, para. IV.1.c).

70   See UNGA (2001, paras. II.7–10, 36–37, III.10).

71   For the report of the GGE, see UNGA (2003).

72   See UNGA (2001, para. IV.1.d).

73   See, for example, UNGA (2010g, para. 3).

74   Standards and procedures for stockpile management are addressed in paragraphs 20, 22, 24, 27b–c, e; the identification of surplus in paragraphs 20, 23, 25–26, 27a; surplus disposal and destruction in paragraphs 22–23, 27e; and resources in paragraphs 21–23, 27d (UNGA, 2008b, outcome).

75   See UNGA (2008b, outcome, paras. 20, 25).

76   Arguably, some of its provisions, such as those relating to stockpile management and surplus disposal, apply to ammunition, but this is contested by some states.

77   See Bevan, McDonald, and Parker (2009, p. 145).

78   For an overview of the report, see Bevan, McDonald, and Parker (2009, pp. 145–47). Note that 'public safety' concerns focus on the prevention of accidental explosion and consequent harm to surrounding populations, whereas 'stockpile security' relates to the risk of diversion to unauthorized groups and individuals (theft or loss). See UNGA (2008a, secs. I.D–E).

79   See UNGA (2008a, paras. 53–54, 61, 72).

80   The PoA devotes an entire section (18 paragraphs) to the topic of 'Implementation, international cooperation and assistance' (UNGA, 2001, sec. III).

81   See also UNGA (2010h, paras. 5–6).

82   For more information, see UNCASA (n.d.).

83   Note that the report provides relatively little information on activities in Asia–Pacific, the Middle East, or North Africa. See UNGA (2010e, paras. 15, 26, 31, 45, 64).

84   See, for example, the review of developments in Asia–Pacific in UNGA (2010e, para. 64).

85   See UNGA (2010e, paras. 50, 52, 60, 65).

86   See UNGA (2010e, para. 9), relating the findings of UN experts groups.

87   See, for example, UNGA (2010g, pmbl. para. 8).

88   Remarks by Sarah Parker, based on research conducted for Parker (2011). UNIDIR workshop on PoA implementation, Geneva, 3 December 2010.

89   See, for example, BtB with IANSA (2006).

90   The Survey study also looks at national coordination agencies and national action plans. See UNGA (2001, paras. II.4–5).

91   As of May 2010, the UNODA website (www.poa-iss.org) listed 151 NPCs, although information for five countries was insufficient to initiate enquiries (Parker, 2010, p. 27).

92   In contrast to the many more complex requirements of the PoA, including the regulation of small arms manufacture, international transfer, brokering, and stockpile management (UNGA, 2001, sec. II), the establishment of a functioning NPC is also quite easy to measure. The existence and identity of an NPC is 'an objectively verifiable fact' (Parker, 2010, p. 32).

93   The UN Secretary-General has identified the PoA's lack of measurability as a factor impeding progress in its implementation (UNSC, 2008, para. 30; see also paras. 58, 63).

94   See Persbo (2010).

95   See McDonald (2004).

96   This estimate is based on the number of functioning points of contact for the PoA (52) and the number of states that reported on their implementation of the ITI in 2010 (around 60, according to information available in January 2011). If, on the other hand, one focuses on the ITI point of contact figures, then the number of states that appear serious about implementation drops to 18 (or 16; see endnote 38).

97   See Borrie et al. (2009, pp. 21–22); Efrat (2010). Efrat notes that '[w]hile global cooperation may be desirable, conflicting government preferences could render it weak'. He concludes that smaller-scale collaboration between like-minded governments, especially at the regional or subregional level, 'offers a second-best alternative' (pp. 127–28).

BIBLIOGRAPHY

Argentina. 2010a. *Chairman's Draft Elements*. 14 July (morning). <http://www.adh-geneva.ch/RULAC/pdf/ChairmanDraftElement-14072010-.pdf>

——. 2010b. *Chairman's Draft Principles*. 14 July (afternoon).<http://www.adh-geneva.ch/RULAC/pdf/ChairDraftPrinc-14072010.pdf>

——. 2010c. *Chairman's Draft Paper*. 22 July. <http://www.reachingcriticalwill.org/legal/att/2010prepcom/docs/ChairDraftPaper.pdf>

Australia. 2010. *Facilitator's Summary on Parameters*. 22 July. <http://www.reachingcriticalwill.org/legal/att/2010prepcom/docs/FS-Parameters.pdf>

Bevan, James. 2009. 'Revealing Provenance: Weapons Tracing during and after Conflict.' In Small Arms Survey. *Small Arms Survey 2009: Shadows of War*. Cambridge: Cambridge University Press, pp. 106–33.

<http://www.smallarmssurvey.org/publications/by-type/yearbook/small-arms-survey-2009.html>

——, Glenn McDonald, and Sarah Parker. 2009. 'Two Steps Forward: UN Measures Update.' In Small Arms Survey. *Small Arms Survey 2009: Shadows of War*. Cambridge: Cambridge University Press, pp. 134–57.

<http://www.smallarmssurvey.org/publications/by-type/yearbook/small-arms-survey-2009.html>

Borrie, John, et al. 2009. 'Learn, Adapt, Succeed: Potential Lessons from the Ottawa and Oslo Processes for other Disarmament and Arms Control Challenges.' *Disarmament Forum*, Nos. 1–2. <http://www.unidir.org/pdf/articles/pdf-art2860.pdf>

BtB with IANSA (Biting the Bullet Project—International Alert, Saferworld, University of Bradford—with the International Action Network on Small Arms). 2006. *Reviewing Action on Small Arms 2006: Assessing the First Five Years of the UN Programme of Action*. London: BtB. <http://www.iansa.org/un/review2006/redbook2006/>

Cattaneo, Silvia and Sarah Parker. 2008. *Implementing the United Nations Programme of Action on Small Arms and Light Weapons: Analysis of the National Reports Submitted by States from 2002 to 2008*. Geneva: United Nations Institute for Disarmament Research. December. <http://www.unidir.ch/bdd/fiche-ouvrage.php?ref_ouvrage=92-9045-008-H-en>

Caughley, Tim. 2009. 'Consensus Rules the Arms Trade Treaty. Or Does It?' *Disarmament Insight*. 10 November. <http://disarmamentinsight.blogspot.com/2009/11/consensus-rules-arms-trade-treaty-or.html>

Čekuolis, Dalius. 2008. 'Tackling the Illicit Small Arms Trade: The Chairman Speaks.' *Arms Control Today*, Vol. 38, No. 8. October, pp. 19–24. <http://www.armscontrol.org/act/2008_10/Cekuolis>

Efrat, Asif. 2010. 'Toward Internationally Regulated Goods: Controlling the Trade in Small Arms and Light Weapons.' *International Organization*, Vol. 64, Iss. 1, pp. 97–131.

<http://journals.cambridge.org/action/displayFulltext?type=1&fid=7093212&jid=INO&volumeId=64&issueId=01&aid=7093204>

Egypt. 2010. *Facilitator's Summary on Implementation and Application*. 22 July. <http://www.reachingcriticalwill.org/legal/att/2010prepcom/docs/FS-Implementation.pdf>

Epps, Kenneth. 2010. 'Arms Trade Treaty Negotiations Begin.' *Ploughshares Monitor*, Vol. 31, No. 3. <http://www.ploughshares.ca/libraries/monitor/mons10b.pdf>

Greene, Owen. 2006. 'Ammunition for Small Arms and Light Weapons: Understanding the Issues and Addressing the Challenges.' In Stéphanie Pézard and Holger Anders, eds. *Targeting Ammunition: A Primer*. Geneva: Small Arms Survey. June.

<http://www.smallarmssurvey.org/publications/by-type/book-series/targeting-ammunition.html>

McDonald, Glenn. 2004. 'Under the Spotlight: Monitoring Implementation of Small Arms Measures.' In Small Arms Survey. *Small Arms Survey 2004: Rights at Risk*. Oxford: Oxford University Press, pp. 248–75.

<http://www.smallarmssurvey.org/publications/by-type/yearbook/small-arms-survey-2004.html>

——. 2006. 'Connecting the Dots: The International Tracing Instrument.' In Small Arms Survey. *Small Arms Survey 2006: Unfinished Business*. Oxford: Oxford University Press, pp. 94–117. <http://www.poa-iss.org/KIT/CH4%20Measures.pdf>

——, Sahar Hasan, and Chris Stevenson. 2007. 'Back to Basics: Transfer Controls in Global Perspective.' In Small Arms Survey. *Small Arms Survey 2007: Guns and the City*. Cambridge: Cambridge University Press, pp. 116–43.

<http://www.smallarmssurvey.org/publications/by-type/yearbook/small-arms-survey-2007.html>

Mexico. 2010. *Consolidated Draft Outcome*. 17 June, 11 p.m. <http://www.poa-iss.org/BMS4/Documents.html>

Parker, Sarah. 2010. *National Implementation of the United Nations Small Arms Programme of Action and the International Tracing Instrument: An Analysis of Reporting in 2009–10*. Working Paper No. 9. Geneva: Small Arms Survey. June.

<http://www.smallarmssurvey.org/fileadmin/docs/F-Working-papers/SAS-WP9-National-Implentation.pdf>

—— 2011. *Improving the Effectiveness of the UN Programme of Action on Small Arms: Implementation Challenges and Opportunities*. Geneva: United Nations Institute for Disarmament Research. <http://www.unidir.ch/html/en/publications.php>

Persbo, Andreas. 2010. 'The Role of Non-governmental Organizations in the Verification of International Agreements.' *Disarmament Forum*, No. 3. <http://www.unidir.org/pdf/articles/pdf-art3003.pdf>

Trinidad and Tobago. 2010. *Facilitator's Summary for Scope.* <http://www.reachingcriticalwill.org/legal/att/2010prepcom/docs/FS-Scope.pdf>

UMass Boston (University of Massachusetts Boston). n.d. 'Boston Symposium on the Arms Trade Treaty.' Boston: John W. McCormack Graduate School of Policy and Global Studies, UMass Boston. <http://www.mccormack.umb.edu/arms_trade_treaty_conference.php>

UN (United Nations). n.d.a. 'Fourth Biennial Meeting of States on Small Arms and Light Weapons.' Website. <http://www.poa-iss.org/BMS4/>

—. n.d.b. 'Fourth Biennial Meeting of States on Small Arms and Light Weapons: Documents.' Website. <http://www.poa-iss.org/BMS4/Documents.html>

—. n.d.c. 'Third Biennial Meeting of States on Small Arms and Light Weapons.' Website.
<http://www.un.org/disarmament/convarms/BMS/bms3/1BMS3Pages/1National%20Reports%202008.html>

UNCASA (United Nations Coordinating Action on Small Arms). 2010. *Project on International Small Arms Control Standards: Phase 2, 1 October 2010–31 December 2011 (15 months).* 4 May. <http://www.un-casa-isacs.org/isacs/Documents_files/ISACS%20Phase%202%20Project.pdf>

——. n.d. 'CASA Project on International Small Arms Control Standards (ISACS).' <http://www.un-casa-isacs.org/isacs/>

UNDP (United Nations Development Programme). 2003. *Assistance Package: Guidelines for Reporting on Implementation of the United Nations Programme of Action to Prevent, Combat and Eradicate the Illicit Trade in Small Arms and Light Weapons in All Its Aspects.*
<http://www.undp.org/cpr/smallarms/docs/PoA_package.pdf>

UNGA (United Nations General Assembly). 2001. Programme of Action to Prevent, Combat and Eradicate the Illicit Trade in Small Arms and Light Weapons in All Its Aspects ('UN Programme of Action'). A/CONF.192/15 of 20 July. <http://www.poa-iss.org/PoA/PoA.aspx>

——. 2003. *Report of the Group of Governmental Experts Established Pursuant to General Assembly Resolution 56/24 V of 24 December 2001, Entitled 'The Illicit Trade in Small Arms and Light Weapons in All Its Aspects'.* A/58/138 of 11 July.
<http://disarmament.un.org/library.nsf/c793d171848bac2b85256d7500700384/e3467dd72a9f4fa285256d88004c099f/$FILE/sg58.138.pdf>

——. 2005. International Instrument to Enable States to Identify and Trace, in a Timely and Reliable Manner, Illicit Small Arms and Light Weapons ('International Tracing Instrument'). A/60/88 of 27 June (annexe). <http://www.poa-iss.org/InternationalTracing/InternationalTracing.aspx>

——. 2006a. Resolution 61/66, adopted 6 December. A/RES/61/66 of 3 January 2007. <http://www.un.org/documents/resga.htm>

——. 2006b. Resolution 61/89, adopted 6 December. A/RES/61/89 of 18 December.
<http://www.un.org/disarmament/convarms/ArmsTradeTreaty/html/ATT.shtml>

——. 2007. *Report of the Group of Governmental Experts Established Pursuant to General Assembly Resolution 60/81 to Consider Further Steps to Enhance International Cooperation in Preventing, Combating and Eradicating Illicit Brokering in Small Arms and Light Weapons.* A/62/163 of 30 August. <http://www.poa-iss.org/BrokeringControls/English_N0744232.pdf>

——. 2008a. *Report of the Group of Governmental Experts Established Pursuant to General Assembly Resolution 61/72 to Consider Further Steps to Enhance Cooperation with Regard to the Issue of Conventional Ammunition Stockpiles in Surplus.* A/63/182 of 28 July.
<http://www.poa-iss.org/DocsUpcomingEvents/a-63-182-e.pdf>

——. 2008b. *Report of the Third Biennial Meeting of States to Consider the Implementation of the Programme of Action to Prevent, Combat and Eradicate the Illicit Trade in Small Arms and Light Weapons in All Its Aspects.* A/CONF.192/BMS/2008/3 of 20 August.
<http://www.poa-iss.org/DocsUpcomingEvents/ENN0846796.pdf>

——. 2008c. *Report of the Group of Governmental Experts to Examine the Feasibility, Scope and Draft Parameters for a Comprehensive, Legally Binding Instrument Establishing Common International Standards for the Import, Export and Transfer of Conventional Arms.* A/63/334 of 26 August.
<http://www.un.org/disarmament/convarms/ArmsTradeTreaty/html/ATT.shtml>

——. 2008d. Resolution 63/72, adopted 2 December. A/RES/63/72 of 12 January 2009. <http://www.un.org/documents/resga.htm>

——. 2008e. Resolution 63/240, adopted 24 December. A/RES/63/240 of 8 January 2009.
<http://www.un.org/disarmament/convarms/ArmsTradeTreaty/html/ATT.shtml>

——. 2009a. 'Provisional Agenda.' A/AC.277/2009/L.2/Rev.1 of 27 February.
<http://www.un.org/disarmament/convarms/ArmsTradeTreaty/html/ATT.shtml>

——. 2009b. 'Provisional Agenda.' A/AC.277/2009/L.3 of 23 June. <http://www.un.org/disarmament/convarms/ArmsTradeTreaty/html/ATT.shtml>

——. 2009c. *Report of the Open-ended Working Group towards an Arms Trade Treaty: Establishing Common International Standards for the Import, Export and Transfer of Conventional Arms.* A/AC.277/2009/1 of 20 July.
<http://www.un.org/disarmament/convarms/ArmsTradeTreaty/html/ATT.shtml>

——. 2009d. Resolution 64/48, adopted 2 December. A/RES/64/48 of 12 January 2010.
<http://www.un.org/disarmament/convarms/ArmsTradeTreaty/html/ATT.shtml>

——. 2009e. Resolution 64/50, adopted 2 December. A/RES/64/50 of 12 January 2010. <http://www.un.org/documents/resga.htm>

——. 2010a. *Annex to the Outcome Document: Implementation of the International Instrument to Enable States to Identify and Trace, in a Timely and Reliable Manner, Illicit Small Arms and Light Weapons.* A/CONF.192/BMS/2010/WP.5 of 15 June.
<http://www.poa-iss.org/BMS4/1workingpapers/WP5-ITI-DraftOutcome/WP5-DraftFinalDoc-ITI-en.pdf>

——. 2010b. *Report of the Fourth Biennial Meeting of States to Consider the Implementation of the Programme of Action to Prevent, Combat and Eradicate the Illicit Trade in Small Arms and Light Weapons in All Its Aspects.* A/CONF.192/BMS/2010/3 of 30 June. <http://www.poa-iss.org/BMS4/Outcome/BMS4-Outcome-E.pdf>

——. 2010c. *Outcome of the Fourth Biennial Meeting of States to Consider the Implementation of the Programme of Action to Prevent, Combat and Eradicate the Illicit Trade in Small Arms and Light Weapons in All Its Aspects.* A/CONF.192/BMS/2010/3 of 30 June (sec. V). <http://www.poa-iss.org/BMS4/Outcome/BMS4-Outcome-E.pdf>

——. 2010d. *Outcome on the Implementation of the International Instrument to Enable States to Identify and Trace, in a Timely and Reliable Manner, Illicit Small Arms and Light Weapons.* A/CONF.192/BMS/2010/3 of 30 June (annexe). <http://www.poa-iss.org/BMS4/Outcome/BMS4-Outcome-E.pdf>

——. 2010e. *Consolidation of Peace through Practical Disarmament Measures; Assistance to States for Curbing the Illicit Traffic in Small Arms and Light Weapons and Collecting Them; The Illicit Trade in Small Arms and Light Weapons in All Its Aspects.* Report of the Secretary-General. A/65/153 of 20 July. <http://www.un.org/Docs/journal/asp/ws.asp?m=A/65/153>

——. 2010f. Resolution 65/50, adopted 8 December. A/RES/65/50 of 13 January 2011. <http://www.un.org/documents/resga.htm>

——. 2010g. Resolution 65/64, adopted 8 December. A/RES/65/64 of 13 January 2011. <http://www.un.org/documents/resga.htm>

——. 2010h. Resolution 65/67, adopted 8 December. A/RES/65/67 of 13 January 2011. <http://www.un.org/documents/resga.htm>

UNODA (United Nations Office for Disarmament Affairs). 2010. 'Template for Use in the Reporting of the Programme of Action.' <http://www.poa-iss.org/PoA/PoA.aspx>

——. 2011. National Points of Contact. Accessed 18 January. <http://www.poa-iss.org/poa/NationalContacts.aspx>

UNSC (United Nations Security Council). 2008. *Small Arms: Report of the Secretary-General.* S/2008/258 of 17 April. <http://www.poa-iss.org/DocsUpcomingEvents/S-2008-258.pdf>

WILPF (Women's International League for Peace and Freedom). 2010. 'Reaching Critical Will: Draft Resolutions and Decisions, Voting Results, and Explanations of Vote from First Committee 2010.' <http://www.reachingcriticalwill.org/political/1com/1com10/resolutions.html>

Wood, Brian and Alberto Estévez. 2010. 'Towards a Bullet-Proof Arms Trade Treaty.' *Disarmament Times.* Fall. <http://disarm.igc.org/index.php?option= com_content&view=article&id=420:towards-a-bullet-proof-arms-trade-treaty&catid=150:disarmament-times-fall-2010&Itemid=2>

Worsnip, Patrick. 2010. 'Progress Seen at Talks on World Arms Trade Treaty.' Reuters. 23 July. <http://www.reuters.com/article/idUSTRE66M52820100723>

Wurche, Dieter. 2010. *Customs and its Role in Combating the Illicit Trafficking in Small Arms.* Presentation to Geneva Process on small arms. Geneva, 27 January.

ACKNOWLEDGEMENTS

Principal author

Glenn McDonald

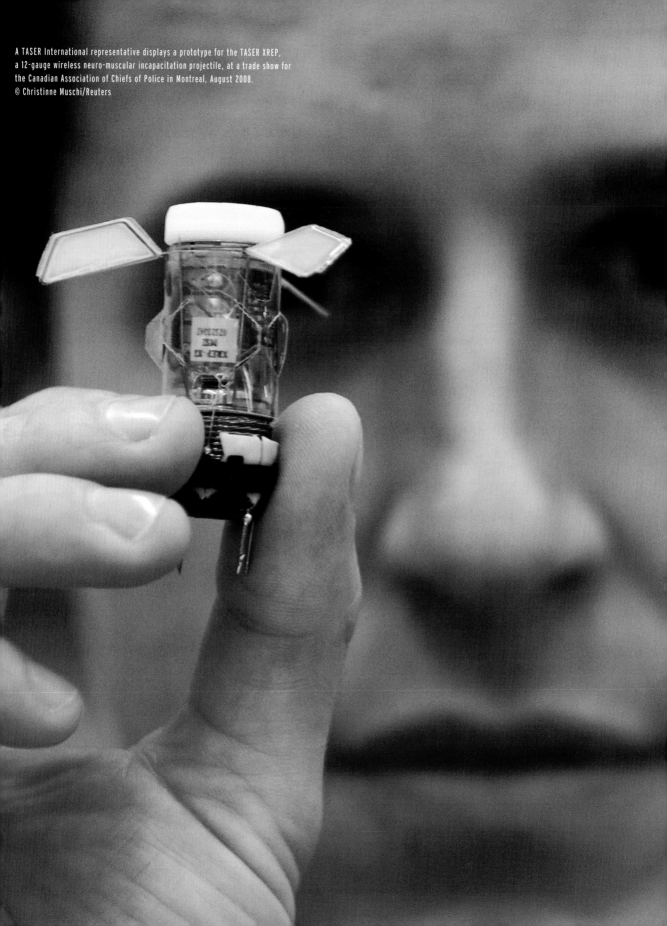

A TASER International representative displays a prototype for the TASER XREP, a 12-gauge wireless neuro-muscular incapacitation projectile, at a trade show for the Canadian Association of Chiefs of Police in Montreal, August 2008.
© Christinne Muschi/Reuters

Procurement and Policy
POLICE USE OF EMERGING WEAPONS TECHNOLOGY

3

INTRODUCTION

In most Western countries, two factors condition the use of force by domestic security agencies: operational guidelines and the type of weaponry available. The relationship between use-of-force policy and weapons procurement is close but complex. The adoption of new weapons may call for new policies, while the need to meet international norms in policing, for example, can affect weapons procurement. Market forces and trends in both civilian and military firearms development and procurement also play influential roles in the adoption of new weapons technology.

On the streets of cities such as Chicago, Manchester, and Marseille, police officers have increasingly adopted so-called 'less-lethal' weapons into their day-to-day activities. A comprehensive review of the use of these weapons by Western police has not yet been undertaken, but police experience in France,[1] the United Kingdom, and the United States suggests that countries are facing similar use-of-force challenges that are drawing them to use these weapons. By examining recent trends in police weapons technology and procurement by law enforcement agencies, as well as their use-of-force policies, this chapter highlights the alignment of policies and practices in a small but important sample of the West's 'leading-edge' states.

Among the chapter's conclusions are the following:

- Aside from the use of new materials to reduce weight and facilitate customization, law enforcement firearms have not recently experienced significant technological development.
- While Western European agencies still predominantly use 9 mm handguns for public order policing, US law enforcement agencies are procuring larger-calibre handguns and semi-automatic rifles to counter armed criminal violence.
- The latest generation of 'less-lethal' weapons allows police officers to engage targets that are farther away and provides them with more flexibility in the use of force across the spectrum from non-lethal to lethal.
- In its effort to adapt police practice and doctrine to new firearms and less-lethal weaponry, the US law enforcement community draws its primary inspiration from the military.
- Use-of-force policies have not kept up with the procurement of some weapons technology by police organizations. In terms of doctrine, practice, and equipment requirements, this discrepancy is accentuated by the absence of consensus among countries and law enforcement agencies.

This chapter begins by reviewing the concepts of the 'use-of-force continuum', 'less-lethal' technology, and the heterogeneity of law enforcement arrangements and needs. The section that follows identifies recent developments and procurement trends in law enforcement firearms and less-lethal weapons. The final section highlights a number of issues concerning the matching of police policies and procedures and emerging weapons technology.

SETTING THE STAGE

Force continuum

The proportionate use of force is key to law enforcement practice, accountability, and perceived legitimacy. Officers learn that the level of force in response to a given situation should match the actual threat. The use of lethal force is usually regarded as a last resort. The UN Code of Conduct for Law Enforcement Officials articulates the basic principles of the proportional, minimum, and discriminate use of force (UNGA, 1979, p. 2). The UN Basic Principles on the Use of Force and Firearms by Law Enforcement Officials call for law enforcement officers to apply non-violent means before resorting to the use of any weapons (UN, 1990, para. 4). If the use of force or armed response is unavoidable, then it should be limited to what is necessary and proportionate to the seriousness of the offence (para. 5a). The Basic Principles also formalize the need for 'various types of weapons and ammunition that would allow for a differentiated use of force and firearms' (para. 2).[2]

Western police organizations illustrate and conceptualize the use-of-force continuum using various charts and schematics, and mainly as a training tool. Such charts present a spectrum of options that reflect increasing levels of threat and indicate what is considered an adequate response. Figure 3.1 shows seven threat levels and matching appropriate police responses, ranging from the officer's mere presence to the use of lethal force. Force continuum models may also seek to match police response to different types of subject behaviour—such as cooperation, passive resistance, active resistance, and assault (Braidwood Commission, 2009, p. 8). In the section on 'Matching weaponry with doctrine' (see below), this chapter studies the factors that can reduce or widen the gap that separates the use of less-lethal weapons from the use of firearms.

Rather than being a fixed, homogeneous concept among police organizations, force continuum models now serve primarily as training tools. Officers in the field cannot reasonably be expected to escalate or de-escalate systematically through every step of a continuum. Use-of-force policies are thus usually upheld nationally by legal standards.

A training officer for the Los Angeles County Sheriff's Department displays an Arwen launcher and the rubber projectile it fires during a demonstration of non-lethal and lethal weaponry at the sheriff's training academy in Whittier, California, November 1995. © Reed Saxon/AP Photo

Figure 3.1 **Example of use-of-force continuum**

Type of situation	Police response
Most serious ▲ ▼ **Least serious**	• Firearm • Intermediate weapons: TASER and impact projectiles • 'Hard' control techniques and compliance • Chemical agents: pepper spray (oleoresin capsicum) • 'Soft' empty-hand control techniques • Verbal commands • Officer presence

In the United States, for instance, the standard of 'objective reasonableness', found in the US Constitution's Fourth Amendment, is more permissive for officers than the force continuum because it addresses the misuse of power but does not require that an officer use the least intrusive means (Peters, 2006). The US Supreme Court has spelled out three general factors for officers to evaluate which degree of force is reasonable and necessary: the severity of the crime, the threat posed by the suspect to surrounding officers or the public, and the suspect's active resistance or attempt to evade arrest (MPD, 2005, p. 10). As such, this standard provides more leeway for officers and more protection against litigation than does the force continuum. The Los Angeles Police Department incorporated the standard of objective reasonableness into its use-of-force policy in 2008 (LAPD, 2009, p. 10; Sargent, 2010).

The diversity of law enforcement

The UN Code of Conduct defines law enforcement officials as:

officers of the law, whether appointed or elected, who exercise police powers, especially the powers of arrest or detention. In countries where police powers are exercised by military authorities, whether uniformed or not, or by State security forces, the definition of law enforcement officials shall be regarded as including officers of such services (UNGA, 1979, art. 1, commentary (a)).

The European Code of Police Ethics, in defining its scope of application, asserts that public police forces are 'empowered by the state to use force and/or special powers for these purposes' (CoE, 2001, appendix).

Each law enforcement agency's characteristics significantly affect its weapons policies and procurement. These include its degree of centralization, whether the force is civilian or military, force size, and structure. An agency may opt for the purchase of a single handgun model for the whole force, or for a system of firearms in full-size, compact, and subcompact models in various calibres but from the same manufacturer (Kaestle and Buehler, 2005). The former option facilitates training and maintenance, while the latter offers more mission flexibility.

Another decisive factor is the police budget. Before 2003, the French police had close to 40 different types of firearms in service, with eight different calibres. The 9 mm SIG SAUER SP2022 was adopted in 2003 by gendarmerie, police, customs, and the penitentiary services to harmonize national procurement and logistics, at significant cost. In 2010, annual firearm maintenance and ammunition procurement cost the French police an estimated EUR 9–10 million (USD 12–13 million), and the gendarmerie an estimated EUR 6–8 million (USD 8–11 million) (DGPN, 2010, p. 10; DGGN, 2010, p. 2). National procurement of new firearms tends to occur in 20-year cycles; officials estimate that the multi-year renewal of a firearm system for the whole police force represents a EUR 25 million (USD 34 million) expenditure (DGPN, 2010, p. 10).

Police organizations are a very difficult market to target because their requirements and doctrine are extremely diverse. Police authorities rarely agree on common weapon requirements, which may explain why some small arms manufacturers mainly focus on military procurement.[3] While NATO standardizes army requirements for infantry small arms, police requirements vary and change frequently. This lack of uniformity makes it difficult for manufacturers to design police-tailored solutions, which, in turn, may explain why some police units follow procurement trends originally observed in the civilian (handgun) or military (rifle) markets.

> Police organizations are diverse and rarely agree on common weapon requirements.

Defining 'less-lethal'

There is little agreement on what constitutes the class of weapons variously called 'non-lethal', 'less-than-lethal', or 'less-lethal'. In the United States and the United Kingdom, the military generally applies the term 'non-lethal' to what law enforcement and criminal justice communities label 'less-lethal' weapons (USDOD, 1996; USDOJ, 2004; UKSG, 2006). Law enforcement organizations often distinguish less-lethal options (such as TASERs and bean bags) from non-lethal ones (such as firm grips, punches, or physical force) (LAPD, 2009, p. 7). These three terms describe what the weapons are *not* supposed to do rather than their specialized effects and the kinds of situations to which they may be suited. They also form a misleading dichotomy, which incorrectly implies that traditional firearms are *systematically* lethal in their effects.

More appropriate names would arguably refer to the ability of these intermediate weapons, when properly used, to neutralize or temporarily incapacitate subjects in medium-threat situations (DGPN, 2010, p. 6). At the same time, manufacturers use marketing terms such as 'conducted energy devices', 'calmatives', 'optical distracters', and 'acoustic hailing devices', which, while descriptive of the technology, may be aimed at softening perceptions and increasing public acceptance of these devices, while playing down their risks (Davison, 2009, pp. 6–7).

The naming of devices may be more than just a question of semantics. If states classify them as firearms, the devices become subject to the same licensing procedures and export control restrictions, which makes them more difficult for manufacturers to market internationally. In France, for instance, the COUGAR and CHOUKA less-lethal projectile launchers manufactured by LACROIX–Alsetex are classified as fourth-category firearms according to national legislation, and can only be exported with an *autorisation d'exportation de matériels de guerre* delivered by the French Ministry of Defence and the Customs administration (France, 2010).[4]

LACROIX-Alsetex's COUGAR and CHOUKA 56 mm calibre less-lethal launchers, on display at the 2010 Eurosatory show.
© Pierre Gobinet

This chapter uses the term 'less-lethal' to reflect the fact that lack of training for or improper use of such weapons can inflict serious or lethal injury on the target.

TRENDS AND DEVELOPMENTS

This section reviews current developments in lethal and less-lethal weaponry, along with several examples of recent procurement. The weapons described here are either 'individual' (assigned to each officer) or 'collective' (shared by officers in the same unit, from one shift to another, or dedicated to special tactical teams and crowd control units). This section presents weapons in a 'lethal' vs. 'less-lethal' dichotomy to highlight the disparity in models procured by Western police units as well as the difficulty of covering both weapons categories in a single, coherent doctrine.

Firearms

Law enforcement firearms have not recently experienced significant technological breakthroughs; improvements have been incremental. The Small Arms Survey reached this conclusion in 2003 and the situation has not changed significantly since then (CAST et al., 2003, pp. 20–25). Research shows that military and civilian markets exert significantly more influence over the procurement of police firearms in the United States than they do in Western Europe. In general, developments in conventional firearm technology emphasize increased compactness, customization, and weight reduction.

The arms race of US law enforcement: handguns and semi-automatic rifles

The .40 S&W cartridge is increasingly procured by US law enforcement agencies.

Confronted with significant armed criminality, the law enforcement community in the United States is spearheading the progressive switch to higher-calibre firearms. In 2009, firearms were used in 67.1 per cent of murders, 42.6 per cent of robberies, and 20.9 per cent of aggravated assaults in the United States (USDOJ, 2010b). In 2008, guns were used in almost 10,000 gun murders (Gun Policy, 2010). National statistics seem to indicate, however, that the annual rate of US firearm homicides has stabilized somewhat between 3 and 5 per 100,000 since 2005, well below the peaks of the 1980s and 1990s (Gun Policy, 2010; Hazen and Stevenson, 2008, p. 282).

Western law enforcement agencies most commonly use four handgun calibres, including the 9 mm Parabellum, .40 Smith & Wesson (S&W), .357 Magnum, and .45 Automatic Colt Pistol (ACP),[5] all of which progressively replaced the traditional .38 Special S&W cartridge more commonly fired by revolvers. Semi-automatic pistols have more recently been favoured by US and European police agencies. Advantages of 9 mm semi-automatic pistols include their smaller size, higher magazine capacities, and lower recoil compared to handguns chambered for the more powerful .45 automatic round, for instance (Kaestle and Buehler, 2005).

In the United States, the .40 S&W cartridge has increasingly been presented as an efficient upgrade for law enforcement use. Its larger bullet, though slower, applies greater force to the target; expanding bullets further heighten the impact through a 'mushrooming' effect, which seems to be an important criterion in the US ammunition selection process.[6] Although firearm instructors advocate the importance of shot placement and training rather than calibre or power (Kaestle and Buehler, 2005), US police departments consistently argue that the switch from 9 mm to .40 S&W is a necessary measure to avoid being 'outgunned' by local gangs and drug traffickers (Matteucci, 2010).

In September 2010, the US Bureau of Alcohol, Tobacco, Firearms, and Explosives (ATF) awarded two ten-year contracts worth USD 40 million each to GLOCK and S&W for .40 calibre service sidearms, namely the GLOCK 22 Gen4, GLOCK 27, and M&P40 pistols (*Atlanta Business Chronicle,* 2010; Sharp, 2010). In July 2010, the Federal Bureau of Investigation also purchased almost 3,000 GLOCK 23 pistols chambered in .40 S&W (FBO, 2010).

This trend may be influencing some police agencies outside of the United States. In October 2010, the Victoria police in Australia announced the replacement of their .38 Special S&W revolvers by semi-automatic S&W M&P pistols in .40 S&W (*Sydney Morning Herald,* 2010). In contrast, Western European countries with more centralized forces still procure 9 mm handguns and are not upgrading to larger-calibre weapon systems. The Dutch police force, for instance, recently chose the 9 mm SIG SAUER PPNL (Politie Pistool Nederland) as its new service pistol (Rijksoverheid, 2011).[7]

Many US police departments are also pushing for the procurement of powerful 'patrol carbines' or 'tactical rifles' (McKenzie, 2010; Spice, 2009). These semi-automatic rifles[8] use the .223 Remington/5.56 mm calibre initially carried by Special Weapons and Tactics (SWAT) teams in large metropolitan police departments. They are now procured by 'regular' police units on the grounds that their field officers encounter increasingly well-equipped gang members and criminals, some of whom have benefited from previous army firearms training (Main, 2010).

The rifles are typically 'collective'; that is, shared by officers from one shift to another. While this approach keeps costs down, semi-automatic rifles such as these require significant training, and large police forces may be unable to provide systematic, standardized training to all their personnel (Levenson and Slack, 2009). Procurement of the weapons, locking racks, ammunition, and training requirements represents a substantial expenditure for any police agency. For example, Pittsburgh's Public Safety Director reported a 2009 state contract order of 46 S&W M&P15 semi-automatic rifles totaling more than USD 37,000; the locking racks added an additional USD 30,000 (Lord, 2009).

Ultimately, the greater distribution of tactical rifles in the field can lead to losses and theft from police departments' patrol vehicles (Matteucci, 2010). Some experts also believe that, by being seen with these weapons, police officers will create 'concomitant pressure within the domestic population for individuals to arm themselves with similar weapons' (DeClerq, 1999, p. 38).

Focusing on ammunition: sub-machine guns, rifles, and sniper rifles

Firearm manufacturers have traditionally developed sub-machine guns, assault rifles, and sniper rifles with the more standardized needs of the military in mind. Consequently, domestic security organizations seeking to upgrade their traditional handguns and shotguns with more effective small arms have turned to the military market (see Table 3.1). For instance, standard bolt-action precision rifles traditionally used by Western police units are now upgraded with

Armed with SIG SG 550 assault rifles, members of a Swiss police intervention team arrive at a house in which an unidentified man barricaded himself, Biel, September 2010. © Pascal Lauener/Reuters

Table 3.1 Selected sub-machine guns, semi-automatic and assault rifles, and sniper rifles

Sub-machine guns	Semi-automatic and assault rifles	Sniper rifles
• Beretta M12	• Bushmaster Model XM15-E2S and Adaptive Combat Rifle (ACR)	• Blaser LRS2
• Brugger & Thomet MP9		• HS Precision Pro Series 2000 HTR
• CZ Scorpion Evo3 A1	• Colt M4 and Modular Carbine Model CM901	• Knight's Armament Company M110 Semi-Automatic Sniper System
• FN Herstal P90 (5.7 x 28)	• FN Herstal Special Operations Forces Combat Assault Rifle (SCAR)	
• Heckler & Koch MP5 and UMP		• PGM Précision Ultima ratio and Hécate II
• Heckler & Koch MP7 A1 (4.6 x 30)	• Heckler & Koch G36, 416 (= HK M27 IAR) and 417	• Remington Model 700 TWS (.308 Win.); M24 Sniper Weapon System
• Israel Weapon Industries UZI PRO (polymer)	• Ruger SR 556	
• Rock River Arms PDS Pistol	• SIG SAUER 516, 551, 556	• Savage 110BA
• SAAB-Bofors CBJ (6.5 x 25)	• S&W M&P15	• Tikka T3T
• ST Kinetics CPW	• Vektor R5	

the free-floated barrels[9] found on their military counterparts. Despite the fact that some manufacturers now capitalize on the law enforcement niche, the technological thrust invariably comes from the military community. The best way to anticipate what Western police units will be showcasing as 'collective' weapons in the near future is thus to look at military small arms and, more specifically, *ammunition* developments.

In this regard, the conflict in Afghanistan has been instrumental in upgrading the standard NATO 5.56 mm SS109/ M855 ammunition, which has been rumoured to suffer from range limitation, erratic target effectiveness, and poor barrier penetration (Arvidsson, n.d.). In 2010, the US Marine Corps announced its intention to upgrade a sizeable proportion of its conventional M855 5.56 mm ammunition stocks in Afghanistan with 5.56 mm Special Operations Science and Technology open-tip[10] rounds. The new round, formerly know as MK318 MOD 0, was specifically developed for short-barrelled rifles and designed to improve accuracy and 'barrier blind' performance against wind-screens, car doors, and other objects (Lamothe, 2010).

Almost simultaneously, the US Army issued more than 200 million rounds of a new 5.56 x 45 M855A1 Enhanced Performance Round to troops in Afghanistan to replace the general-purpose M855 ammunition. The M855A1 is referred to as 'green ammo' because it uses a lead-free projectile (Cox, 2010; PPAO, 2010). If the Marines and Army evaluations of both rounds remain positive, the rounds will probably draw the attention of the law enforcement community as well.

Experts hint that the Afghan conflict has also highlighted a gap between the 5.56 mm and the 7.62 mm, and that a new, versatile family of general-purpose small arms should be conceived around an intermediate calibre (Williams, 2010). Many of the features and requirements envisioned for these next-generation military small arms would also make them ideal for special law enforcement applications.

Some recently marketed products attempt to bridge the demands for interchangeability and versatility by combining the technical strengths of short-barrelled 5.56 mm rifles and long-range 7.62 mm rifles. Remington's Adaptive Combat Rifle, developed in cooperation with Bushmaster and Magpul Industries, boasts modular components and ambidextrous controls. The shooter can change calibres (5.56 mm, 6.5 mm, 6.8 mm Special Purpose Cartridge), barrel lengths, and stock configurations at will (Remington Military, 2010). FN Herstal emphasizes similar interchangeability capacity in marketing both versions of the Special Operations Forces Combat Assault Rifle (SCAR), the MK16 in 5.56 mm and the MK17 in 7.62 mm (Military.com, 2010). According to an industry expert, rifle modularity is still purely a military demand.[11] Yet, in the coming years, Western police units may procure some of these new rifles in their favoured calibre.

Regardless of the calibre issue, a reduction in weight is also appealing for law enforcement use. Nevertheless, the format of short-barrelled 5.56 x 45 rifles has considerable drawbacks for police use and training, such as excessive power, muzzle flash, blast, and recoil. Alternatives exist mainly in the form of pistol-calibre sub-machine guns, which are usually variants of compact military personal defence weapons (PDWs). Small-calibre, high-velocity rounds, presented as possible alternatives to the 9 x 19 mm calibre, have been developed for these weapons.

The 5.7 x 28 mm SS190 cartridge used in FN's P90 PDW and Five-seveN pistol is significantly slimmer and lighter than the 9 x 19 mm ball ammunition. This high-velocity, full metal jacket round was designed to increase short-range lethality. The round is meant to deposit most of its energy in the target, thereby limiting risks of ricochets in populated areas. In the 1990s, Heckler & Koch introduced a similar concept with the 4.6 x 30 mm cartridge for the MP7 machine pistol. The latest newcomers to the 'intermediate round' market appear to be Sweden's CBJ Tech with the 6.5 x 25 CBJ cartridge, and Knight's Armament Company's PDW in the new calibre of 6 x 35 mm (Williams, 2009).

Police tactical teams procure limited numbers of small-calibre, high-velocity systems. A small arms inventory provided by the French police administration mentions that its intervention unit procured a small but undisclosed number of FN P90 rifles (DGPN, 2010, p. 4). The costs of replacement and the requirements of calibre standardization prevent any mass procurement of these cartridges by large police organizations. In this context, it may not be clear why law enforcement agencies would need these intermediate calibres, especially since the use of soft-nosed and expanding bullets[12] by law enforcement officers in a necessary and proportionate use-of-force context has become common.[13] Many manufacturers market an extensive array of expanding ammunition specifically for law enforcement use. Examples include RUAG's Sintox Action 4 and Federal Premium's LE Tactical HST round—which features a copper jacket that folds out of the way during penetration while the round's petals fully expand (ATK, n.d.).

> Manufacturers have developed firearms with the more standardized needs of the military in mind.

Inputs from the civilian market: polymers, accessories, and customization

While soldiers are the primary source of inspiration for firearm manufacturers, it is clear that the customization niche, which bolstered the development of optical sights, light mounts, polymer frames, and ancillary equipment, has been greatly influenced by the sports shooting and hunting communities.

Law enforcement firearms have benefited from incremental developments leading to size and weight reduction. Handguns come in three categories—full-size, compact, and subcompact—with decreasing magazine capacities. Agencies typically choose the first two categories for general issue while specialized units, such as undercover or criminal investigation squads, prefer subcompact models as backup or concealed-carry firearms. Balancing between weight reduction, magazine capacity, and overall compactness has resulted in the development of models such as the SIG P238 (in .380 ACP), the S&W M&P compact, and the Springfield Armory XD(M) 3.8 (see Table 3.2).

Semi-automatic rifles will eventually benefit from similar developments. Experts increasingly refer to the 'bullpup' design—in which the barrel is moved back into the stock and the action and magazine are placed behind the trigger system—as the way forward. The concept is not new and can be combined with polymer technology and heat-resistant alloys to develop lighter, more manoeuvrable firearms while maintaining barrel length, and thus accuracy. For instance, Austrian and French law enforcement have been using the AUG Steyr and MAS FAMAS, respectively, for a number of years. It also appears that China has chosen this type of technology for its new service rifle, the QBZ-95G in 5.8 x 42 mm calibre. Further advances will most probably enhance magazine capacity (Crane, 2010).

Modern mouldings and polymer technology enable manufacturers to produce guns that combine various materials, for instance steel for the slide and aluminium or polymer for the frame. These materials allow handguns to be tailored to various uses, benefiting from the rust-resistant properties of stainless steel and from the tactical advantage provided by matte black, non-reflective finishes. The market is mostly populated by polymer-framed handguns featuring specific metal inserts, a combination favoured because of its durability and ease of maintenance (see Table 3.3). In addition, the use of polymers—which can be translucent—in the production of assault rifle magazines allows for rapid checks of ammunition supply.

Modern sights, developed to enhance target acquisition capabilities, are the driving force behind firearm development technology. Although advanced rifle sights are mainly procured by tactical teams, collective patrol rifles now feature Picatinny accessory rails that allow for the mounting of flashlights and laser sights to complement the basic iron sights (see Table 3.4). Low-powered 1X to 4X optic sights are being upgraded to variable-power optics with higher magnification in the 5–25X range. Modern riflescopes can now assist long-range target engagement without sacrificing the ability to deal with close threats. Not only do new systems magnify vision, but they can also accurately determine ranges using built-in laser range finders and image intensifiers to aim in low-light conditions. The US Defense Advanced Research Projects Agency is working on the 'One Shot' sniper scope programme, whose stated aim is to develop a ballistic calculation system prototype to fully assess ballistic trajectory and 'enable snipers to hit targets with the first round, under crosswind conditions, up to the maximum effective range of the weapon' (DARPA, 2010, p. 4; Calamia, 2010). Crosswind detection and compensation is in effect the cornerstone of future sniper scope developments. In this context, the Israeli Soreq Nuclear Research Center is developing 'Laser Identification Detection and Ranging' to detect the direction and velocity of any crosswind (Williams, 2010).

Table 3.2   **Selected compact handguns**

- Beretta Px4 Storm Type F subcompact
- Boberg Arms XR-9S pistol
- GLOCK 26
- Heckler & Koch USP Compact
- Ruger LCP (.380 ACP)
- S&W M&P compact pistol
- SIG P238 series (.380 ACP) & P290
- Springfield Armory XD(M) 3.8 Compact

Table 3.3   **Selected polymer handguns**

- Beretta 92FS, Px4 Storm
- Caracal
- FN Herstal NP9 and FNP40
- GLOCK series
- Israel Weapon Industries Jericho B
- Ruger LCR Double-Action Revolver and SR40
- SIG SAUER SP 2340, 2022, P226
- S&W M&P compact pistol
- Springfield Armory XD

Table 3.4   **Selected ancillary equipment**

- Aimpoint Concealed Engagement Unit
- Eotech Holographic Sights
- Horus Vision 1.5-8 x 24 mm BlackBird
- Corner Shot
- Leupold 1.1-8 x 24 mm CQBSS
- Premier Reticles V8 1.1-8 x 24 Tactical

Safety features: opportunities for 'police-specific' developments

Safety features are among the few technological developments specifically tailored for law enforcement use. Traditional safety options are external devices, such as cable locks. For the more popular semi-automatic handguns, integral safety features can be as diverse as standard manual hammer blocks, de-cocking levers, and trigger or magazine disconnect safeties. For manufacturers, reliable safety options represent important value added in firearm design, since law enforcement agencies rely on them to prevent or minimize unintentional discharges. Procurement agencies have the choice between single-action, double-action/single-action (DA/SA), or double-action-only handguns, depending on mission requirements and the level of training of personnel. Single-action handguns require the hammer to be cocked manually before the first round is fired; the slide cocks the hammer on subsequent shots. In handguns with double-action-only triggers, the trigger cocks the hammer and releases it; the trigger pull is thus longer and requires more weight, which can prevent negligent discharges. The hammer returns to its de-cocked position after each shot. Double-action/single-action handguns combine double-action for the first shot and single-action for subsequent shots. Some semi-automatic handguns now feature complex triggers with pre-set strikers or hammers, intended to reduce the hard trigger pull associated with the double-action mechanism. They also have a number of safety features that require specific training and familiarization (Kastle and Buehler, 2005).

Despite these systems, preventable injuries and deaths do occur, including of police officers. Between 2000 and 2009 in the United States, at least 41 officers were killed with their own weapon, and 69 had their weapon stolen (USDOJ, 2010a, table 13). Friendly fire accidents and shootings of police officers with their own service firearm periodically revive debates about the feasibility of 'smart', or personalized, gun systems (Long, 2009; Main et al., 2010). In the mid-1990s a number of US manufacturers received federal grants to explore and test personalized safety options. The US National Institute of Justice provided research funds to develop a safe gun system that would:

Safety features are among the few technological developments specifically tailored for law enforcement use.

- operate reliably in all environments;
- have all the capabilities of a current firearm;
- be able to be fired by other police officers;
- be easy to operate and maintain;
- verify and approve the user in the time it takes to draw and aim;
- only work when the transponder is behind the gun;
- include an indicator that tells the user if the system is enabled;
- fire even if the electronics fail (Taylor, 2000).

In response, Colt developed a wristwatch containing a transponder chip; Smith & Wesson explored a fully electronic locking system using a PIN code; and FN Herstal looked into microelectronic and ultrasonic wave technology. Industry interest soon faded, however. The main issue was the requirement for total reliability. Systems that use radio frequencies, for instance, could potentially be vulnerable to interference. Second, the stringent weapon requirements probably doomed the project from the start, not least because police consultants wanted the weapon to fail 'live'— that is, still be able to fire in case of system failure. This requirement is, according to an industry official, technically impossible to meet[14]. Moreover, the private sector preferred the weapon to fail 'dead' in order to make it more acceptable to the public (Taylor, 2000, p. 19). Furthermore, stakeholders of the project expressed uncertainties as to how this technology would influence civilian gun ownership—possibly creating additional unnecessary demand.[15]

In 2003, Australian gun maker Metal Storm briefly collaborated with the New Jersey Institute of Technology's Dynamic Grip Recognition programme. The system used biometric technology to match a single user to a single gun (NJIT, 2003; Di Justo, 2005). The resulting O'Dwyer variable lethality personalized smart gun prototype was designed to feature lethal and non-lethal capacities, as well as an aural signal to indicate readiness.[16] However, a lack of funding stopped further development of this technology.[17]

A man holds a prototype of a smart gun by Armatix during an international gun exhibition in Nuremberg, Germany, March 2009.
© Michaela Rehle/Reuters

Law enforcement and industry interest in personalized gun technology may again be growing. The US states of New Jersey and Maryland anticipated these trends and tasked oversight bodies to monitor the development of personalized handgun technology, and to report on its availability on the retail market (LCAV, 2008). One company investing in personalized gun technology is Germany's Armatix. Its iP1 pistol design incorporates most of the initial requirements of the US National Institute of Justice. Because the weapon is a .22 LR calibre, typically subject to fewer user restrictions, it may become widely known. According to the company's head of development, Armatix is currently approaching gun manufacturers worldwide to promote the iP1's Smart System technology. Armatix partnered with Anschütz to present a biathlon rifle with user and target control at the World Shooting Championship in Munich in July 2010.[18]

Less-lethal weapons

Less-lethal weapons have rapidly become the public and private sectors' primary technological response to perceived gaps in the force continuum. The defence industry trade show Eurosatory 2010 catalogue lists 17 exhibitors as either manufacturers or retailers of 'reduced lethality or incapacitating weapons and munitions', excluding lasers (barred from the exhibition) and anti-riot vehicle manufacturers (Eurosatory, 2010).

Much of today's less-lethal technology attempts to meet the military and law enforcement demand for longer-range weapons with variable effects. This section covers less-lethal technology that Western law enforcement agencies currently use or may use in the near future. It focuses on kinetic (impact), electric-shock, acoustic, and directed-energy weapons that a trained officer can use *discriminately* and *proportionately*. It omits incapacitating chemical weapons, which are distinct from riot control agents and have not been endorsed for police use by Western governments.

Increasing 'stand off' and accuracy

First-generation less-lethal weapons were designed for close engagement (less than ten metres). Some newer less-lethal systems, however, feature increased 'stand-off' range and discrimination capability; they allow security personnel to accurately engage or mark targets at a greater distance. As described below, they are commonly marketed as both individual targeting weapons and crowd control force multipliers; police use-of-force regulations do not always emphasize the doctrinal difference.

Kinetic energy weapons deliver blunt or penetrating trauma impact to the target. Some law enforcement agencies have been deploying very short-range, smoothbore, oversized-calibre launchers for a number of years, including the French 44 x 83 mm Verney-Carron Flash-Ball munitions (180 to 200 joules), with a limited stand-off range of about 12 metres (Verney-Carron, n.d.; Martin, 2008). Similarly, the 56 mm COUGAR and CHOUKA launchers manufactured by LACROIX–Alsetex fire 'oversized'-diameter impact projectiles but also dye-marking canisters, CS gas canisters, or combined-effects rounds. The COUGAR's 'Bliniz' round is filled with talcum and designed to maximize projectile deformation upon impact at very short distances (7–10 m). The round's diameter reportedly prevents any eye penetration. Nevertheless, the projectile delivers a significant 157 joules of energy at five metres.[19]

Most kinetic launchers now feature smaller diameters, 37 or 40 mm (see Table 3.5). Projectiles are either flexible or rigid, the latter increasingly fin-stabilized, full-bore diameter rubber projectiles used for long-range, accurate direct-fire shots. US companies such as Defense Technology Corporation, Combined Tactical Systems (which sells Combined Systems, Inc., products), and NonLethal Technologies, Inc., are major exporters of this type of ammunition (Mispo.org, 2010). The latest approach taken up by manufacturers and law enforcement agencies involves the use of single-shot, purpose-built *rifled* grenade launchers with break-open configurations that provide greater accuracy than traditional smoothbore launchers. Police agencies favour the single-shot configuration for weight reduction (see Table 3.5).[20]

In addition to extending the range, the obvious advantage of having a rifled tube is that, thanks to the gyroscopic effect, the projectile arrives nose first rather than tumbling. In the UK, the soft nose Attenuating Energy Projectile replaced the L21A1 round in June 2005. The rounds were submitted to the Defence Ordnance Safety Group for testing and to the Defence Scientific Advisory Council subcommittee on the Medical Implications of Less-Lethal Weapons for medical evaluation trials. The projectile was designed in 37 mm calibre to be used with the Heckler & Koch L104A1 launcher. The authorities that oversaw its development also argued that its non-standard calibre would reduce the likelihood of ammunition confusion (UKSG, 2006). Similarly, the French police have deployed the 40 x 46 Brügger & Thomet GL-06 single-shot launcher since 2008. Brügger & Thomet's technical sheet for the GL-06 advertises

Table 3.5 **Selected purpose-built less-lethal launchers**	
Single-shot	**Multiple-shot**
• Brügger & Thomet LL-06 & GL-06 (40 mm)	• FN Herstal FN 303 (.68 calibre = 18 mm)
• Condor AM 640 (40 mm)	• Police Ordnance Company Inc. ARWEN (37 mm)
• Defense Technology Corp. Rifled Barrel Launcher (40 mm)	• Sage Control Ordnance SL-6 (37 mm)
• LACROIX-Alsetex COUGAR/CHOUKA (56 mm)	• Verney-Carron Flash-Ball Super Pro & Compact (44 mm)
• Police Ordnance Company Inc. ARWEN ACE (37 mm)	
• Sage Control Ordnance SL-1 (37 mm)	

a 112-joule energy output capacity at five metres and 86 joules at 50 metres (Brügger & Thomet, n.d.). It is often fitted with an EOTech 512 electronic sight (DGPN, 2010, p. 6).

Modern kinetic energy weapons still suffer from deteriorating accuracy at longer distances (Kenny, Heal, and Grossman, 2001, p. 1). Officers thus face a conundrum whenever they have to balance range, accuracy, and kinetic energy if the impact is to remain less-lethal. According to US military research from the early 1970s, kinetic impact weapons that deliver energy greater than 122 joules inflict severe damage and should be considered lethal (Omega Foundation, 2000, p. 26, citing Egner, 1973). Recent research provides a more nuanced assessment, highlighting numerous criteria for blunt trauma depending on the type of injury one wants to avoid—skin penetration, head injury, or chest injury (Paulissen, 2010). It is therefore difficult to quantify effectiveness universally and impose a single threshold value for the muzzle velocity of impact projectiles (Papy and Pirlot, 2007). Energy delivery is not the only technical criterion worth considering: stand-off range, calibre, and projectile structure (gradient of hardness or softness) are equally critical, and the law enforcement community has yet to agree on common criteria for selection.

Police officers demonstrate the LRAD portable loud hailer—which was used to communicate with crowds during the G8 and G20 summits—during a technical briefing for media in Toronto, June 2010. © Mike Cassese/Reuters

Another domain that has directly benefited from the demand for increased 'stand-off' distances is acoustic technology. Acoustic devices are becoming increasingly accepted in US military and law enforcement circles thanks to a well-targeted marketing campaign. These devices use audible sound technology to deliver warning messages such as speeches, recordings, or 'alert' warning tones across a distance of up to 3 km (LRAD Corporation, n.d.b). Sceptics have been quick to point out that 'hailing devices' such as the ones manufactured and sold by the LRAD Corporation (specifically, the Long Range Acoustic Device) can produce tones of 150 decibels at one metre (Altmann, 2008, p. 45). Very short exposures to levels of 120 decibels can result in hearing loss (Lewer and Davison, 2005, p. 41). Considering the risk of permanent hearing damage, some analysts argue that these devices should be qualified as 'weapons' so that their procurement and deployment by government agencies may be properly monitored (Altmann, 2008, p. 53).

Acoustic weapons remain expensive and cumbersome to use; they have yet to be miniaturized in order to be of any significant interest for most law enforcement agencies. Nevertheless, the US police departments of Boston, New York, Pittsburgh, San Diego, and Santa Ana advertise their use of LRAD systems for a range of public order applications (LRAD Corporation, n.d.a). According to Mispo.org, the LRAD was used against protesters in 2007 in Tiblisi, Georgia, and in the US city of Pittsburgh during the 2010 G20 Summit (Mispo.org, 2010). In June 2010, a ruling by Ontario's Superior Court of Justice restricted the use of the 'alert' function (as opposed to the loudspeaker mode) on the 100X and 300X models purchased by the Toronto Police Service (Ontario, 2010). At the time of writing, there was no evidence of these devices being extensively procured or used by Western European law enforcement agencies. Nor does it appear that this family of less-lethal weapons has substantially affected use-of-force doctrine.

> To date, no single weapons technology has enabled users to switch at will from non-lethal to lethal.

The dream of rheostatic

The second axis of technological development for less-lethal weapons involves the search for a device that can deliver variable—or 'rheostatic'—effects along a continuum from lethal to non-lethal. Harkening back to science-fiction novels and comic books, the idea inspired the development of the new less-lethal products and 'combinations' found off-the-shelf today (see Box 3.1). Yet to date, no single weapons technology has enabled users to switch at will from non-lethal to lethal. As a result, practitioners have sought to fill the gap by combining less-lethal weapons with firearms.

Kinetic energy weapon developers are taking two distinct approaches to meet the perceived need for variable weapons. The first assumes that less-lethal launchers should be versatile, dual-purpose firearms or military grenade launchers that can be used either as stand-alone or mixed (under-barrel) systems (see Table 3.6). Standard 12-gauge pump-action shotguns, such as the Remington 870 and Mossberg 500, can fire a range of impact projectiles and bean bags. The advantage of using a shotgun for riot control is that the user can switch quickly from less-lethal projectiles to lethal shells; yet such switching increases the risk that the wrong ammunition will be used.

Modern dual-purpose shotguns, however, seem to be gaining in popularity. Metal Storm's semi-automatic MAUL shotgun prototype, for example, is actively marketed to law enforcement agencies. It fires a mix of blunt impact, frangible-nose chemical, and marker munitions, as well as lethal and door-breaching 12-gauge shells. It is sold either as a stand-alone weapon or as a module on a combat rifle. According to Metal Storm, the shotgun is still undergoing certification in the United States. In July 2010, the company announced a USD 3.4 million production contract to supply 500 MAUL shotguns and 50,000 rounds of non-lethal ammunition for the Correctional Service of Papua New Guinea. The first 50 weapons were due for delivery by February 2011.[21]

Box 3.1   Science fiction and less-lethal weapons

Science fiction and public relations campaigns have influenced policy and technology for law enforcement and military weapons, creating unrealistic expectations about less-lethal capabilities and colouring public perceptions.

In the 1960s and 1970s, private US companies sought to capitalize on the search for new unconventional law enforcement weapons. One idea that would later have a great impact on police use of force around the world was contained in a patent filed by Jack Cover in 1970 and awarded in 1974 for a 'weapon for immobilization and capture' (US Patent, 1974). In naming the weapon, Cover, an aerospace engineer who worked for NASA, was inspired by the 1911 children's adventure novel, *Tom Swift and His Electric Rifle,* in which the protagonist invents an 'electric rifle' that is adjustable from stun to kill. The adapted expression 'Thomas A. Swift's Electrical Rifle' provided the acronym he used: TASER (Woo, 2009; Kroll and Ho, 2009, p. 11).

Greater investment in more exotic weapons technology began in the early 1990s, in response to support in US military policy circles. Several of the key advocates in this process were science fiction enthusiasts. Janet and Chris Morris are science fiction authors who, in the late 1980s, worked at an influential think tank in Washington, DC, the US Global Strategy Council, led by former CIA Deputy Director Ray Cline. In 1990 Janet Morris authored *Nonlethality: A Global Strategy,* which presented an argument for particular policy on 'non-lethal' weapons and set out proposed weapons technology to fulfil it (Morris, 1990). Cline took up the issue with US President George H.W. Bush and Secretary of Defense Dick Cheney, and a strategy group was established, although initial policy proposals did not gain traction. The Global Strategy Council generated publicity along with others at the national laboratories, where the majority of research and development was taking place (Swett, 1993). Janet and Chris Morris joined an Independent Task Force set up by another Washington think tank, the Council on Foreign Relations, whose 1995 report on 'non-lethal' weapons contributed to the institutionalization of this policy in the US Department of Defense (CFR, 1995). The Morris team then began receiving military contracts to help develop these weapons (M2 Technologies, n.d.). They were also members of subsequent Council on Foreign Relations panels on the subject, publishing reports in 1999 and 2004 (CFR, 1999; Allison, Kelley, and Garwin, 2004).

A key individual at the Los Alamos National Laboratory, and also a science fiction enthusiast, was John Alexander, a former US Army colonel who has remained a high-profile proponent of these weapons. His 1980 article in *Military Review*–'The New Mental Battlefield: "Beam Me Up, Spock"'–called for US Army attention to 'psychotronic' weapons 'effecting sight, sound, smell, temperature, electromagnetic energy or sensory deprivation' as well as 'parapsychological' techniques such as remote viewing and extrasensory perception (Alexander, 1980). Alexander's involvement in US Army exploration of these fields was portrayed in Jon Ronson's lighthearted book, *The Men Who Stare at Goats,* adapted into a feature film in 2009 (Ronson, 2004). Alexander was closely connected with policy discussions and weapons development via the national laboratories and the military. He subsequently published two books advocating increased investment in these weapons (Alexander, 1999; 2003).

Proceeding alongside efforts to gain internal policy support in military and law enforcement organizations, concerted public relations campaigns, reflected in the media, have often exaggerated the capability of laser, microwave, acoustic, and chemical weapons technology. In 2002 George Fenton, then director of the Joint Non-Lethal Weapons Directorate, the US military body tasked with coordinating related weapons research and development, described a prototype 'pulsed energy projectile' weapon to a military industry magazine as 'the closest thing we have right now to the phasers on the television series *Star Trek*' (Kennedy, 2002). The programme has since been cancelled, although basic research continues on pulsed lasers with the aim of developing a weapon to cause pain without permanent damage (Hambling, 2008a). Fenton, now retired from the military, is a vice president for government and military programmes at TASER International (TASER International, n.d.a).

In late 2005 some details emerged about a previously secret laser weapons programme called the Personnel Halting and Stimulation Response (PHaSR), co-funded by US law enforcement and military agencies (Knight, 2005; Hambling, 2008b). It was specifically named by the developers as a tribute to *Star Trek* (Burgess, 2005). The weapon, which in its most recent prototype form had two lasers (one to temporarily blind, and the other to heat the skin), has not been fielded. According to Davison, the numerous claims about new weapons capabilities and the public relations efforts to persuade people that the real-life *Star Trek* phaser has arrived have created a demand for weapons that do not yet exist (Davison, 2010).

The myth of harmlessness is arguably perpetuated through the portrayal of certain weapons in fictional films. In a scene from *The Hangover* (2009), for instance, a police officer invites a group of children in a classroom to shoot a TASER at the groin and face of the two main characters. In real life, probe deployments targeting the eyes or genitals present serious risks of injury. In another comedy, *Meet the Fockers* (2004), a police officer uses a TASER in 'drive stun' mode directly against the chest of the lead character. Again, in real life, this type of use would be dangerous. Even TASER International advises against aiming the weapon at the chest (CNN, 2009). These new weapons are also portrayed and promoted in non-fiction. Perhaps the most popular television documentary series in this area is *Future Weapons* on the Discovery Channel. The documentary popularizes all manner of weapons, and the website invites viewers to 'test out' these weapons virtually (Discovery Channel, 2010).

Source: Davison (2010)

Table 3.6 Selected dual-purpose launchers

Shotguns	Single-shot grenade launchers	Multiple grenade launchers
• Metal Storm MAUL (semi-automatic) • Mossberg 500 • Remington 870	• Česká Zbrojovka 805 G1 • Colt M203 (40 mm) • Heckler & Koch AG36, AG-C/GLM, HK69A1 • Springfield Armory M79 (40 mm)	• Metal Storm 3GL • Milkor M32, MGL-140, MGL-105 • Rippel Effect MSGL40

The second approach is taken by manufacturers who develop purpose-built, less-lethal launchers (see Table 3.5) that can be combined with firearms in under-barrel configurations. The best representative of this category is FN Herstal's FN 303 in .68 calibre. Since its initial development in 1994, US police and military units have procured the FN 303; in 2004, the weapon was involved in a controversial incident in Boston (see below). The FN 303's innovation came from the fact that it was developed using paintball marker technology; it can be used to 'mark' aggressive individuals in a crowd, and single them out prior to arrest. The semi-automatic weapon uses compressed air to fire a range of less-lethal ammunition, such as paint capsules filled with oleoresin capsicum, which fracture on impact. Manufacturer specifications state that the kinetic impact at 15 metres is 29 joules ($15J/cm^2$)—enough to stop any suspect (Jacobs, n.d.). The FN 303 can be integrated as a separate module under a rifle, a characteristic that is marketed as giving the police officer or soldier the ability to use a less-lethal system as well as a conventional firearm. There is no consensus on requirements among manufacturers or police agencies on this issue. France's LACROIX, for instance, has chosen *not* to develop modular launchers.[22]

'Rheostatic' implies variable electrical resistance, and electric-shock weapons are direct descendants of this concept. In the past ten years, TASER International, Inc., has had a significant presence in the less-lethal market, with an increased diversification of products that cater to the military, law enforcement agencies, and the civilian self-defence niche. Other companies have capitalized on this success, such as Stinger Systems with its S-200 AT, an alternative two-dart projectile stun device. The technology used in the TASER Models M26 and X26 is therefore not new;[23] in the United States, it has arguably influenced police use-of-force doctrine like no other weapon in the less-lethal category (see below).

TASERs are actively marketed as ideal alternatives to impact projectiles and hand-held CS sprays. Yet electric-shock weapons follow development paths surprisingly similar to their kinetic energy siblings: range extension, and the possibility of combination with firearms. The X-Rail System can be used to clip a TASER X26 to assault rifles, mimicking the modularity of 40 mm grenade launchers. More significantly, the new eXtended Range Electro-Muscular Projectile (XREP) can be fired from any 12-gauge shotgun to a range of 30 metres. The projectile thus combines the kinetic energy of blunt impact with an electric shock that causes the target's muscles to contract involuntarily. The recently marketed X3 model replicates the spread of multiple-projectile kinetic ammunition. Finally, the TASER Shockwave Remote Area Denial system fulfils the needs of area denial by juxtaposing multiple units to extend area coverage and allow multiple salvos (TASER International, n.d.b).

Directed-energy weapons are another offshoot of the US military's research into rheostatic capability. Such research has produced weapon prototypes, including high-power microwave weapons designed to destroy the circuits of electronic equipment. The best-documented example is the Active Denial System, which uses millimetre waves to heat up water and fat molecules in the subcutaneous layers of the skin. Raytheon Corp. has been marketing the product to military and law enforcement agencies since 2007 under the name Silent Guardian. While military directed-energy weapon prototypes are generally too bulky for individual law enforcement use, the US National Institute of Justice and Raytheon Corp. are reportedly developing smaller versions (Davison, 2009, p. 170). In mid-2010, the Los Angeles County Sheriff's Department equipped the Pitchess Detention Center with a 2.3-metre, tripod-mounted version of the Active Denial System, dubbed the Assault Intervention Device, for a six-month trial (Hadhazy, 2011; Shachtman, 2010).

Today's most appropriate, off-the-shelf options for law enforcement applications are light-emitting diode (LED)-based systems dubbed 'illuminators' or 'dazzlers', such as the LED Incapacitator. Developed by Intelligent Optical Systems, Inc., in collaboration with the US Department of Homeland Security and the Defense Advanced Research Projects Agency, LED Incapacitators are marketed as:

> diversionary devices [that] utilize high powered, multicolored light emitting diode (LED) clusters and complex optical concentrators to produce a temporary high level of visual impairment and potential disorientation of the targeted subject(s), while optical intensities remain at eye-safe levels (Intelligent Optical Systems, n.d.).

Directed-energy weapons are an offshoot of the US military's research into 'rheostatic' capability.

Laser-based weapons, such as the Dissuader Laser Illuminator, have been in use for a number of years. This weapon is described in a US Department of Justice report as a handheld laser 'flashlight' that produces a powerful variable-width beam of red light 'that forces adversaries to signal their intent to retreat, surrender, or continue aggressive behavior' (USDOJ, 2004, p. 35). Modern laser 'dazzlers' use green lasers because the human eye is more sensitive to their wavelength (Davison, 2009, p. 151). The US military's deployment of dazzlers in Iraq and Afghanistan is likely to inspire law enforcement units to adopt them in the near future.

MATCHING WEAPONRY WITH DOCTRINE

Police and military overlap

In the United States, the militarization of police firearms described above also applies to less-lethal weaponry. The cross-fertilization of police and military procurement is even clearer in this category because it highlights the fact that agencies with supposedly very different operational cultures are nevertheless using the same arsenal. The US law enforcement community's strong shift towards militarization is characterized by the development, procurement, and use of weapons that tend to blur law enforcement use of force and military doctrine. In contrast, European doctrine emphasizes the specificity of law enforcement approaches to the use of force.

In October 1999, NATO approved a new Non-Lethal Doctrine to facilitate their future peacekeeping and peace-enforcement deployments (NATO, 1999). Interestingly, transfers of non-lethal technology from the military to state and local police forces in the United States also began to be institutionalized in the late 1990s (USDOJ and USDOD, 1997, p. 8). In fact, the development of less-lethal weaponry involved both military and law enforcement input, usually in the form of joint research and funding from the National Institute of Justice—the research, development, and evaluation arm of the US Department of Justice—and the Department of Defense's Joint Non-Lethal Weapons Program (Davison, 2010). Thus many less-lethal weapons programmes were initiated with both military and domestic law enforcement applications in mind.

Despite these research and development efforts, most of the less-lethal weapons recently adopted by the US military and law enforcement are 'off-the-shelf', commercially available products developed by the private sector (such as the TASER X26 and FN 303). This may explain why police departments and infantry units actually procure very similar less-lethal arsenals. The *Department of Defense Nonlethal Weapons and Equipment Review* enumerates a series of commercially produced impact projectiles (such as Defense Technology Corp.'s 40 mm Area Round) that it considers suited for both law enforcement and military use (USDOJ, 2004, p. 28). Similarly, Non-Lethal Capability Sets used by the US Marines and Army include off-the-shelf kinetic (FN 303) and electric (M26 and X26) TASERs, along with chemical, optical, and flash-bang systems (Davison, 2007, p. 19; Calloway, 2008).

Police and military applications are not the same, however. The US military sees 'non-lethal' weapons primarily as an adjunct to lethal force (USDOD, 1996, para. 4.7). Recent military documentation refers to these weapons as battlefield force multipliers that can be used to coerce enemies 'to move from hiding, and thereby be exposed to lethal effects' (TRADOC, 2008, p. 84). This approach emphasizes less-lethal weapons as a *prelude* to lethal force, not an alternative. It may also explain why, in the force continuum, the gap between the use of less-lethal weapons and firearms is relatively small in US law enforcement, whereas it is more significant in European policing practice.[24]

Yet the military approach ignores the constraints and weapon requirements specific to law enforcement, such as:

- portability;
- affordability;
- shorter stand-off ranges;
- high safety standards;
- compatibility with domestic use-of-force legislation;
- 'acceptability' to local and national media; and
- the presence of civil society and democratic control mechanisms.

> **The military approach ignores the constraints and weapon requirements specific to domestic law enforcement.**

Since police officers traditionally work within a strict legal framework based on minimum, discriminate, and proportionate use of force, they should arguably be setting the norms and technical development standards for less-lethal weapons to be used both domestically and abroad, such as in peacekeeping settings. The French approach, for instance, sees law enforcement as the primary provider of crowd and riot control know-how to the military. French Army infantry units are trained for crowd control by law enforcement personnel; they procure less-lethal weapons that have been tried, tested, and used domestically for years.[25] French infantry units serving in Kosovo in early 2000 were first trained by gendarmerie instructors and provided with domestic crowd control equipment before being sent to the Balkans (de Laforcade, n.d., p. 20; Merchet, 2008). The military clearly attributes the techniques and equipment they are using abroad to the gendarmerie:

> *Obviously, during such out-of-area crowd control actions, the committed units apply, at the subordinate levels (company-taskforce, platoon) the know-how used by Gendarmerie on the national territory* (de Laforcade, n.d., p. 6).

In 2008, the South African Police Service asked the French Compagnies Républicaines de Sécurité, the riot control body of the civilian *police nationale,* and the gendarmerie to train their public order and rapid response teams in crowd management techniques in preparation for the 2010 FIFA World Cup. The training focused on platoon manoeuvrability in a simulation of clashes between supporters and stadium stewards, intimidation of players, and the evacuation of VIPs (Gabara, 2008). This cooperation undoubtedly created business opportunities for riot control equipment manufacturers. Moreover, the South African police clearly chose to upgrade their doctrine as much as their gear. French crowd control methods have been described as deliberately passive and defensive (France24, 2009; Mispo.org, 2010); these had already been proven effective during the 1998 Football World Cup and the 2007 Rugby World Cup. Every year, the Centre national d'entraînement des forces de gendarmerie in Saint-Astier trains close to 10,000 officers from various French and European security organizations (CNEFG, n.d.). The centre has traditionally promoted *passive* riot control techniques in which weapons are displayed by troops at the very last moment in order to promote a high threshold for the use of force (Gendarmerie Nationale, 2007, p. 8).

Discriminate targeting or crowd control?

The proliferation of off-the-shelf less-lethal products sometimes encourages police agencies to procure weapons without considering whether the equipment is intended for individual targeting or crowd control requirements. Whereas CS grenades are tacitly understood by police as a means to break up unruly crowds *indiscriminately,* there is no official international standard on how other, more recent weapons systems, such as kinetic launchers, should be used. In many cases, manufacturer recommendations substitute for doctrine-led procurement.

A case in point is the controversial death of Victoria Snelgrove in Boston during baseball's 2004 American League Championship Series. When police resolved to clear a crowd celebrating a Boston Red Sox victory, an officer fired two pepper ball rounds from an FN 303 launcher, allegedly aiming for a specific individual moving in the crowd. Although Victoria Snelgrove was not close to the group, one of the rounds hit her left eye, entered the brain, and killed her. The 2005 Stern Commission report notes that the Boston Police Department procurement officials knew very little about the FN 303's technical characteristics, maintenance, or operational deployment prior to purchasing the weapons. Furthermore, the weapons had been deployed in the field without a solid training programme or accompanying rules for use. The report also highlights significant gaps in the operational planning of the event, which further contributed to the FN 303 launchers being deployed improperly due to a blurry chain of command (Stern et al., 2005, pp. 3–5; Conley, 2005).

Other incidents involving extended-range impact weapons in crowd control were reported in Seattle during the 1999 World Trade Organization protests, in Los Angeles during the Democratic National Convention in 2000, in Cincinnati during demonstrations following the controversial police shooting of Timothy Thomas in 2001, and during anti-war protests at the Port of Oakland in April 2003 (Stone, Buchner, and

Crowd control training at the Centre national d'entraînement des forces de gendarmerie in Saint-Astier, France, 2001. Unable to use less-lethal launchers to accurately target fast-moving rioters through curtains of smoke, the troops use extended batons in barrage formation. © Pierre Gobinet

Dash, 2005). The issue was also raised in Paris when a 16-year-old student was hit in the eye and wounded, this time at short range, by a 44 x 83mm Verney-Carron Flash-Ball projectile during the October 2010 pension reform demonstrations (*Le Figaro,* 2010).

These events highlight the pitfalls of using long-range impact weapons for crowd control without a strict, specific use-of-force policy and proper personnel training. Among other things, use-of-force doctrine should allow for clear distinctions between discriminate targeting and crowd control. A 2006 report by the UK Steering Group in consultation with the Association of Chief Police Officers clearly states that the new Attenuating Energy Projectile:

> *has not been designed for use as a crowd control technology but has been designed for use as a less lethal option in situations where officers are faced with individual aggressors whether such aggressors are acting on their own or as part of a group* (UKSG, 2006, p. 7).

Distinguishing lethal from less-lethal in the field

A common argument in academic and advocacy circles is that the wide availability of less-lethal weapons has led to the relaxed use of such weapons as well as to an erosion of proportionality in the use of force. The weapons are criticized as making the use of force more acceptable and, overall, more widespread. Since the early 1990s, progress has been made by the scientific community to provide independent biomedical research, and by less-lethal weapons manufacturers to improve quality control at the production level. Yet developments are hampered by a glaring lack of internationally agreed tests and standards (Paulissen, 2010). While it is clear that less-lethal weapons have important weaknesses, in considering their overall utility it is important to weigh the alternatives, including the use of firearms.

The recent generation of less-lethal weapons does not *replace* lethal weapons but is deployed *in addition* to them.

Electric-shock weapons exemplify some of the uncertainties inherent in use-of-force decision-making. The police community generally endorses the proportionate use of TASERs to save lives, reduce injuries to officers and suspects, and ultimately contribute to a positive outcome (Hawk, 2010). In 2010, for instance, the San Francisco Police Department undertook an elaborate and comparative statistical analysis of annual officer-involved shootings to advocate, in the end, for the procurement of TASERs to reduce the use of firearms even further (Wyllie, 2010). In the minds of some practitioners, there is thus no question that TASERs have provided police with a much needed use-of-force option.

Opponents of the use of TASERs point out that the alleged flexibility they offer can conflict with norms governing the proportionate use of force. General Provision 2 of the UN Basic Principles on the Use of Force and Firearms by Law Enforcement Officials advocates the development and use of non-lethal incapacitating weapons (1) 'with a view to increasingly restraining the application of means capable of causing death or injury to persons', but also (2) 'to decrease the need to use weapons of any kind' (UN, 1990). It is not clear that either has happened.

The recent generation of less-lethal weapons, and specifically TASERs, is not *substituted* for lethal weapons, but deployed *in addition* to them, with officers who already carry a sidearm (South Yorkshire Police, n.d.).[26] This approach is deeply rooted in police practice. In September 2007, the UK Home Secretary provided an exception by allowing non-firearms officers to use TASERs under a controversial pilot scheme; the initiative was refused by some of the country's largest forces, including the Metropolitan Police (Whitehead, 2009).

Other examples cast doubt as to whether less-lethal weapons have effectively reduced the police use of firearms. A recent audit of the Houston Police Department reported no significant reduction in the number of police shootings of civilians since the introduction of electric-shock weapons (Parker and Schoonover, 2008, p. 12). A 2010 report on the use of TASER weapons by the Western Australia Police found that despite an overall 25 per cent increase in TASER use between 2007 and 2009, the use of firearms by police doubled during the same period (CCC, 2010, p. 13, para. 7).

Secondly, academic, police, and media reports suggest that the increased availability of less-lethal weapons for routine police operations may have lowered the threshold for use-of-force against passive resisters, and widened the net of people against whom the police has used coercion, including minorities (Davison, 2009, p. 5; CNN, 2010; AI, 2008, p. 22).

According to a 2010 company press kit, more than 15,500 law enforcement agencies in 40 countries have purchased at least 499,000 TASER devices since 1998 (TASER International, 2010). The company website states that more than 86 per cent of these agencies place the use of a TASER device at the level of pepper spray deployment. Yet police agencies have not come to any consensus with respect to where TASERs should be deployed within a use-of-force continuum. In Figure 3.1, the use of less-lethal weapons is just one step below firearms, meaning that it ranks as a responding officer's final attempt to avoid the use of lethal force.

Inversely, agency policy can place TASERs *early* in the use-of-force continuum, notably at active resistance level, in order to *prevent* the conflict from escalating into assault. However, if TASER deployment is placed too close to the beginning of the continuum, for example as an accessory to verbal action, their use may be seen as trivialized, with verbal exchanges quickly escalating to the use of physical force. Indeed, TASERs have

Wearing crowd-control gear, police hold shotguns during a demonstration in Oakland, California, after Johannes Mehserle was found guilty of involuntary manslaughter in the shooting death of Oscar Grant, July 2010. © Paul Sakuma/AP Photo

reportedly been used as compliance tools on restrained individuals or passive resisters (Cook, 2010).

A number of cases have seen officers immediately deploying and firing a TASER on the grounds that the suspect was visibly disturbed, delusional, or prone to 'excited delirium', and thus not receptive to verbal efforts to de-escalate the situation. This argument was provided by the Royal Canadian Mounted Police officers who used TASERs against Robert Dziekanski at Vancouver International Airport in October 2007. Dziekanski, who did not speak English, had spent ten hours at the airport and had begun throwing furniture in the international terminal because he was unable to find his mother or communicate with anyone. Dziekanski was 'tased' a total of five times by four officers; he subsequently died. An enquiry into the event concludes that, in the case of emotionally disturbed subjects, 'the best practice is to de-escalate the agitation, which can best be achieved through the application of recognized crisis intervention techniques' (Braidwood Commission, 2009, p. 15).

Nevertheless, the placement of TASERs further up the continuum, as an *accessory* to firearms, can facilitate confusion regarding weapons use among improperly trained personnel.[27] In the United States, one of the most controversial recent cases of weapons confusion involving a TASER and a handgun is the shooting of Oscar Grant by police officer Johannes Mehserle on a Bay Area Rapid Transit platform in Oakland, California, on 1 January 2009. Following a fight on a train involving New Year's Eve revellers, police proceeded to handcuff and detain a number of people, including Grant. A bystander's cell-phone video footage shows the latter being taken to the ground to be handcuffed, and physically resisting. The use-of-force expert involved in the trial testified that Mehserle, visibly under stress, attempted to undo what he thought was a TASER holster safety strap, and instead ended up pulling out his SIG SAUER P226. He performed all the steps consistent with a TASER dart deployment and fired once in Grant's back (Meyer, 2010).

The expert explained the role that holster configurations may have played in Mehserle's mistake and advocated for TASERs to be placed in weak-side holsters requiring a weak-hand draw to reduce the risk of weapon confusion. His testimony also highlights the department's lack of realistic, stress-inducing training scenarios to practice force-options decision-making. A few months later, the local media reported that the Bay Area Rapid Transit Police Department had removed TASERs from the field after an officer had fired the darts at a 13-year-old boy fleeing from police in Richmond on his bicycle. Police staff declared that the force would be retrained in updated TASER use policies (Bulwa, 2010).

Police do not agree on where TASERs should be deployed within a use-of-force continuum.

The private sector is actively working to bridge the divide between lethal and less-lethal weapons. The Portugal-based company Inventarium Security, Research & Development is apparently looking to commercialize the Biggun, or 'Begane', a weapon that can alternately be used as a .38 Special-calibre firearm, a baton, an electrical discharge weapon, or a pepper spray launcher (Begane, 2010). Whatever their tactical advantages, such products will probably present increased risks of weapons confusion.

Doctrine defines the tools

In 2007, the Boston Police Department melted down their FN 303 semi-automatic less-lethal launchers and refashioned them into sewer caps (Shea, 2007). According to local media, in 2006 the family of Victoria Snelgrove settled a USD 10 million wrongful death suit against FN Herstal. They argued that the FN 303 was inherently inaccurate and lethal, thus shifting the responsibility from the Boston police to the launcher's manufacturer (Murphy, 2006). Yet the Stern Commission determined that the lack of proper tactical judgement, training, and doctrine contributed as much to Snelgrove's death as did the launchers (Stern et al., 2005. pp. 3–5).

It is clear that standard operating procedures and tactical rules of engagement, which together define the implementation of police use of force, need to take precedence over the procurement of hardware. This rule of thumb explains the UK Home Office's decision to revoke the licence of Pro-Tect Systems, a company that had supplied X12 TASERs and XREP ammunition directly to the Northumbria Police during a six-hour stand-off with a gunman. At that time, these TASER models had only been available for supply to the Home Office science and development branch for testing (BBC News, 2010).

In countries where law enforcement is fragmented and decentralized, each police department may use different use-of-force guidelines. In some cases, doctrine defines the tools. Wisler describes the differences in crowd control doctrines employed by Switzerland's French- and German-speaking police, and how the doctrines have dictated the use of different weapons. Geneva police units train for riot control with their French counterparts; they thus use more traditional methods, allowing proximity and physical contact with the crowd. In contrast, police in Zurich systematically *avoid* direct contact. Zurich riot police therefore employ CS grenades and impact projectiles on a regular basis. Their colleagues in Geneva do not, although they are legally allowed to (Wisler, 1997, p. 7).

In cases where a police agency's weapons procurement has overtaken its doctrine, two important measures can reverse the trend. The first is provincial, national, or federal harmonization. In Canada, the Braidwood Commission highlighted a lack of consistency regarding electric-shock weapon use and training policy, as well as a lack of direction from provincial governments (Braidwood Commission, 2009, p. 9). In the United States, the final recommendations of the Stern Commission call for the federal government to issue more guidance on national standards, testing, and certification for new less-lethal weapons, regardless of the technology being used (Stern et al., 2005, p. 42). Centralized police organizations seem structurally better suited to the harmonization of use-of-force doctrine, as evidenced by a national directive regarding the standard use of electric-shock weapons by gendarmerie units (DGGN, 2006).

The second factor is accountability and independent civilian oversight (Coginta, 2010). The UN Basic Principles assert that:

> Governments and law enforcement agencies shall ensure that an effective review process is available and that independent administrative or prosecutorial authorities are in a position to exercise jurisdiction in appropriate circumstances (UN, 1990, art. 22).

The Stern Commission echoes this concern by urging independent civilian oversight of police actions (Stern et al., 2005, pp. 41–42; Stone, Buchner, and Dash, 2005, p. 6). Similarly, the Braidwood Commission recommends the creation of a civilian-based criminal investigative body in British Columbia, the Independent Investigation Office (Braidwood Commission, 2010, p. 24).

The private sector is well aware that the need for accountability introduces market opportunities. Most of the electric-shock weapons procured by Western law enforcement agencies already feature miniature cameras and the bar-coded serialization of each cartridge to improve traceability and officer accountability. Another development in this field comes from the Brazilian manufacturer Condor, which produces the I-REF line of traceable grenades. All of the company's non-lethal, tear gas, flash-bang, and smoke grenades are assigned a reference number that can be read electronically. This allows traceability and accountability throughout a product's life cycle—even after detonation, as each grenade's components can be traced back to the user.[28] This initiative could catch on with other kinetic energy projectiles, which are often criticized for their lack of ballistic traceability.

CONCLUSION

This chapter identifies the trends that are likely to influence the development and procurement of firearms and less-lethal weapons by Western law enforcement agencies.

In general, the procurement of new police weaponry reflects military trends and precepts. In the case of US law enforcement, this pattern is exemplified by the increased use of .40 S&W handguns, and tactical rifles to match criminal firepower. Observing military small arms development is the most reliable way to predict what police officers will be issued in the near future. Conversely, the only area to benefit from direct police input is the development of firearm safety features, which has yet to attract significant research funds and continued manufacturer interest.

Since firearms have not undergone any evolution significant enough to modify the police approach to use of force, all hopes have turned to less-lethal weapons to provide much needed force-option flexibility in the field. Less-lethal weapons have not replaced firearms but have been added to the police officer's tools for intermediate coercive means. Whereas first-generation less-lethal weapons originally permitted only close engagements, new technology increasingly allows officers to target discriminately from a distance. Moreover, the private sector appears intent on progressively marketing fully scalable, lethal-to-non-lethal weapons. To date, for lack of a truly 'rheostatic' option, manufacturers are addressing the demand for technology by combining less-lethal systems with firearms.

Decentralized, market-driven procurement of some intermediate weapons technology can prove detrimental to the development of a coherent use-of-force doctrine and related police training. There is a risk, as in the United States, that new technology—and associated marketing efforts—influence police procurement and use-of-force doctrine, not the other way around. Police forces may be acquiring new products and operational policies that are not suited to their structure, doctrine, or environment. This trend is not as present in Western Europe, where more centralized police structures have helped strengthen and harmonize use-of-force policies. Nevertheless, it remains to be seen whether marketing and 'peer' pressure will prompt European law enforcement agencies to adopt procurement patterns similar to those in the United States.

In the end, the notion of a force continuum, designed to represent proportionate responses to specific threats, can only serve as a guide to decision-making in the field. It does not account for the complexity of circumstances faced by individual officers. Nor does it imply that there should be a weapon capable of covering the whole range of options. Despite the fact that a number of less-lethal weapons have already proven effective, it is clear that these weapons cannot, on their own, improve police use of force, or replace proper doctrine and training. ▰

LIST OF ABBREVIATIONS

ACP	Automatic Colt Pistol
ATF	US Bureau of Alcohol, Tobacco, Firearms and Explosives
LED	Light-emitting diode
LRAD	Long Range Acoustic Device
M&P	Military & Police
PDW	Personal defence weapon
S&W	Smith & Wesson
XREP	eXtended Range Electro-Muscular Projectile

ENDNOTES

1   France has a dual law enforcement structure, composed of the civilian *police nationale* and the military *gendarmerie nationale*—both of which have public order, judicial policing, and crowd control duties. For convenience, these two components are described jointly as 'law enforcement' or 'police' in this chapter. See Gobinet (2008) for more on dual police structures.

2   For further information on the UN Basic Principles, see Wood and McDonald (2004).

3   Author interview with a marketing representative from a Western small arms and light weapons manufacturer, July 2010.

4   Author interview with Bertrand Colin and Marc-Antoine Galzin, LACROIX–Alsetex, Paris, 3 September 2010.

5   European calibre notations use projectile diameter multiplied by length of casing in millimetres with bullet mass expressed in grams (g). US notations use projectile diameter in decimal fractions of an inch, with bullet mass expressed in grains (gr).

6   Author interview with a marketing representative from a Western small arms and light weapons manufacturer, July 2010.

7   See also the above reference to France's purchase of 9 mm SIG SAUER SP2022 handguns.

8   On semi-automatic rifles, the trigger must be pulled each time a shot is fired. Fully automatic rifles continue firing when the trigger is pulled. The term 'assault rifle' usually includes the requirement that the weapon can be switched from semi-automatic to fully automatic ('selective fire').

9   The barrel never touches the stock material. When fired, the bullet creates a whipping action in the barrel, enhancing consistency of the shots.

10   In the past, US authorities have stressed the structural and legal distinctions between open-tip and hollow-point bullets to legalize the use of sniper ammunition such as the M852 Sierra MatchKing or M118 rounds (Parks, 1990).

11   Author interview with a marketing representative from a Western small arms and light weapons manufacturer, July 2010.

12 Soft-nosed ammunition applies most of its kinetic energy to the target by deforming on impact, thus producing a larger wound cavity. Consequently, the projectile is less susceptible to ricochet and less likely to cause collateral casualties in a densely populated area.

13 Law enforcement agencies (and hunters alike) use considerably more effective expanding bullets that are otherwise explicitly forbidden for warfare by the Hague Declaration concerning expanding bullets of 29 July 1899 (Coupland and Loye, 2003).

14 Author interview with a marketing representative from a Western small arms and light weapons manufacturer, July 2010.

15 Author interview with a marketing representative from a Western small arms and light weapons manufacturer, July 2010.

16 Author correspondence with Arthur Schatz, vice president of business development, Metal Storm, 14 June 2010.

17 Author correspondence with Arthur Schatz, vice president of business development, Metal Storm, 14 June 2010.

18 Author correspondence with Karl Giebel, head of development, Armatix, 11 August 2010.

19 Author interview with Bertrand Colin and Marc-Antoine Galzin, LACROIX–Alsetex, Paris, 3 September 2010.

20 Author correspondence with Massilon Miranda, international business manager, Condor Naoletal, 10 August 2010.

21 Author correspondence with Arthur Schatz, vice president of business development, Metal Storm, 10 July 2010.

22 Author interview with Bertrand Colin and Marc-Antoine Galzin, LACROIX–Alsetex, Paris, 3 September 2010.

23 The M26 advanced TASER fires two probes up to 6 m from a gas cartridge. These probes are connected to the weapon by high-voltage insulated wire. The weapon uses electro-muscular disruption to override the central nervous system and trigger an uncontrollable contraction of the muscle tissue.

24 This view was also expressed by a marketing representative from a Western small arms and light weapons manufacturer, July 2010.

25 Author interview with Bertrand Colin and Marc-Antoine Galzin, LACROIX–Alsetex, Paris, 3 September 2010.

26 The situation is different for crowd control technologies. Aggressive rioters in the first half of the 20th century were often contained by regular police troops (or even army units) using firearms. In most Western countries, the introduction of specialized, non-lethal riot control equipment, such as CS grenades, undeniably allowed many police organizations to professionalize their riot control doctrine and practice, and thus replace the use of firearms altogether.

27 Author interview with a business representative from a Western small arms and light weapons manufacturer, July 2010.

28 Author correspondence with Massilon Miranda, International Business Manager, Condor Naoletal, 10 August 2010.

BIBLIOGRAPHY

AI (Amnesty International). 2008. *'Less Than Lethal?' The Use of Stun Weapons in U.S. Law Enforcement*. 16 December.
   <http://www.amnesty.org/en/library/asset/AMR51/010/2008/en/530be6d6-437e-4c77-851b-9e581197ccf6/amr510102008en.pdf>

Alexander, John. 1980. 'The New Mental Battlefield: "Beam Me Up, Spock."' *Military Review*, Vol. LX, No. 12. December.
   <http://www.icomw.org/documents/alexander.pdf>

—. 1999. *Future War: Non-Lethal Weapons in Twenty-First-Century Warfare*. New York: St. Martin's Press.

—. 2003. *Winning the War: Advanced Weapons, Strategies, and Concepts for the Post-9/11 World*. New York: St. Martin's Press.

Allison, Graham, Paul Kelley, and Richard Garwin. 2004. *Nonlethal Weapons and Capabilities*. New York: Council on Foreign Relations.
   <http://www.cfr.org/content/publications/attachments/Nonlethal_TF.pdf>

Altmann, Jürgen. 2008. *Millimetre Waves, Lasers, Acoustics for Non-Lethal Weapons? Physics Analyses and Inferences*. Osnabrück: Deutsche Stiftung Friedensforschung. <http://www.bundesstiftung-friedensforschung.de/pdf-docs/berichtaltmann2.pdf>

Arvidsson, Per. n.d. 'Is There a Problem with the Lethality of the 5.56 NATO Caliber?' Presentation for the NATO Army Armaments Group.
   <http://www.dtic.mil/ndia/2010armament/WednesdayLandmarkBPerArvidsson.pdf>

ATK. n.d. 'ATK: Law Enforcement.' Company website. <http://le.atk.com/general/federalproducts/pistol/tacticalhst.aspx>

Atlanta Business Chronicle. 2010. 'ATF Gives GLOCK $40M Contract.' 8 September. <http://www.bizjournals.com/atlanta/stories/2010/09/06/daily21.html>

BBC News. 2010. 'Raoul Moat Taser Firm Licence "Revoked."' 28 September. <http://www.bbc.co.uk/news/uk-england-tyne-11425262>

Begane. 2010. Company website. <http://www.begane.com/html/home.html>

Braidwood Commission (Braidwood Commission on Conducted Energy Weapon Use). 2009. *Restoring Public Confidence: Restricting the Use of Conducted Energy Weapons in British Columbia*. June. Victoria, British Columbia: Braidwood Commission. <http://www.braidwoodinquiry.ca/report/>

—. 2010. *Why? The Robert Dziekanski Tragedy*. Victoria, British Columbia: Braidwood Commission. 20 May.
   <http://www.braidwoodinquiry.ca/report/P2_pdf/00-TitlePage.pdf>

Brügger & Thomet. n.d. 'Technical Specifications: B&T GL-06.' <http://www.bt-ag.ch/pdf/TS-2889.pdf>

Bulwa, Demian. 2010. 'BART Cops Lose TASER Privileges: Announcement Comes after Officer Shocked 13-year-old.' *San Francisco Chronicle*. 16 April.
   <http://www.policeone.com/police-products/less-lethal/articles/2048391-BART-cops-lose-TASER-privileges/>

Burgess, Lisa. 2005. 'PHaSRs May Soon Make "Trek" to Battle.' *Stars and Stripes*. November. <http://www.military.com/features/0,15240,80290,00.html>

Calamia, Joseph. 2010. 'DARPA's New Sniper Rifle Offers a Perfect Shot Across 12 Football Fields.' *Discover*. 25 May.
   <http://blogs.discovermagazine.com/80beats/2010/05/25/darpas-new-sniper-rifle-offers-a-perfect-shot-across-12-football-fields/>

Calloway, Audra. 2008. 'Army Fields First Brigade Nonlethal Capability Set.' Army.mil. 4 August.
   <http://www.army.mil/-news/2008/08/04/11422-army-fields-first-brigade-nonlethal-capability-set/index.html>

CAST (Centre for Analysis of Strategies and Technologies) et al. 2003. 'Workshops and Factories: Products and Producers.' In Small Arms Survey. *Small Arms Survey 2003: Development Denied*. Oxford: Oxford University Press, pp. 9–55.

CBC News. 2010. 'Facts about Stun Guns and Their Use in Canada.' 8 December. <http://www.cbc.ca/canada/story/2009/03/18/f-taser-faq.html>

CCC (Corruption and Crime Commission). 2010. *The Use of TASER Weapons by the Western Australia Police*. October. <http://www.ccc.wa.gov.au/Publications/Reports/Published%20Reports%202010/Full%20Report%20-%20Use%20of%20Taser%20Weapons%20by%20WAPOL.pdf>

CFR (Council on Foreign Relations). 1995. *Non-Lethal Technologies: Military Options and Technologies*. Washington, DC: CFR Press. January.
   <http://www.cfr.org/americas/non-lethal-technologies/p131>

—. 1999. *Nonlethal Technologies Progress and Prospect*. New York: CFR.
   <http://www.cfr.org/content/publications/attachments/Non-lethal_Tech_Progress.pdf>

CNEFG (Centre national d'entraînement des forces de gendarmerie). n.d. 'Public Order.' <http://cnefg.org/02_training/03_public_order.html>

CNN. 2009. 'Taser Makers Say Don't Aim at Chest.' 22 October. <http://edition.cnn.com/2009/CRIME/10/22/taser.advice/>

—. 2010. 'Police Chief: Multiple Taser Incident "Unacceptable."' 6 October.
   <http://edition.cnn.com/2010/WORLD/asiapcf/10/05/australia.tasers/index.html?hpt=T2>

CoE (Council of Europe). 2001. Recommendation Rec(2001)10 of the Committee of Ministers to Member States on the European Code of Police Ethics. Adopted 19 September. Strasburg: CoE. <https://wcd.coe.int/wcd/com.instranet.InstraServlet?command=com.instranet.CmdBlobGet&InstranetImage=1277578&SecMode=1&DocId=212766&Usage=2>

Coginta. 2010. 'La gouvernance civile des agences policières en Europe.' August. <http://coginta.org/pdf/MANUALS/gouvernance%20civile.pptx>

Conley, Daniel. 2005. *Investigation into the Death of Victoria Snelgrove and Other Uses of the FN 303 on Landsdowne Street on October 20–21, 2004*. 12 September. <http://www.mass.gov/dasuffolk/docs/091205a.html>

Cook, Rhonda. 2010. '2 Officers out of Jobs in Wake of Repeated Tasering of Woman.' *Atlanta Journal-Constitution*. 13 July.
   <http://www.ajc.com/news/2-officers-out-of-568967.html>

Coupland, Robin and Dominique Loye. 2003. 'The 1899 Hague Declaration concerning Expanding Bullets: A Treaty Effective for More than 100 Years Faces Complex Contemporary Issues.' *International Review of the Red Cross*, Vol. 85, No. 849.
   <http://www.icrc.org/eng/assets/files/other/irrc_849_coupland_et_loye.pdf>

Cox, Matthew. 2010. '"Green" Ammo Shipped to Afghanistan.' *Army Times*. 23 June.
   <http://www.armytimes.com/news/2010/06/army_green_ammo_062310w/>

Crane, David. 2010. 'MagPul Industries Developing 4×2 Quad Stack Magazine (5.56mm NATO): Enhanced-Capacity AR Mag (4179 STANAG).' *Defense Review*. 8 June.
   <http://www.defensereview.com/magpul-files-patent-application-for-quad-stack-magazine-4179-stanag-enhanced-capacity-ar-ar-15-mag/>

DARPA (Defense Advanced Research Projects Agency). 2010. 'One Shot Phase 2 Enhanced (2E).' DARPA-BAA-10-67. Arlington, VA: Adaptive Execution Office, DARPA. <https://www.fbo.gov/download/1c8/1c871251d4f47d3cf30d40d64c7cc251/DARPA-BAA-10-67_One_Shot_Phase_2E_Final_For_Posting_21May10.pdf>

Davison, Neil. 2007. *The Contemporary Development of 'Non-Lethal' Weapons*. Occasional Paper No. 3. May. Bradford: Bradford Non-Lethal Weapons Research Project, University of Bradford. <http://www.brad.ac.uk/acad/nlw/research_reports/docs/BNLWRP_OP3_May07.pdf>

—. 2009. *'Non-Lethal' Weapons*. Hampshire, UK: Palgrave Macmillan.

—. 2010. *The Influence of Science Fiction on the Research and Development, Marketing, and Public Perception of New Law eEnforcement Weapons*. Unpublished background paper. Geneva: Small Arms Survey.

DeClerq, David. 1999. *Trends in Small Arms and Light Weapons (SALW) Development: Non-Proliferation and Arms Control Dimensions*. Ottawa: Department of Foreign Affairs and International Trade, Canada.
   <http://www.international.gc.ca/arms-armes/isrop-prisi/research-recherche/nonproliferation/declerq1999/index.aspx?lang=eng>

de Laforcarde. n.d. 'Land Forces Employment in Crowd Control'. *Objectif Doctrine*, No. 30. <http://ids.nic.in/FRENCH%20DOCTRINE/objdoc_30.pdf>

DGGN (Direction générale de la gendarmerie nationale). 2006. *Circulaire relative à l'emploi du pistolet à impulsions électriques (PIE) au sein de la gendarmerie nationale*, No. 13183. 25 January. Paris: DGGN.

—. 2010. *Elaboration du prochain rapport Small Arms Survey*. Unpublished background paper. Geneva: Small Arms Survey.

DGPN (Direction générale de la police nationale). 2010. *Elaboration du prochain rapport Small Arms Survey*. Unpublished background paper. Geneva: Small Arms Survey.

Di Justo, Patrick. 2005. 'A Smart Shooter?' *Popular Science*. 20 December. <http://www.popsci.com/scitech/article/2005-12/smart-shooter>

Discovery Channel. n.d. 'Future Weapons.' <http://dsc.discovery.com/tv/future-weapons/future-weapons.html>

Egner, Donald, et al. 1973. *A Multidisciplinary Technique for the Evaluation of Less Lethal Weapons*, Vol. 1. Aberdeen, MD: United States Army Land Warfare Laboratory.

Eurosatory. 2010. 'List of Exhibitors.' <http://www.eurosatory.com/media/downloads/2010-List-of-Exhibitors.pdf>

FBO (Federal Business Opportunites). 2010. 'Glock Weapons—Solicitation Number: RFQ0200727.' 30 July.
   <https://www.fbo.gov/index?s=opportunity&mode=form&id=c2f363bc858841aa74c756688c5905cc&tab=core&_cview=1>

France. 2010. 'Les matériels de guerre, armes et munitions et matériels assimilés, les biens et technologies à double usage, les produits explosifs.' <http://www.douane.gouv.fr/page.asp?id=246>

France24. 2009. 'South African Police Trained in Crowd Control Techniques.' 12 June. <http://www.france24.com/en/20090612-south-african-police-trained-crowd-control-techniques-2010-world-cup>

Gabara, Nthambeleni. 2008. 'Saps Anti-Riot Water Truck to Manage Crowds During 2010.' AllAfrica.com. 31 October. <http://allafrica.com/stories/200810310173.html>

Gendarmerie Nationale. 2007. *La Gendarmerie au maintien de l'ordre: généralités, formation, équipement—Fiche N° 36-05.* Intranet document.

Gobinet, Pierre. 2008. 'The Gendarmerie Alternative: Is There a Case for the Existence of Police Organisations with Military Status in the Twenty-first Century European Security Apparatus?' *International Journal of Police Science and Management,* Vol. 10, No. 4.

Gun Policy. 2010. 'United States: Gun Facts, Figures and the Law.' <http://www.gunpolicy.org/firearms/region/united-states>

Hadhazy, Adam. 2011. 'New Laserlike Weapon Could Break Up Prison Rumbles.' TechNewsDaily.com. 3 February. <http://www.technewsdaily.com/laser-gun-could-break-up-prison-fights-2094/>

Hambling, David. 2008a. 'Pain Laser Finds New Special Forces Role.' Wired.com. 15 September. <http://www.wired.com/dangerroom/2008/09/pulsed-laser-fi/>

—. 2008b. 'US Police Could Get "Pain Beam" Weapons.' *New Scientist.* 24 December. <http://www.newscientist.com/article/dn16339-us-police-could-get-pain-beam-weapons.html>

Hawk, Matt. 2010. 'Local Police Still Trust in Tasers: Others Increasingly Skeptical of Misuse, Point to Rare Cases of Death by Taser.' *Times Standard.* 19 July. <http://www.times-standard.com/ci_15549303?IADID=Search-www.times-standard.com-www.times-standard.com>

Hazen, Jennifer and Chris Stevenson. 2008. 'Targeting Armed Violence: Public Health Interventions.' In Small Arms Survey. *Small Arms Survey 2008: Risk and Resilience.* Cambridge: Cambridge University Press, pp. 275–302.

Intelligent Optical Systems. n.d. 'Nonlethal Countermeasures.' Company website. <http://www.intopsys.com/nonlethal.html>

Jacobs, Thierry. n.d. 'Less Lethal System: The FN303 Approach.' FN Herstal. <http://www.non-lethal-weapons.com/sy02abstracts/P49.pdf>

Kaestle, Chad and Jon Buehler. 2005. 'Selecting a Duty-Issue Handgun.' *FBI Law Enforcement Bulletin,* Vol. 74, No.1. January. <http://www2.fbi.gov/publications/leb/2005/jan2005/jan2005.htm>

Kennedy, Harold. 2002. 'U.S. Troops Find New Uses For Non-Lethal Weaponry.' *National Defense.* March. <http://www.nationaldefensemagazine.org/archive/2002/March/Pages/US_Troops4127.aspx>

Kenny, John, Sid Heal, and Mike Grossman. 2001. *The Attribute-Based Evaluation (ABE) of Less-Than-Lethal, Extended-Range, Impact Munitions.* State College, PA: Applied Research Laboratory, Pennsylvania State University, and Los Angeles Sheriff's Department. 15 February. <http://nldt.arl.psu.edu/documents/abe_report.pdf>

Knight, Will. 2005. 'US Military Sets Laser PHASRs to Stun.' *New Scientist.* November. <http://www.newscientist.com/article/dn8275>

Kroll, Mark and Jeffrey Ho, eds. 2009. *TASER® Conducted Electrical Weapons: Physiology, Pathology, and Law.* New York: Springer.

Lamothe, Dan. 2010. 'Corps to Use More Lethal Ammo in Afghanistan.' *Marine Corps Times.* 15 February. <http://www.marinecorpstimes.com/news/2010/02/marine_SOST_ammo_021510w/>

LAPD (Los Angeles Police Department). 2009. *2008 Use of Force: Year End Report.* 2 December. <http://lapdonline.org/assets/pdf/2008-year-end-review-uof.pdf>

LCAV (Legal Community against Violence). 2008. 'Personalized Firearms.' February. <http://www.lcav.org/content/personalized_firearms.pdf>

Le Figaro. 2010. 'Le préfet de Paris recadre l'utilisation du flashball.' 15 October. <http://www.lefigaro.fr/actualite-france/2010/10/15/01016-20101015ARTFIG00433-le-prefet-de-paris-recadre-l-utilisation-du-flash-ball.php>

Levenson, Michael and Donovan Slack. 2009. 'Boston Mayor Says No to M-16 Patrol Units.' *Boston Globe.* 1 June. <http://www.policeone.com/police-products/firearms/articles/1838021-Boston-mayor-says-no-to-M-16-patrol-units/>

Lewer, Nick and Neil Davison. 2005. 'Non-Lethal Technologies: An Overview.' *Disarmament Forum,* Iss. 1. <http://www.unidir.org/pdf/articles/pdf-art2217.pdf>

Long, Colleen. 2009. 'NYPD Looking at Futuristic Weapons Technology.' Associated Press. 6 June. <http://www.policeone.com/police-products/firearms/handguns/articles/1839643-NYPD-looking-at-futuristic-weapons-technology/>

Lord, Rich. 2009. 'Semiautomatic Rifles Ordered for Pa. Patrol Officers.' *Pittsburgh Post-Gazette.* 29 April. <http://www.policeone.com/police-products/firearms/articles/1815685-Semiautomatic-rifles-ordered-for-Pa-patrol-officers/>

LRAD Corporation. n.d.a 'Public Safety/Law Enforcement.' <http://www.lradx.com/site/content/view/285/110>

—. n.d.b 'Product Overview.' <http://www.lradx.com/site/content/view/15/110/>

M2 Technologies. n.d. 'Leadership.' Company website. <http://www.m2tech.us/company/leadership.html>

Main, Frank. 2010. '"Scary" Growth of Gangs in War Zones.' *Chicago Sun-Times.* 18 July. <http://www.freerepublic.com/focus/f-news/2554375/posts>

— et al. 2010. 'Chicago Cop Killed with Own Gun.' *Chicago Sun-Times.* 8 July. <http://www.policeone.com/police-heroes/articles/2093668-Chicago-cop-killed-with-own-gun/>

Martin, Cyriel. 2008. 'La police présente sa nouvelle arme de dissuasion.' *Le Point.* 21 March. <http://www.lepoint.fr/actualites-politique/2008-03-21/la-police-presente-sa-nouvelle-arme-de-dissuasion/917/0/231380>

Matteucci, Megan. 2010. 'Ga. Police Beef up Sidearms to Counter Criminals'. *Atlanta Journal-Constitution.* 25 January. <http://www.policeone.com/police-products/firearms/articles/1995106-Ga-police-beef-up-sidearms-to-counter-criminals/>

McKenzie, Kevin. 2010. 'West Memphis Police to Carry Semiautomatic Rifles.' *Commercial Appeal*. 29 June.
<http://www.commercialappeal.com/news/2010/jun/29/w-memphis-police-to-carry-semiautomatic-rifles/>

Merchet, Jean-Dominique. 2008. 'Contrôler la foule, un job de fantassins.' Secret Défense blog. 30 March.
<http://secretdefense.blogs.liberation.fr/defense/2008/03/contrler-la-fou.html>

Meyer, Greg. 2010. 'The BART Shooting Tragedy: Lessons to Be Learned.' PoliceOne.com. 12 July.
<http://www.policeone.com/less-lethal/articles/2095072-The-BART-shooting-tragedy-Lessons-to-be-learned/>

Military.com. 2010. 'SOCOM Cancels Mk-16 SCAR.' 25 June. <http://kitup.military.com/2010/06/socom-cancels-mk-16-scar.html>

Mispo.org. 2010. 'South Africa 2010: World Cup Special Issue—A Guide to Police and Security Equipment in South Africa.' June.

Morris, Janet. 1990 (revised edn., 2009). *Nonlethality: A Global Strategy*. Washington, DC: United States Global Strategy Council.
<http://m2.cabem.com/images/upload/Nonlethality-A%20Global%20Strategy.pdf>

MPD (Madison Police Department). 2005. *Taser Report Submitted to Chief of Police Noble Wray by Lieutenant Victor Wahl*. 31 January.
<http://www.taser.com/research/statistics/Documents/Madison_WI_TASER_Report_2_05.pdf>

Murphy, Shelley. 2006. 'Snelgrove Family Settles Lawsuit—Kin Sought $10m from Gun Maker.' *Boston Globe*. 14 July.
<http://www.boston.com/news/local/massachusetts/articles/2006/07/14/snelgrove_family_settles_lawsuit/>

NATO. 1999. 'NATO Policy on Non-lethal Weapons.' Press statement. 13 October. <http://www.nato.int/docu/pr/1999/p991013e.htm>

NJIT (New Jersey Institute of Technology). 2003. 'New Jersey Institute of Technology Moves Ahead to Get Smart Gun on Market.' 5 September.
<http://www.njit.edu/news/2003/2003-089.php>

Omega Foundation. 2000. *Crowd Control Technologies: An Appraisal of Technologies for Political Control*. Working document for the European
Parliament's Science and Technology Options Assessment Panel. June.
<http://www.europarl.europa.eu/stoa/publications/studies/19991401a_en.pdf>

Ontario. 2010. *Canadian Civil Liberties Association v. Toronto Police Service*. 2010 ONSC 3698.
<http://www.torontopolice.on.ca/media/text/20100625-CCLA_v_TPS_G20_Injunction_LRAD_supplemental_june_25_10.pdf>

Papy, Alexandre and Marc Pirlot. 2007. *Evaluation of Kinetic-Energy Non-Lethal Weapons*. <http://blog.apapy.com/publications/nlw07.pdf>

Parker, Annise and Steve Schoonover. 2008. *Conducted Energy Device Program Audit*. City of Houston Report No. 2009-09. 8 September.
<http://www.houstontx.gov/controller/audit/2009-09.pdf>

Parks, W. Hays. 1990. 'Sniper Use of Open-Tip Ammunition.' 12 October. <http://www.thegunzone.com/opentip-ammo.html>

Paulissen, Pascal. 2010. 'Identifying the Right Non-lethal Weapons Needs.' Presentation prepared for the IQPC Infantry Weapons Conference 2010.
London, 20 September.

Peters, John. 2006. 'Force Continuums: Three Questions.' *Police Chief*, Vol. 73, No. 1. January.
<http://www.policechiefmagazine.org/magazine/index.cfm?fuseaction=display_arch&article_id=791&issue_id=12006>

PPAO (Picatinny Public Affairs Office). 2010. 'Army Begins Shipping Improved 5.56mm Cartridge.' Army.mil. 23 June.
<http://www.army.mil/-news/2010/06/23/41283-army-begins-shipping-improved-556mm-cartridge/>

Remington Military. 2010. 'Carbines: ACR™.' Company website. <http://www.remingtonmilitary.com/Firearms/Carbines/ACR.aspx>

Rijksoverheid. 2011. 'Nieuw pistool voor Nederlandse politie.' 27 January.
<http://www.rijksoverheid.nl/nieuws/2011/01/27/nieuw-pistool-voor-nederlandse-politie.html>

Ronson, Jon. 2004. *The Men Who Stare at Goats*. London: Picador. <http://www.jonronson.com/goats_04.html>

Sargent, Scott. 2010. 'Ensuring Relevant Training after a Use-of-Force Incident.' PoliceOne.com. 28 June.
<http://www.policeone.com/less-lethal/articles/2088597-Ensuring-relevant-training-after-a-use-of-force-incident/>

Shachtman, Noah. 2010. 'Pain Ray, Rejected by the Military, Ready to Blast L.A. Prisoners.' Wired.com. 24 August.
<http://www.wired.com/dangerroom/2010/08/pain-ray-rejected-by-the-military-ready-to-blast-l-a-prisoners/>

Sharp, Liz. 2010. 'Smith & Wesson Awarded ATF Firearms Contract.' 13 September.
<http://ir.smith-wesson.com/phoenix.zhtml?c=90977&p=irol-newsArticle&ID=1470110&highlight=>

Shea, Paddy. 2007. 'BPD: Pellet Guns to be Destroyed.' BerkeleyBeacon.com. 1 March.
<http://media.www.berkeleybeacon.com/media/storage/paper169/news/2007/03/01/News/Bpd-Pellet.Guns.To.Be.Destroyed-2755628.shtml/>

South Yorkshire Police. n.d. 'Use of Less Lethal Technologies: TASER (Superseded by Statement of Agreed Policy Use of Conductive Energy Device:
Taser, November 2009). 28 November. <http://www.southyorks.police.uk/print/foi/publicationscheme/policiesandprocedures/archive/392005>

Spice, Linda. 2009. 'Milwaukee Police Trade Shotguns for Tactical Rifles.' *Milwaukee Journal Sentinel*. 24 April.
<http://www.policeone.com/police-products/firearms/articles/1814387-Milwaukee-police-trade-shotguns-for-tactical-rifles/>

Stern, Donald, et al. 2005. *Report of the Commission Investigating the Death of Victoria Snelgrove*. Boston: Boston Police Department.
<http://www.cityofboston.gov/police/pdfs/report.pdf>

Stone, Christopher, Brian Buchner, and Scott Dash. 2005. *Crowd Control That Can Kill: Can American Police Get a Grip on Their New, 'Less-Lethal'
Weapons Before They Kill Again?* Policy Brief 2005-6. 24 October.
<http://www.parc.info/client_files/Articles/2%20-%20Less%20Lethal%20Policy%20Brief%20(Oct.%202005).pdf>

Swett, Charles. 1993. *Strategic Assessment: Non-Lethal Weapons*. 9 November. Washington, DC: Pentagon.
<http://www.dod.gov/pubs/foi/reading_room/823.pdf>

Sydney Morning Herald. 2010. 'Vic Police Roll out Semi-automatic Guns.' 14 October.
<http://news.smh.com.au/breaking-news-national/vic-police-roll-out-semiautomatic-guns-20101014-16kmm.html>

TASER International. 2010. 'TASER Technology.' <http://www.taser.com/company/pressroom/Documents/Press%20Kit%2012%2005%2010.pdf>

—. n.d.a. 'Military Points of Contact.' Company website. <http://www.taser.com/Pages/military_04.aspx>

—. n.d.b. 'TASER Products.' <http://www.taser.com/products/Pages/default.aspx>

Taylor, Lauren. 2000. 'Getting Smarter: Making Guns Safer for Law Enforcement and Consumers.' *National Institute of Justice Journal*. July.
<http://www.ncjrs.gov/pdffiles1/jr000244d.pdf>

TRADOC (Training and Doctrine Command). 2008. *Force Operating Capabilities*. TRADOC Pamphlet 525-66. Fort Monroe, Virginia: Department of the Army Headquarters. 7 March. <www.tradoc.army.mil/tpubs/pams/p525-66.pdf>

UKSG (United Kingdom Steering Group). 2006. *Patten Report Recommendations 69 and 70 Relating to Public Order Equipment*.
<http://www.nio.gov.uk/less_lethal_weaponry_steering_group_phase_5_report.pdf>

UN (United Nations). 1990. Basic Principles on the Use of Force and Firearms by Law Enforcement Officials.
<http://www2.ohchr.org/english/law/pdf/firearms.pdf>

UNGA (United Nations General Assembly). 1979. Code of Conduct for Law Enforcement Officials. Adopted by General Assembly Resolution 34/169 of 17 December. <http://www2.ohchr.org/english/law/pdf/codeofconduct.pdf>

USDOD (United States Department of Defense). 1996. 'Policy for Non-Lethal Weapons.' Directive Number 3000.3. 9 July.
<http://www.dtic.mil/whs/directives/corres/pdf/300003p.pdf>

USDOJ (United States Department of Justice). 2004. *Department of Defense Nonlethal Weapons and Equipment Review: A Research Guide for Civil Law Enforcement and Corrections*. Washington, DC: National Institute of Justice. <http://www.ncjrs.gov/pdffiles1/nij/205293.pdf>

—. 2010a. *Law Enforcement Officers Killed and Assaulted 2009: Uniform Crime Reports*. Washington, DC: US Department of Justice, Federal Bureau of Investigation. October. <http://www.fbi.gov/about-us/cjis/ucr/leoka/2009>

—. 2010b. *Crime in the United States: Violent Crime*. September. Washington, DC: Federal Bureau of Investigation, US Department of Justice.
<http://www2.fbi.gov/ucr/cius2009/offenses/violent_crime/index.html>

— and USDOD (United States Department of Defense). 1997. *Joint Technology Program: Second Anniversary Report*. Washington, DC: USDOJ. February.
<http://www.ncjrs.gov/pdffiles/164268.pdf>

US Patent (United States Patent). 1974. 'Weapon for Immobilization and Capture.' US Patent 3,803,463. April.
<http://www.freepatentsonline.com/3803463.pdf>

Verney-Carron. n.d. 'Flash-Ball.' <http://www.flash-ball.com/pages-us/i_fb_tech.htm>

Whitehead, Tom. 2009. 'Taser Roll-out Snubbed by Police.' *Telegraph*. 22 June.
<http://www.telegraph.co.uk/news/uknews/law-and-order/5595375/Taser-roll-out-snubbed-by-police.html>

Williams, Anthony. 2009. 'Where Next For PDWs?' 29 August. <http://www.quarry.nildram.co.uk/PDWs.htm>

—. 2010. 'Future Infantry Small Arms.' <http://www.quarry.nildram.co.uk/future%20small%20arms.htm>

Wisler, Dominique. 1997. 'Variation et impact des pratiques policières: le cas de la Suisse.' *Les cahiers de la sécurité intérieure*, No. 27, pp. 58–85.

Woo, Elaine. 2009. 'Jack Cover Dies at 88—Scientist Invented the Taser Stun Gun.' *Los Angeles Times*. 13 February.
<http://www.latimes.com/news/printedition/california/la-me-jack-cover13-2009feb13,0,4344955.story>

Wood, Brian and Glenn McDonald. 2004. 'Critical Triggers: Implementing International Standards for Police Firearms Use.' In Small Arms Survey. *Small Arms Survey 2004: Rights at Risk*. Oxford: Oxford University Press, pp. 212–47.

Wyllie, Doug. 2010. 'Calif. Study on Officer-involved Shootings Underscores Need for Cops to Have TASERs.' PoliceOne.com. 2 July. <http://www.policeone.com/officer-shootings/articles/2090691-Calif-study-on-officer-involved-shootings-underscores-need-for-cops-to-have-TASERs/>

ACKNOWLEDGEMENTS

Principal author

Pierre Gobinet

Contributors

Neil Davison and Anna Alvazzi del Frate

A private security guard looks out into a shopping centre in Quito, Ecuador, 2001. © Rhodri Jones/Panos Pictures

A Booming Business

PRIVATE SECURITY AND SMALL ARMS

4

INTRODUCTION

In August 2010, President Hamid Karzai issued a decree requiring private security companies (PSCs) to cease all operations in Afghanistan by December 2010, calling them unwelcome 'parallel structures' and a 'cause for insecurity' (Afghanistan, 2010; Rubin, 2010). With billions of dollars in Afghan-based development programmes that require constant protection, donor governments reacted by placing intense pressure on Karzai to withdraw the decree. The deadline was ultimately extended, and some PSCs were exempted from the ban, but the president stood by his decision. The case illustrates how deeply embedded PSCs have become in some contexts.

PSCs have come under increased international scrutiny in the 2000s due to the central roles they have been granted in the conflicts of Afghanistan and Iraq, as well as concerns over the perceived lack of accountability for action taken by private personnel. Incidents such as the killing of 17 civilians by Blackwater personnel in September 2007 in Nisoor Square, Baghdad, have significantly tarnished the industry's image (Glanz and Lehren, 2010).

The highly publicized involvement of international PSCs in contemporary conflicts tends to overshadow the much wider trend of security privatization across society as a whole, particularly in non-conflict settings. Around the globe, individuals, communities, local businesses, government agencies, large corporations, and powerful militaries are increasingly outsourcing aspects of their security to private entities. The growing reliance on PSCs in conflict is just one aspect of a global phenomenon that must be assessed in its entirety to be properly understood.

This chapter attempts to shed light on a poorly documented aspect of the global private security industry: its use of arms. While much attention has been devoted to debating the legitimacy of PSCs undertaking what may be considered state functions, less effort has gone into documenting the types of small arms used by PSCs and potential gaps in their control. The chapter examines the scale of the private security industry at the global level, calculates the extent to which it is armed, and asks whether PSC equipment contributes to or threatens security.

Main findings include:

- Based on a review of 70 countries, this study estimates that the formal private security sector employs between 19.5 and 25.5 million people worldwide. The number of PSC personnel has grown at a fast pace since the mid-1980s and exceeds the number of police officers at the global level.
- PSCs hold between 1.7 and 3.7 million firearms worldwide, an estimate based on extrapolations from reported inventories. If undeclared and illegally held weapons were to be included, the global PSC stockpile would undoubtedly be higher.
- Globally, PSC firearm holdings are just a fraction of the stockpiles held by law enforcement agencies (26 million) and armed forces (200 million).

- While several states ban the use of small arms by PSCs, private security stockpiles in some conflict-affected areas amount to more than three weapons per employee.
- Outside of armed conflict settings, PSCs are most armed in Latin America, with ratios of arms per employee about ten times higher than in Western Europe.
- PSCs working in Afghanistan and Iraq have been equipped with fully automatic assault rifles, machine guns, sniper rifles, and, in some cases, rocket-propelled grenade launchers (RPGs), raising questions about their stated 'defensive' roles.
- Some PSCs have been involved in illegal acquisition and possession of firearms, have lost weapons through theft, and have used their small arms against civilians although they were unprovoked. Available information remains anecdotal, however, and makes it challenging to measure PSC performance over time or compare it to that of state security forces.
- The rapid growth of the private security sector has outpaced regulation and oversight mechanisms. International initiatives to tackle regulatory gaps remain in their infancy.

Several states ban the use of small arms by PSCs. This chapter focuses on PSCs, using the term in its widest possible sense to include all legally registered business entities that provide, on a contractual basis, security or military services, regardless of whether they operate in situations of conflict. Security and military services may include protecting persons, guarding objects (such as convoys or buildings), the maintenance and operation of weapons systems, prisoner detention, the provision of advice or training for security forces and personnel, and associated surveillance and intelligence operations.[1]

The chapter begins by providing an overview of factors that contribute to the growing role of PSCs and documents the scale of the phenomenon worldwide. The second section focuses on the weaponry used by PSCs worldwide, with reference to both quantity and type. The last section assesses the extent to which the existing regulatory regime as well as ongoing initiatives can prevent incidents of small arms misuse by private security personnel. In addition to desk research and interviews with industry representatives and other stakeholders, the chapter relies on a number of original expert contributions commissioned by the Small Arms Survey.

THE PRIVATIZATION OF SECURITY

The private security spectrum is extremely broad and diverse. While the media spotlight has focused on international PSCs operating in the conflict zones of Afghanistan and Iraq, private security is employed in virtually all societies.[2] PSCs are often portrayed as protecting property and people, in contrast to private military companies (PMCs), which provide offensive services meant to have military impact,[3] yet analysts argue that such a distinction is misleading (Holmqvist, 2005, p. 5). Indeed, a single company can perform a variety of services encompassing both defensive and offensive support. Furthermore, what can be termed protective services in peacetime—such as the protection of public institutions—can have military and offensive implications in situations of conflict. Additional analysis of the sector according to company size, level of compliance with standards, and proximity to the state would undoubtedly move the discussion forward. Yet since this chapter is a first attempt to shed light on the small arms used by the industry as a whole, it refers to PSCs in a broad sense.

Scale

The private security sector has been booming since the mid-1980s and continues to grow steadily (van Dijk, 2008, p. 217). Recent estimates show that the security market is worth about USD 100–165 billion per year, and that it has been growing at an annual rate of 7–8 per cent.[4] The scale of growth is further illustrated by significant increases in the number of personnel employed over time and across regions:

- In France, the sector expanded from just over 100,000 employees in 1982 to 160,000 in 2010 (Ocqueteau, 2006, p. 65; CoESS and APEG, 2010, p. 12).
- Japanese PSC personnel increased from just over 70,000 guards in 1975 to nearly 460,000 in 2003 (Yoshida and Leishman, 2006, p. 232).
- In South Africa, the number of registered security officers more than tripled in the space of 13 years, from about 115,000 in 1997 to nearly 390,000 in 2010 (Berg, 2007, p. 5; PSIRA, 2010, p. 4).

The main impediment to accounting for the total number of PSC employees in the world is the lack of global data collection and monitoring systems. Nevertheless, this chapter is able to present recent figures on PSC personnel in 70 countries (see Table 4.1); the sources for this data are various, including regional reviews of the industry, academic articles examining the industry at the country level, and media reports.[5] While different sources may rely on varying definitions of PSC personnel, this study focuses on active PSC employees registered by a national government body or a private security industry association. Where possible, multiple and multi-year sources have been cross-checked to obtain the most plausible figure.

PSC size varies from a dozen to several hundred thousand employees.

Table 4.1 shows that the private security sector employs a reported 19.5 million people in the 70 countries. An extrapolation from this figure yields a global range of registered PSC personnel of 19.5–25.5 million.[6] The size of individual companies varies greatly, ranging from a dozen employees to several hundred thousand. For example, G4S has 530,000 staff in 115 countries, while Securitas employs 260,000 people in 40 countries (Abrahamsen and Williams, 2009, p. 2; Securitas, n.d.). Countless smaller firms are also active; about 30,000 companies are registered in the Russian Federation, while South African PSCs numbered nearly 7,500 in 2010 (Modestov, 2009; PSIRA, 2010, p. 4).

Taken together, PSC personnel employed in the 70 countries covered in Table 4.1 outnumber police officers by a ratio of 1.8 to 1. These countries employ a combined 19.5 million PSC personnel (a rate of 435 per 100,000) compared with fewer than 11 million police officers (240 per 100,000), suggesting an even greater imbalance than previously thought.[7] Global private security dominance in terms of personnel does not apply systematically across countries, however. More than half (39) of the countries listed in Table 4.1 actually employ more police officers than PSC personnel, but their effect on global numbers is negated by the situation in larger PSC markets, such as China, India, and the United States.

It is beyond the scope of this chapter to document the number of people participating in informal security arrangements; however, the figures reportedly hover around 50,000 in Argentina, between 670,000 and 1,000,000 in Brazil, and from 240,000 to 600,000 in Mexico (Godnick, 2009; Arias, 2009, pp. 26–27). In Francophone African countries, some communities seek to fill the state security vacuum by establishing informal neighbourhood militia groups, while young men faced with economic hardship provide free bodyguard services to businessmen in exchange for food—activities that are reported by neither industry nor governments (Kougniazondé, 2010, pp. 6, 8). Informal security schemes, ranging from neighbourhood watch to armed vigilante groups, can be found across the globe and provide additional evidence of a global demand for security that exceeds what states can offer.

Table 4.1 Private security personnel in 70 countries

Country	Year	Private security personnel	Police officers	Population	Ratio of private security to police	Private security per 100,000	Police per 100,000
Afghanistan	2010	26,000	115,500	24,507,000	0.23	106	471
Albania	2004	4,092	11,987	3,111,000	0.34	132	385
Angola	2004	35,715	17,000	16,618,000	2.10	215	102
Argentina	2007	150,000	120,000	38,732,000	1.25	387	310
Australia	2008	114,600	52,400	20,395,000	2.19	562	257
Austria	2009	11,200	20,500	8,372,930	0.55	134	245
Belgium	2009	18,609	47,000	10,827,519	0.40	172	434
Bolivia	2002	500	19,365	9,182,000	0.03	5	211
Bosnia and Herzegovina	2009	4,207	10,589	4,590,310	0.40	92	231
Brazil	2005-07	570,000	687,684	186,075,000	0.83	306	370
Bulgaria	2009	56,486	47,000	7,576,751	1.20	746	620
Chile	2008	45,020	35,053	16,297,000	1.28	276	215
China	2010	5,000,000	2,690,000	1,312,253,000	1.86	381	205
Colombia	2005-07	190,000	119,146	43,049,000	1.59	441	277
Costa Rica	2008	19,558	12,100	4,328,000	1.62	452	280
Côte d'Ivoire	2009	50,000	32,000	19,245,000	1.56	260	166
Croatia	2009	13,461	19,000	4,697,548	0.71	287	404
Cyprus	2009	1,700	3,000	801,851	0.57	212	374
Czech Republic	2009	51,542	46,000	10,512,397	1.12	490	438
Denmark	2009	5,250	10,000	5,547,088	0.53	95	180
Dominican Republic	2008	30,000	29,357	9,533,000	1.02	315	308
Ecuador	2005-07	40,368	42,610	13,063,000	0.95	309	326
El Salvador	2008	21,146	16,737	6,059,000	1.26	349	276
Estonia	2009	4,283	6,000	1,340,274	0.71	320	448
Finland	2009	10,000	8,000	5,350,475	1.25	187	150
France	2009	160,000	250,000	64,709,480	0.64	247	386
Germany	2009	170,000	250,000	81,757,600	0.68	208	306
Greece	2009	30,000	50,000	11,306,183	0.60	265	442

Country	Year	Private security personnel	Police officers	Population	Ratio of private security to police	Private security per 100,000	Police per 100,000
Guatemala	2008	120,000	19,974	12,710,000	6.01	944	157
Honduras	2005-07	60,000	12,301	6,893,000	4.88	870	178
Hungary	2009	105,121	40,000	10,013,628	2.63	1,050	399
India	2010	7,000,000	1,406,021	1,130,618,000	4.98	619	124
Iraq	2008	35,000	153,000	28,238,000	0.23	124	542
Ireland	2009	21,675	12,265	4,450,878	1.77	487	276
Italy	2009	49,166	425,000	60,397,353	0.12	81	704
Jamaica	2010	15,000	8,441	2,668,000	1.78	562	316
Japan	2003	459,305	246,800	127,449,000	1.86	360	194
Kenya	2005	48,811	36,206	35,817,000	1.35	136	101
Kosovo	2005	2,579	6,282	2,000,000	0.41	129	314
Latvia	2009	8,000	10,600	2,248,961	0.75	356	471
Lithuania	2009	10,000	20,000	3,329,227	0.50	300	601
Luxembourg	2009	2,200	1,573	502,207	1.40	438	313
Macedonia, former Yugoslav Republic of	2009	5,600	14,500	2,114,550	0.39	265	686
Malta	2009	700	1,904	416,333	0.37	168	457
Mexico	2005-07	450,000	495,821	105,330,000	0.91	427	471
Moldova	2000	10,000	13,431	3,386,000	0.74	295	397
Montenegro	2005	1,900	4,227	660,000	0.45	288	640
Morocco	2010	20,000	48,394	30,495,000	0.41	66	159
Netherlands	2009	30,936	49,000	16,576,800	0.63	187	296
Nicaragua	2008	19,710	9,216	5,455,000	2.14	361	169
Nigeria	2005	100,000	360,000	140,879,000	0.28	71	256
Norway	2009	6,700	8,500	4,854,824	0.79	138	175
Panama	2008	30,000	15,255	3,232,000	1.97	928	472
Peru	2005-07	50,000	90,093	27,836,000	0.55	180	324
Poland	2009	165,000	100,000	38,163,895	1.65	432	262
Portugal	2009	38,874	50,000	10,636,888	0.78	365	470
Romania	2009	107,000	55,000	21,466,174	1.95	498	256

▶▶

Country	Year	Private security personnel	Police officers	Population	Ratio of private security to police	Private security per 100,000	Police per 100,000
Russian Federation	2009	800,000	601,000	143,170,000	1.33	559	420
Serbia	2009	28,500	34,000	10,100,000	0.84	282	337
Sierra Leone	2005	3,000	9,300	5,107,000	0.32	59	182
Slovakia	2009	17,200	21,500	5,424,057	0.80	317	396
Slovenia	2009	7,554	7,500	2,054,119	1.01	368	365
South Africa	2010	387,273	150,513	48,073,000	2.57	806	313
Spain	2009	86,000	227,250	46,087,170	0.38	187	493
Sweden	2009	13,500	19,000	9,347,899	0.71	144	203
Switzerland	2009	13,075	16,000	7,760,477	0.82	168	206
Trinidad and Tobago	2010	5,000	6,500	1,318,000	0.77	379	493
Turkey	2009	257,192	201,064	74,816,000	1.28	344	269
United Kingdom	2009	120,000	140,000	62,041,708	0.86	193	226
United States	2007	2,000,000	883,600	302,741,000	2.26	661	292
Total		19,545,308	10,799,059	4,496,715,554	1.81	435	240
Median					0.83	298	311

Source: Annexe 4.1

Table 4.2	Public perception of private security providers in seven African countries

Percentage of survey respondents who answered 'yes' to the question, 'Do you think that policing functions performed by private security is a good development?'

	Year	Percentage	Survey sample size
Ghana	2009	93	1,560
Uganda	2007	88	2,147
Tanzania	2008	81	1,888
Rwanda	2008	65	2,100
Egypt	2008	64	3,126
Cape Verde	2008	62	1,844
Kenya	2010	57	2,777

Source: Small Arms Survey elaboration of unpublished UNODC victimization survey data, 30 June 2010

Reasons for growth

The global trend towards downsizing government, including public security institutions, has contributed to the growth of the private security sector. Previously core state functions—such as prison surveillance, immigration control, and airport security—have increasingly been outsourced in order to save financial and human resources within government agencies (Abrahamsen and Williams, 2009, pp. 3, 4).

The gap left behind by downsized public sectors is being felt across the globe, and PSCs represent one of the ways to fill it. As Table 4.2 illustrates, the involvement of PSCs in policing is rather well accepted by the majority of the public in seven African countries, reflecting local demand for the services—and possibly for the employment opportunities—offered by PSCs. Multinational corporations, international organizations, peacekeeping missions, non-governmental organizations, and the general population, in addition to government, are among the clients (Holmqvist, 2007, p. 8; Baker and Pattison, 2010; MULTINATIONAL CORPORATIONS).

It would be too simplistic to claim that shortcomings of the public security sector alone are responsible for the growth and scale of private security. Analysts have shown that per population rates of PSC personnel are not statistically related to rates of police officers, and that more complex political and economic factors contribute to the size of private security in a given context (van Dijk, 2008, p. 216).

Industry leaders attribute the continued growth of the sector to clients' greater awareness of security risks as well as their increased demand for technology. Alarm and electronic

Security cameras for China's closed-circuit television system in Beijing, China.
© Stewart Cohen/Getty Images

surveillance systems have permitted costs to drop and the reliability of private security services to increase by allowing constant surveillance and better incident recording (Securitas, 2009, pp. 28–29). Western armies' increasing use of high-tech weaponry has made them reliant on levels of technological expertise that appear impossible to maintain within the ranks, pushing them to outsource aspects of maintenance and training to PSCs (Cusumano, 2009, p. 2). This is especially true with respect to 'robotic' weapons such as unmanned drones.[8]

Some major Western militaries and government agencies, such as the US Department of Defense, have gradually institutionalized the outsourcing of functions other than combat in order to free up uniformed personnel for fighting (USDOD, 2001, p. 53). Some states contracting PSCs argue that the private sector can be hired and fired faster than uniformed personnel and can therefore be deployed more flexibly, which is more affordable in the long run than maintaining a permanent in-house capability (Schwartz, 2010, p. 2). As a result, the proportion of non-military personnel contracted by the US military has increased over time; while it represented 1/20 of the size of regular US forces during World War I, this ratio grew to 1/7 during World War II and 1/6 in Vietnam, to reach and even exceed parity in the conflicts of the Balkans, Afghanistan, and Iraq (Fontaine and Nagl, 2010, p. 9).[9]

A side effect of reductions in state security personnel has been the creation of a vast supply of available and trained individuals, many of whom secured jobs in PSCs or created their own. An estimated 5–6 million soldiers were demobilized worldwide between 1985 and 1996 (Renou, 2005, p. 289; Holmqvist, 2007, fn. 17). If reservists are included, military downsizing from the 1980s to 2007 resulted in more than 30 million trained personnel leaving military positions worldwide (Karp, 2008). A number of demobilized public security personnel and fighters in post-conflict societies such as Sierra Leone found employment as PSC employees (Abrahamsen and Williams, 2005b, p. 12). Companies such as Military Professional Resources, Inc., reportedly maintained a list of 12,500 'on-call' recruits, and Blackwater (now known as Xe Services) had its own database of 21,000 names (Scahill, 2007, p. xviii; Singer, 2003, p. 120).

Plainclothes Blackwater contractors take part in a firefight as demonstrators loyal to Muqtada al-Sadr attempt to advance on a facility defended by US and Spanish soldiers, Najaf, Iraq, 4 April 2004. © Gervasio Sanchez/AP Photo

The perils of growth[10]

One of the principal concerns regarding the private security sector is that, like other commercial services, only those who are able and willing to pay will benefit from it (Holmqvist, 2005, p. 12). This dynamic runs the risk of exacerbating disparities between the wealthy—protected by increasingly sophisticated systems—and the poorest, who may need to resort to informal and sometimes illegal means to secure their safety.

Another crucial question concerns the legitimacy of outsourcing activities that some consider an inherently governmental function (Cusumano, 2009, p. 18). The use of PSCs redistributes the control over the use of force, and drawing a line on the types of services that PSCs can perform has been the subject of continuing debate. Reports that the Central Intelligence Agency hired Blackwater to carry out a plan to assassinate al-Qaeda operatives caused significant controversy (Marlowe, 2010). The possible use of PSCs to conduct internationally mandated peacekeeping operations and humanitarian interventions is similarly contentious (Baker and Pattison, 2010). While very few firms currently undertake offensive combat missions, PSCs generally do not have policies ruling out this possibility. A voluntary industry code of conduct, for instance, does not exclude taking on offensive missions if 'mandated by a legitimate authority under international law' (ISOA, 2009, art. 8.2.).

Insufficient oversight of PSC performance and a lack of accountability in cases of alleged abuse represent a third set of concerns. Privileged links between private security personnel and current or former government and law enforcement agencies

Box 4.1 PSCs in armed conflict: debates in international law

Considerable debate surrounds the legal implications of the use of PSCs in areas affected by armed conflict. Yet the view that PSCs operate in a 'legal vacuum'[11] is somewhat misleading.[12] In situations of armed conflict, international humanitarian law (IHL) and international criminal law govern the activities of PSC employees. Serious violations they commit or order to be committed may be prosecuted in national or international courts, such as the International Criminal Court (ICC).[13] Both IHL and international human rights law also apply to states that hire PSCs (contracting states), states where they operate, and those where they are incorporated.[14]

Much of the discussion surrounding private contractors and their relationship to IHL has focused on determining whether these individuals have status as combatants or civilians. As combatants, PSC personnel would represent legitimate targets of attacks at all times,[15] but they would also have the right to directly participate in hostilities. If captured, they would be entitled to prisoner-of-war status and would not be prosecuted for having taken part in hostilities.

Various criteria must be met for an individual to qualify as a legal combatant, most of which arguably would not apply to PSCs as they are currently structured. The great majority of private contractors and civilian employees active in armed conflicts have not been incorporated into state armed forces and assume functions that clearly do not involve their direct participation in hostilities on behalf of a party to a particular conflict. Accordingly, under IHL, PSC personnel are generally defined as civilians and are (legally) protected against direct attack, except if and when they directly participate in hostilities (Melzer, 2009, pp. 39, 49).

The notion of direct participation in hostilities has, in fact, been the subject of ongoing debate among members of academia, government, and industry, specifically with reference to the type of work PSC personnel should be permitted to perform. For a specific act to qualify as 'direct' participation in hostilities, some scholars maintain that it must have a close causal relation to the resulting harm (Melzer, 2009, p. 52). Legal experts have argued that PSC participation in combat operations can include guarding military bases against attacks from the enemy,[16] gathering tactical military intelligence,[17] and operating weapons systems in combat operations (Heaton, 2005, p. 202). While participating in these activities, contractors would lose their protection against enemy attack. But as the acts that constitute direct participation are not yet codified, PSC employee participation in hostilities must be examined on a case-by-case basis (Gillard, 2006, p. 539).

International human rights law, applicable to situations of armed conflict (with limited scope for derogation),[18] is also relevant to PSC activity. It imposes an obligation on states to ensure that private parties, including PSCs, not infringe on the human rights of persons in any state's territory or within its jurisdiction. For this purpose, states are required to adopt appropriate legislative and other measures that serve to prevent, investigate, and provide remedies for human rights abuses.

Despite the existence of clear legal obligations and a well-established network of national and international courts with potential jurisdiction over serious IHL violations, proceedings against PSC employees are rare (Gillard, 2006, pp. 542-43). The problem lies less with the applicable norms, although some aspects of the law require clarification, than with a lack of oversight, accountability, and enforcement, including the inherent difficulties associated with gathering evidence of abuses in settings affected by conflict.

Sources: Richard (2010); Bushnell (2010)

can contribute to reducing oversight of PSC activities (Richards and Smith, 2007, p. 4). The possibility of links between the private security sector and criminal networks also worries analysts (Godnick, 2009). A large PSC firm in Tanzania, for example, found that as many as 30 per cent of its employees had criminal records.[19]

There is particular concern over perceived gaps in the accountability of PSC personnel operating in conflict situations. While aspects of international law apply to PSC personnel operating in contexts of warfare (see Box 4.1), enforcement is often difficult because of the specific features of PSC contracting and operation. In cases such as Iraq, where PSCs were granted immunity from Iraqi law between 2004 and 2009, accountability rested with the contracting states. Bringing to justice private security personnel operating overseas also entails obtaining evidence and initiating proceedings in the theatre of operations (Bailes and Holmqvist, 2007, p. iii). Furthermore, conflicts of interest can emerge if a contracting state takes on the roles of both client and watchdog (Cockayne and Speers Mears, 2009, p. 3). For these reasons and others, very few cases of alleged PSC abuse against civilians in Iraq have been prosecuted.[20]

Trainees take aim at each other during an anti-piracy drill aboard a ship in Haifa, Israel, June 2009.
© Baz Ratner/Reuters

THE PRIVATE SECURITY ARSENAL

The quantities and types of firearms at the disposal of PSCs vary greatly across settings, depending largely on the activities they perform and on national legislation. This section reviews available information on the quantities and categories of small arms available to PSCs in different situations.

Estimating arms holdings[21]

National legislation is a major factor influencing the extent to which PSCs arm themselves. A number of countries prohibit—at least on paper—the use of firearms by PSCs operating on their territory, including the Bahamas,[22] Denmark, Japan, Kenya, the Netherlands, Nigeria, Norway, and the UK.[23] Elsewhere, PSCs are allowed to use firearms only for very specific activities. In China and France, for instance, PSC personnel may legally carry firearms only when escorting money to and from banks ('cash-in-transit') (CoESS, 2008; Trevaskes, 2008, p. 38).

Restrictions on the transfer of arms to PSCs as non-state actors appear to be relatively common in countries that are in the midst of, or have recently emerged from, conflict. For example, the Sierra Leone National Security and Intelligence Act 2002 allows PSCs to hold arms in principle; however, the 1998 UN arms embargo prevented the sale of arms to non-state actors until 2010 (Abrahamsen and Williams, 2005b, p. 7). Yet Sierra Rutile, a rutile and bauxite mine in Sierra Leone, obtained permission by way of a specific decree to operate the only armed private security force in the country, despite the embargo on sales (p. 10). In Afghanistan, only the Afghan government, foreign military, and embassies are permitted to import a limited number of firearms for use by their international staff. As a result, there is no official weapons market in Afghanistan for PSCs to legally access firearms. PSCs can circumvent these restrictions by hiring local people who have their own weapons, and turning a blind eye to how they were obtained (Joras and Schuster, 2008, p. 14; Karimova, 2010a).

In practice, PSCs provide a number of services that do not require the use of firearms, such as risk analysis and advisory services. In non-conflict settings, PSCs are most likely to use arms when guarding sensitive industrial, government, and bank sites, performing mobile patrols and emergency interventions (in case an alarm system is activated), or protecting convoys (such as cash-in-transit) and people (acting as bodyguards).[24] In areas affected by conflict, PSCs may need weapons when escorting military supply convoys, protecting government and expatriate personnel, guarding military and government facilities, and training local security forces.[25] Maritime protection—of both ships and ports—may also require armed guards.[26]

Table 4.3 Reported armed PSC personnel in selected settings				
Location or company	Total PSC personnel	Personnel authorized to carry firearms*	Armed vs. total personnel ratio	Source
Croatia	16,000	300	0.02	CoESS (2008)
G4S in India	141,488	2,912	0.02	Author correspondence with a G4S representative, 12 October 2010
Sweden	13,500	300	0.02	CoESS (2008)
Germany	173,000	10,000	0.06	CoESS (2008)
One PSC in the Canton of Geneva, Switzerland	860	85	0.10	Author interview with private security representative 1, Geneva, 19 August 2010
Slovenia	4,500	1,000	0.22	CoESS (2008)
Turkey	158,839	35,263	0.22	CoESS (2008)
Russian Federation	850,000	196,266	0.23	Abrahamsen and Williams (2009, p. 2), citing Volkov (2002)
Spain	83,000	20,000	0.24	CoESS (2008)
Bulgaria	58,700	23,400	0.40	CoESS (2008)
Dominican Republic	30,000	24,000	0.80	Godnick (2009)
Colombia	200,000	170,000	0.85	Arias (2009, p. 48)

Note: * The number of personnel authorized to carry firearms in Bulgaria is calculated based on the country's reported total PSC personnel and its reported ratio of armed vs. total personnel.

PSC personnel are therefore not all licensed or authorized to be armed, as reflected by variations in the proportion of armed guards vs. total PSC personnel across settings. Table 4.3 illustrates that as few as two per cent of PSC personnel are armed in Croatia and in an international firm with significant presence in India, while more than 80 per cent of employees are armed in the Dominican Republic and Colombia. In Bosnia and Herzegovina, national legislation states that one-fifth of personnel may carry short-barrel firearms in the Federation of Bosnia and Herzegovina, while one-half of employees may do so in Republika Srpska (Page et al., 2005, p. 22).

PSC personnel who are authorized to carry firearms often do not each have their own weapon, nor do they always carry one. Guns may be stored in a central armoury and shared by employees from shift to shift. A PSC operating in the Canton of Geneva in Switzerland, for instance, explained that while ten per cent of personnel were licensed to carry firearms, the number of firearms in inventory amounted to just six per cent of the total number of employees.[27]

Reported PSC firearm stockpiles in 16 situations are presented in Table 4.4. They illustrate a wide range of PSC stockpile levels, starting at less than one firearm for ten employees in the above-mentioned Geneva company, to

Table 4.4 **Reported number of firearms held by PSCs in selected settings**				
Location or company	**PSC personnel**	**PSC firearms**	**Firearms per PSC personnel**	**Source**
One PSC in the Canton of Geneva, Switzerland	860	50	0.06	Author interview with private security representative 1, Geneva, 19 August 2010
Serbia	28,000	2,395	0.09	CoESS (2008); Page et al. (2005, p. 93)
Moscow	157,138	22,294	0.14	Falalyev (2010); Karimova (2010b, pp. 1-2)
Russian Federation	800,000	116,000	0.15	Modestov (2009); Karimova (2010b, p. 1)
Albania	4,093	938	0.23	CPDE and Saferworld (2005, p. 38)
South Africa	248,025	58,981	0.24	Gould and Lamb (2004, p. 185)
Bosnia and Herzegovina	4,207	1,075	0.26	Krzalic (2009, p. 34, fn. 38)
Angola	35,715	12,087	0.34	Joras and Schuster (2008, p. 46)
Nicaragua	19,710	6,799	0.34	Godnick (2009)
Costa Rica	19,558	8,884	0.45	Godnick (2009)
Brazil	570,000	301,526	0.53	Dreyfus et al. (2010, p. 100); Carballido Gómez (2008, slide 9)
Colombia	120,000	82,283	0.69	UNODC (2006, p. 59)
São Paolo	330,000	255,000	0.77	Wood and Cardia (2006, p. 156)
El Salvador	21,146	18,125	0.86	Godnick (2009)
35 PSCs in Afghanistan	1,431	4,968	3.47	Joras and Schuster (2008, p. 15)
Sandline operation in Papua New Guinea	42	160	3.81	PNG and Sandline (1997, pp. 8-9)

almost four small arms for every Sandline International employee in the 1997 Papua New Guinea operation. Together with Table 4.3, this information makes it possible to establish broad estimates of the level of PSC armament according to region and context (for example, exposure to armed conflict). Applying these ratios to reported numbers of PSC personnel contained in Table 4.1 generates a first global estimate of PSC firearm stockpiles (see Table 4.5).

It should be noted that any estimate risks under-representing actual levels of armament of PSCs as reports on PSC weapons are scarce and unlikely to take into account personnel who carry personal, or illegal, weapons on duty. For instance, while Kenya currently prohibits PSC firearm use, industry sources admit that some companies arm small elite units responsible for protecting important people and high-value facilities (Mbogo, 2010). In countries that prohibit the arming of private personnel, PSCs are nevertheless able to provide an armed service through arrangements with the public security forces. This is the case in Nigeria, where Mobile Police officers are permanently seconded to most PSCs and equipped with fully automatic weapons, usually AK-47s or FN assault rifles (Abrahamsen and Williams, 2005a, p. 11). Improved reporting, data collection, and transparency on PSC firearm holdings are therefore required to fully understand its scope.

Overall, and based on available information, Latin America stands out as the region where PSCs are the most armed, with ratios of arms to personnel ranging from 0.34 firearms in Nicaragua to 0.86 in El Salvador (see Table 4.4). A range of 0.3 to 0.8 firearms per PSC employee is therefore applied to other known PSC staff in the region in Table 4.5.

Even though data on African countries is scarce, industry representatives argue that Angola's 0.34 ratio of arms to personnel and South Africa's 0.24 rate (see Table 4.4) should not differ greatly from the situation in other African countries that allow PSC firearm use. PSCs probably have fewer weapons elsewhere on the continent, however.[28] For these reasons, a 0.05–0.30 range is applied to reported African PSC personnel.

Despite high rates of personnel, Eastern European PSCs are less equipped than their Latin American counterparts, with less than 0.1 firearm per employee in Serbia and up to 0.26 in Bosnia and Herzegovina (see Table 4.4). A 0.05–0.20 range is therefore applied to documented PSC personnel in the region.

Western European rates are believed to be particularly low. Countries such as Norway and the United Kingdom do not allow PSCs to possess weapons at all (CoESS, 2008). The Geneva PSC's rate of 0.06 firearms per employee[29] and information revealing that only two per cent of Swedish PSC employees are authorized to use firearms (CoESS, 2008) point to low levels of PSC armament even in countries where the use of firearms by PSCs is allowed. Some countries in the region may be home to larger PSC stockpiles, however. In Spain, for instance, more than 20 per cent of PSC personnel may be armed (see Table 4.3). As a result, 0.02–0.15 is the ratio applied to reported PSC personnel in Western European states.

Patterns of armament among PSCs in China, India, and the United States, with combined PSC personnel of more than 14 million, have a significant impact on a global estimate. Very little research exists on China's PSC industry. While Chinese PSC personnel can carry firearms only when escorting cash-in-transit (Trevaskes, 2008, p. 38), experts argue that up to several hundred thousand guards may be armed, although often illegally.[30] A minimal ratio of 0.01–0.05 is therefore applied to China to reflect low PSC arming.

Most private security guards in India are unarmed or carry only batons or long sticks (lathis) (Karp, 2010b). So equipped, they are able to perform little more than surveillance roles (Thottam and Bhowmick, 2010). While the total number of legally armed private security guards cannot be estimated systematically, it appears to be relatively low, in the range of one to three per cent (Karp, 2010b).[31] Similarly, about two per cent of the roughly 140,000 G4S guards in India are authorized to be armed (see Table 4.3). For these reasons, a low range of 0.01–0.05 is also applied to India's seven million PSC staff.

Latin America stands out as the region where PSCs are the most armed.

Table 4.5   Estimated global PSC firearm holdings

Group of countries	Combined PSC personnel (see Table 4.1)	Low firearm per employee ratio	High firearm per employee ratio	Low PSC firearms estimate	High PSC firearms estimate
Countries with reported PSC personnel and firearm holdings (see Table 4.4): Albania, Angola, Bosnia and Herzegovina, Brazil, Colombia, Costa Rica, El Salvador, Nicaragua, Russian Federation, Serbia, South Africa	2,080,201	0.29	0.29	609,093	609,093
Countries with reported PSC personnel and estimated firearms ratios in Latin America: Argentina, Bolivia, Chile, Dominican Republic, Ecuador, Guatemala, Honduras, Jamaica, Mexico, Panama, Peru, Trinidad and Tobago	995,888	0.30	0.80	298,766	796,710
Countries with reported PSC personnel and estimated firearms ratios in Africa: Côte d'Ivoire, Morocco, Sierra Leone	73,000	0.05	0.30	3,650	21,900
Countries with reported PSC personnel and estimated firearms ratios in Eastern Europe: Bulgaria, Croatia, Czech Republic, Estonia, Hungary, Kosovo, Latvia, Lithuania, Macedonia (former Yugoslav Republic of), Moldova, Montenegro, Poland, Romania, Slovakia, Slovenia	565,726	0.05	0.20	28,286	113,145
Countries with reported PSC personnel and estimated firearms ratios in Western Europe: Austria, Belgium, Cyprus, Finland, France, Germany, Greece, Ireland, Italy, Luxembourg, Malta, Portugal, Spain, Sweden, Switzerland	626,699	0.02	0.15	12,534	94,005
Australia: reported PSC personnel and estimated firearms ratio	114,600	0.02	0.15	2,292	17,190
China: reported PSC personnel and estimated firearms ratio	5,000,000	0.01	0.05	50,000	250,000
India: reported PSC personnel and estimated firearms ratio	7,000,000	0.01	0.05	70,000	350,000
United States: reported PSC personnel and estimated firearms ratio	2,000,000	0.20	0.30	400,000	600,000
Turkey: reported PSC personnel and estimated firearms ratio	257,192	0.15	0.20	38,579	51,438
Afghanistan and Iraq: reported PSC personnel and estimated firearms ratios	61,000	3.00	4.00	183,000	244,000
Countries where PSC employees are not allowed to carry firearms: Denmark, Japan, Kenya, Netherlands, Nigeria, Norway, United Kingdom	771,002	0	0	0	0
Rest of the world: estimated PSC personnel and firearms ratios	Between 0 and 6,000,000	0	0.10	0	600,000
World total				**1,696,200**	**3,747,481**

Source: estimates and calculations based on Tables 4.1, 4.3, and 4.4

Among countries with large numbers of PSC personnel, the United States appears to stand out with a relatively high proportion of armed guards. Typical functions for US PSCs include patrolling businesses and protecting gated communities. But there is no official information on what percentage of personnel normally carry a gun. Most US security guards do not carry a firearm; their functions are essentially those of watchmen and gatekeepers, with instructions to call the police in case of danger. A reasonable estimate of the proportion of PSC personnel armed while on duty would be one-quarter to one-third (Karp, 2010a). Since guards may share firearms between shifts, a ratio of 0.2–0.3 is thus applied to the two million US private security personnel.

A range of 0.15–0.20 is applied to Turkey, given information that 22 per cent of its private guards are armed (see Table 4.3). Western Europe's ratio of 0.02–0.15 is also applied to Australia, given that the proportion of armed PSC personnel in that country has dropped from 10–30 per cent in 2003–04 to 4–5 per cent in 2010 (Prenzler, 2005, p. 61).[32] Finally, a conservative ratio of 0.0–0.1 is applied to countries for which PSC personnel figures are estimated but not reported.

While the ratio of arms per PSC employee is usually lower than 1:1 in societies not affected by armed conflict, it is common for PSC personnel to carry more than one firearm in more hostile settings. PSC staff in Afghanistan and Iraq are typically equipped with two weapons: a handgun and an automatic rifle, with additional weaponry kept in vehicles and company armouries.[33] As illustrated by Table 4.4, individual PSC employees had access to more than three firearms each in Afghanistan and Sandline International's 1997 operation in Papua New Guinea. A high ratio of 3–4 firearms per employee is therefore applied to reported PSC staff in Afghanistan and Iraq.

Based on the above assumptions, it appears that PSCs worldwide hold somewhere between 1.7 and 3.7 million legal firearms. While the dearth of information explains such a broad range, this estimate remains significant in that PSCs hold only a small proportion of the global firearm stockpile of at least 875 million units. PSC holdings are comparable to the quantities of small arms held worldwide by gangs and armed groups (2 to 11 million units), but much lower than those of law enforcement (26 million), armed forces (200 million), and civilians (650 million) (Small Arms Survey, 2010, p. 103).

A private contractor guards a NATO convoy armed with a machine gun in Ghazin, Afghanistan, October 2010. © Rahmatullah Naikzad/AP Photo

Types of firearms[34]

National legislation usually leaves very little discretion to PSCs when it comes to the types of weapons they can use.[35] A survey of the industry across 34 European states reveals, for instance, that the vast majority of PSCs are only allowed to use handguns (pistols and revolvers) (CoESS, 2008). Smoothbore firearms (such as shotguns) are authorized in few countries, and almost all European countries prohibit PSCs from using automatic firearms. Fully automatic firearms and other types of military weapons are also generally banned from PSC use in other settings, including in Argentina, Brazil, Guatemala, Peru,[36] and South Africa.[37] In the Philippines, PSCs are not allowed to possess:

> *high caliber firearms considered as military-type weapons such as M16, M14, cal .30 carbine, M1 Garand, and other rifles and special weapons with bores bigger than cal .22, to include pistols and revolvers with bores bigger than cal .38 such as cal .40, cal .41, cal .44, cal .45, cal .50, except cal .22 centerfire magnum and cal .357 and other pistols with bores smaller than cal .38 but with firing characteristics of full automatic burst and three-round burst* (RoP, 2005, rule VII, sec. 2).

Many exceptions exist, however. In Turkey, for instance, PSCs may use MP5 sub-machine guns and G3 rifles for the protection of oil refineries, oil wells, and power plants (CoESS, 2008). Although Russian law seems to only allow PSCs to use pistols, revolvers, and other self-defence weapons, some company websites list sub-machine guns among the weapons available to their staff (Karimova, 2010b).[38] In some cases, legislation does not provide clear definitions of the weapons that PSCs may not use, resulting in broad interpretation and application. For example, under Angolan law, PSC staff are allowed to use and bear only 'defensive' firearms, for which they are required to undertake regular arms training. In practice, however, PSCs continue to use AK-47s and similar 'weapons of war', seen by the population as especially intimidating (Joras and Schuster, 2008, pp. 40, 56).

PSCs operating in hostile conflict environments rely on a greater variety of weapons, with Afghanistan and Iraq representing extreme examples. Although PSCs operating in these two countries procure mainly 9 mm handguns and assault rifles of calibre 7.62 mm or smaller,[39] reports show access to a broad range of small arms and light weapons, including general-purpose machine guns, sniper rifles, and, in some cases, RPGs (see Table 4.6).[40] Sandline International personnel, controversially recruited by the government

Table 4.6   Examples of small arms and light weapons reportedly held by PSCs in Afghanistan and Iraq

Weapon category	Afghanistan	Iraq
Handguns	• GLOCK (9 x 19 mm) • Smith & Wesson Sigma (9 x 19 mm)	• Beretta (9 x 19 mm) • Browning (9 x 19 mm) • Colt M1911 (.45) • CZ (9 x 19 mm) • GLOCK 17 (9 x 19 mm) • GLOCK 19 (9 x 19 mm) • Walther PPK (9 x 17 mm/.380 ACP)
Shotguns	• Remington 12-gauge	• 12-gauge
Sniper rifles	• Unspecified type	• Dragunov (7.62 x 54 mm R)
Semi- and fully automatic rifles	• AK-47 (7.62 x 39 mm) • AMD-65 (7.62 x 39 mm) • HK G36 and G36K (5.56 x 45 mm) • M4 (5.56 x 45 mm) • SIG 556 (5.56 x 45 mm)	• AK-47 (7.62 x 39 mm) • AR-M9 (5.56 x 45 mm) • HK G3 (7.62 x 51 mm) • HK G36 (5.56 x 45 mm) • M4 (5.56 x 45 mm) • M16 (5.56 x 45 mm) • SIG 552 (5.56 x 45 mm)
Machine guns	• PKM (7.62 x 54 mm R) • RPK (7.62 x 39 mm)	• Beretta M12S SMG (9 x 19 mm) • FN Minimi/M-249 (5.56 x 45 mm) • HK MP5 (9 x 19 mm) • M-240 (7.62 x 51 mm) • PKM (7.62 x 54 mm R) • RPK (7.62 x 39 mm) • SMG Sterling (9 x 19 mm or 7.62 x 51 mm)
Portable anti-tank weapons	• Unspecified RPG	• Unspecified RPG • AT4 (84 mm)

Sources: Isenberg (2009); JASG (2008); Joras and Schuster (2008, p. 14); Miller and Roston (2009); USASC (2010); USHR (2007, pp. 3, 8); author interviews with private security representatives 2, 3, 4, 5, 6, and 8

of Papua New Guinea to quell the Bougainville secessionist movement in 1997, were equipped with 60 mm and 80 mm mortars as well as AGS-17 30 mm automatic grenade launchers, in addition to pistols, AK-47 assault rifles, and PKM light machine guns (PNG and Sandline, 1997, p. 9).

Few companies have internal policies that specify restrictions on the arms their personnel may carry. Responsible PSCs undertake risk assessments to determine the level of threat involved in each operation; they adapt their equipment accordingly. The risk of collateral damage can be part of such assessments. One British company, for instance, systematically advises clients against using armed guards on ships, arguing that the presence of arms can only increase the likelihood of use of force by potential hijackers.[41]

Reported PSC use of sniper rifles, machine guns, and, in some cases, RPGs in Afghanistan and Iraq seems contradictory to PSC and contracting states' claims that private security personnel play an essentially protective, defensive role, and do not get involved in combat operations.[42] While light weapons and fully automatic assault rifles clearly give PSCs offensive capabilities, industry representatives argue that maintaining weapon capabilities at least equal or superior to potential attackers' is crucial for the purpose of suppressing enemy fire in case of attacks.[43] Rate of fire is particularly important when responding to an ambush while in a moving vehicle, and machine guns are commonly deployed for this purpose during convoy escorts.[44] The choice of weapon is also driven by the environment and 'local norms' where PSCs operate. The widespread availability of the AK-47 in Afghanistan and Iraq means that PSCs seek to carry similar or more advanced weapons systems in order to repel attacks. The type of weapon and its calibre will usually be determined and authorized by the host government.[45]

Contractual arrangements with clients sometimes specify the types of weapons PSCs may use. Standard operating procedures (SOPs) agreed by PSCs and clients usually indicate the allocation of firearms, ammunition, and magazines for each function, including the team leader, personnel protection officer, shooter, and driver.[46] Western-made weapons were reportedly popular at the outset of war among diplomatic outposts in Iraq, as proof that PSC equipment was in line with that of coalition forces rather than that of insurgents.[47] In Iraq, clients could sometimes be identified solely based on the type of arms carried by PSC personnel.[48] As Iraq progressively moved into a post-conflict phase, some PSCs preferred the AK-47 to the M4 as a symbol of return to normalcy and adherence to local norms.[49]

> Contracts with clients sometimes specify the types of weapons PSCs may use.

PSCs in Afghanistan and Iraq use standard ball, full metal jacket ammunition; expanding and exploding bullets are not permitted.[50] The amount of ammunition carried depends on the threat level a PSC team expects to encounter. Operators often carry smoke grenades, used to provide a screen behind which personnel can withdraw to safety. Industry sources explain that PSCs may use incendiary grenades only to destroy their own vehicles, such as when these are disabled by roadside improvised explosive devices, and to deny insurgents access to their contents.[51]

TACKLING MISUSE

Incidents of armed violence against civilians perpetrated by PSC personnel, particularly in Afghanistan and Iraq, have come under intense international scrutiny. Less attention has been devoted to the role that weapons, and gaps in regulations covering them, have played in such situations. This section reviews apparent loopholes in controls over PSC acquisition, management, and use of firearms and discusses the extent to which current initiatives may help address them.

Arms misuse by PSCs[52]

Arms acquisition

In most countries where the rule of law prevails, PSCs purchase their weapons locally through a registered dealer.[53] If firearms are not available locally, PSCs work with government arms procurement agencies or dealers to obtain an import licence from their country of operation, as well as an export licence from the country from which the arms are to be shipped.[54]

Reports of illicit firearm acquisition and use by PSCs suggest that such procedures are either not systematically followed or do not exist in all countries. In Brazil, for instance, the federal police recorded 760 cases of illicit arms possession by PSC personnel from January 2001 to September 2003 (FPB, 2009). In Tanzania, illegally produced 'homemade' guns called *magobori* feature among PSC weapons.[55] In 2010 in North Bengal, Indian intelligence seized illegal firearms and forged licences from PSC personnel, who had reportedly bought them from former soldiers (Das, 2010).

Due to increased media and government monitoring, several cases of illicit arms acquisition and possession by PSCs in Afghanistan and Iraq have been documented. One company was found to have procured firearms from US Armyguarded Afghan National Police stockpiles without proper authorization, for instance (USASC, 2010). In February 2009, US and Iraqi government officials found unauthorized 9 mm hollow-point ammunition, as well as unregistered MP5s, during random inspections of PSC armouries (MNF–I, 2009). In a separate inspection, Iraqi authorities raided the headquarters of a foreign security firm in Baghdad and seized unregistered arms and ammunition, including 20,000 rounds of ammunition and 400 rifles (al-Ansary, 2010). On 18 August 2010, Xe Services (formerly Blackwater) entered into a civil settlement with the US Department of State for 288 alleged violations of the International Traffic in Arms Regulations involving the unauthorized export of defence articles and provision of defence services to foreign end users in several countries between 2003 and 2009 (USDOS, 2010).

While negligence and criminal intent may explain several cases, it appears that regulatory constraints sometimes lead PSCs to break laws to acquire firearms. In the early days of the operations in Iraq, for instance, the time required to obtain the necessary authorization to import weapons into Iraq was such that some PSCs chose to procure arms illegally on the local market in order to be able to execute their contracts on time (Bergner, 2005; Miller and Roston, 2009). Faced with similar constraints, some companies in Afghanistan hired staff that already possessed weapons, turning a blind eye to the origins of their firearms (USHR, 2010, p. 2). Bureaucratic delays are no excuse for breaking laws, but improving procedures for the legal acquisition of arms by PSCs, including enhanced transparency and oversight, might have prevented some of the above-mentioned incidents.

A Pakistani officer inspects unlicensed weapons confiscated from a local security firm, Islamabad, September 2009.
© Anjum Naveed/AP Photo

Stockpile management

National legislation rarely provides details on how PSCs should secure firearm stockpiles from theft or diversion, or how to account for ammunition issue and expenditure (da Silva, 2010). When it does, the law tends to focus on whether personnel may keep their weapons at home when off duty. In Europe, for instance, PSC weapons must usually be secured in armouries (CoESS, 2008).

Stockpile security is crucial to preventing PSC arms from leaking to criminal networks through theft or loss. In Australia in 2007, for instance, gangs repeatedly targeted armed PSC employees in at least 11 attacks to seize not just the money they were escorting, but also their firearms (Gee and Jones, 2007). In South Africa, criminals have reportedly attacked—and killed—armed PSC personnel for the sole purpose of stealing their weapons (Gould and Lamb, 2004, pp. 192–93). Accountability of PSC small arms seems particularly problematic in maritime security operations. Some armed guards protecting ships from Somali pirates, for instance, reportedly dump weapons offshore before reaching countries' territorial waters in order to evade arms transfers regulations, save time, and cut costs (Hope, 2011).

In practice, the specifics of managing and securing PSC stockpiles are usually left to the companies themselves. Some large international PSCs have developed lengthy SOPs—up to several hundred pages—that contain detailed firearm policies and procedures for arms management.[56] Partly because SOPs are often required in client tenders, companies usually consider these documents proprietary information and keep them confidential. Making SOPs public would allow smaller, less well-resourced companies to simply reproduce existing written procedures and compete unfairly without necessarily being able to implement such regulations.[57] While large companies argue that their arms management procedures are strict and based on military standards,[58] lack of transparency makes an objective evaluation difficult. Controls over ammunition appear particularly critical. As industry sources admit, it is virtually impossible for PSCs—and state armed forces—to account for every round, even when every effort is made to do so.[59]

> A lack of transparency regarding internal PSC procedures makes objective evaluation difficult.

Where detailed regulations on PSC stockpile management are in place, setting up monitoring and enforcement is critical for these measures to be effective. Examples suggest that governments have been reactive rather than active in enforcing regulations and imposing oversight. In Iraq, for instance, despite the existence of detailed firearm-related regulations since the early days of operations, effective enforcement mechanisms were only put in place following Blackwater's killing of 17 Iraqi civilians at Nisoor Square in Baghdad in 2007 (Glanz and Lehren, 2010; Isenberg, 2010b). The Armed Contractor Oversight Division, for instance, was only established in November 2007. The Division has since carried out random inspections of PSC personnel and compounds, confiscating unrecorded weapons and ammunition from several companies (MNF–I, 2009).

Another issue concerns the disposal of firearms once a PSC no longer uses them. In most countries, PSCs have a long-term presence and simply renew their licences periodically.[60] For PSCs operating in conflict environments, however, weapons are often procured only for the duration of specific contracts. At the end of an assignment, PSCs may destroy their stockpiles and produce a government-issued destruction certificate, transfer weapons to their operations in another country, or return weapons to the original procurement agent or dealer.[61] The latter two options require PSCs to obtain the relevant export and import licences and are rarely implemented in practice. Resale to the host government or other PSCs operating locally is generally the favoured option.[62]

Use of force and firearms

Abusive use of force by PSCs has been the most controversial and publicized aspect of their activities, especially in Iraq and Afghanistan. Human Rights First documents that contractors in Iraq have discharged their weapons thousands of times, and hundreds of times against civilians, without facing investigation (HRF, 2008, p. 3). A RAND Corporation study also finds that more than one-fifth of US Department of State personnel in Iraq had first-hand knowledge of armed contractors mistreating civilians (Cotton et al., 2010, p. xv). The US Department of Defense reports that from May 2008 to February 2009, PSC personnel in Iraq discharged their weapons 109 times, of which more than one-third were categorized as 'negligent' (Isenberg, 2010b, citing CONOC, 2010).

The extent to which a PSC team will use its weapons also depends greatly on the type of operation. One company providing close protection services to government officials in Iraq reported that personnel only fired weapons five times in more than six years of operations.[63] In contrast, PSCs entrusted with protecting military convoys may fire their weapons on a daily basis, as their roles render them much more exposed to enemy attack.[64]

PSC use of force is regulated by international and national law (see Box 4.1). According to the Swiss criminal code, for instance, personnel can only use firearms in self-defence, and each firearm discharge must trigger a police investigation.[65] Moreover, standard rules for the use of force are an integral part of contracts with clients such as US government agencies.[66]

Some large PSCs develop their own rules, which they then validate with national authorities and clients.[67] The level of threat required to legitimize the use of force can vary greatly from company to company. Some PSCs require an imminent threat to life to justify the use of force by employees[68] (see Box 4.2). Other PSCs reportedly legitimize the use of force to protect not only life, but also infrastructure and materiel they are hired to guard.[69]

While regulations on the use of force and firearms do exist, their effectiveness is difficult to evaluate. Data on weapons discharge incidents by PSC personnel is improving in Iraq, but such progress is far

---

Box 4.2 Excerpts from internal PSC rules for opening fire in Iraq[70]

General rules

- In all situations you are to use the minimum force necessary. Firearms must only be used as a last resort.
- Your weapon must always be made safe; that is, no live round is to be carried in the breech [. . .] unless you are authorized to carry a live round in the breech or are about to fire.

Challenging

- A challenge must be given before opening fire unless:
  - To do so would increase the risk of death or grave injury to you, the client or other [company] personnel.
  - You, the client or other [company] personnel in the immediate vicinity are being engaged by hostile forces.
- You are to challenge by shouting: 'Security: Stop or I fire' or words to that effect.

Opening fire

- You may only open fire against a person:
  - If s/he is committing or about to commit an act likely to endanger life to you, the client or other [company] personnel and there is no other way to prevent the danger. The following are some examples of acts where life could be endangered, dependent always upon the circumstances:
    - Firing or being about to fire a weapon.
    - Planting, detonating or throwing an explosive device.
    - Deliberately driving a vehicle at a person [. . .] where it is assessed there is no other way of stopping him/her.
  - If you know that s/he has just killed or injured the client or other [company] personnel by such means as s/he does not surrender if challenged and presents a clear and hostile threat to you, the client or other [company] personnel.
- If you have to open fire you should:
  - Fire only aimed shots.
  - Fire no more rounds than are necessary.
  - Take all reasonable precautions not to injure anyone other than your target.

from universal. Furthermore, existing data provides no basis for assessing the performance of PSC personnel compared with state security officers, for instance. Complicating matters further, even the best-intentioned firms keep their internal rules on the use of force confidential, which prevents any external assessment or monitoring of their implementation.

Training requirements

Training of PSC personnel in the use of firearms is another area that appears not to be systematically controlled. Some countries do not require any level of training or competence for individuals employed in the private security sector. For example, in Sierra Leone, governmental regulations relating to the qualifications and training of security personnel are non-existent, and there are no minimum training standards specified for PSCs, nor any requirements relating specifically to firearms (Abrahamsen and Williams, 2005b, p. 11). In the Democratic Republic of the Congo, there is no training requirement for PSCs at all (de Goede, 2008, p. 48). In the United States, there are no federal laws governing the domestic PSC industry. State laws with regard to training of PSC guards vary: 16 US states do not require background checks before someone can be hired by a PSC; 30 states do not require training; 20 states provide for mandatory training, but the requirements vary between 1 and 48 hours; in 22 states, private security services do not have to be licensed (da Silva, 2010).

Armed guards from a private security company practice firing 9 mm pistols at a shooting range, Johannesburg, South Africa, June 1997.
© Reuters

Even when legal requirements exist regarding the vetting and training of PSC sector workers, they often merely indicate that the PSC is responsible for ensuring that employees are properly trained (da Silva, 2010). Under the Private Guards Act in Nigeria, for instance, the training syllabus and instruction notes of every licensed PSC must be submitted to and approved by the Minister of Internal Affairs. These are not, however, assessed against a set of common standards. As a consequence, the quality and duration of training varies greatly among PSCs (Abrahamsen and Williams, 2005a, p. 8).

Specific requirements for training in the use of arms are rare. For example, Colombian Decree 356 of 1994 states that responsibility for the training of personnel lies with the PSC, but it makes no specific mention of training in the use of arms (Colombia, 1994, art. 64). In Angola, PSC employees are legally required to undertake regular arms use training (RoA, 1992, art. 11); however, Angolan law does not establish training standards. Few states actually require accredited firearms training. South Africa appears to be an exception. The Firearms Control Act 2000 requires that security industry employees produce a competence certificate before a firearm can be issued to them. In order to acquire such a certificate, the individual must already have been trained at an accredited training facility (South Africa, 2000, ch. 5, sec. 9.1).

Overall, training in firearms for PSC personnel lacks standardization and accreditation. As a result, designing the content of training modules is often left to companies, resulting in disparate standards. Training programmes used by large international firms are often based on recognized systems, such as the British Army small arms instructors' course. They sometimes require personnel to practice on ranges more frequently than the military—more than once every three months.[71] Poor weapons handling performance can result in additional training to the satisfaction of a weapons instructor.[72] Little is known, however, about any training packages that may be available to the employees of the many other PSCs.

International initiatives[73]

Several international initiatives have emerged in recent years to increase accountability of PSCs and establish standards against which to measure their performance. Initiated by the Swiss government and the International Committee of the Red Cross, the Montreux Document on Pertinent International Legal Obligations and Good Practices for States Related to Operations of Private Military and Security Companies during Armed Conflict[74] was adopted in 2008 and had the support of 35 countries at this writing. Responding to a need for clarification, it summarizes contracting and hosting states' legal obligations

Tim Spicer, representing the Aegis Group, signs the International Code of Conduct for Private Security Providers in Geneva, Switzerland, on 9 November 2010.
© Anja Niedringhaus/AP Photo

under international humanitarian and human rights law with respect to PSCs, while also compiling good practices. Although the Montreux Document applies primarily to the activities of PSCs in contexts of armed conflict, it contains several firearm-specific recommendations that are relevant to the broader operations of the private security industry (see Box 4.3).

Building on the Montreux Document, the Swiss government, with support from the Geneva Centre for the Democratic Control of Armed Forces and the Geneva Academy of International Humanitarian Law and Human Rights, has worked with industry, civil society, private sector clients, and governments—principally the UK and United States—to develop an International Code of Conduct for Private Security Providers (ICoC). Like the Montreux Document, the ICoC is based on international human rights and humanitarian law, but it speaks directly to the private security industry by establishing common international principles that will guide PSC work.

The ICoC was formally adopted in Geneva on 9 November 2010 by 58 companies, including market leaders Aegis, G4S, DynCorp, Triple Canopy, and Xe Services (FDFA, 2010).[75] Significantly, key contracting government agencies—including the US Department of Defense and the British Foreign Office—have announced their intent to favour companies that sign up to the ICoC when allocating contracts, providing important incentives for companies to comply with it in practice.[76] The next step in the Swiss-led process involves the creation of governance and oversight mechanisms that

---

Box 4.3 Firearm-specific recommendations contained in the Montreux Document

States, when hiring a PSC, should take into account:

- the past conduct of the PSC and its personnel, including whether any of its personnel, particularly those who are required to carry weapons as part of their duties, have a reliably attested record of not having been involved in serious crime or have not been dishonourably discharged from armed or security forces (part two, paras. 6, 32);
- whether the PSC maintains accurate and up-to-date personnel and property records, in particular with regard to weapons and ammunition, available for inspection on demand (paras. 9, 34);
- whether the PSC's personnel are adequately trained, including with regard to rules on the use of force and firearms (paras. 10(a), 35(a));
- whether the PSC:
  - acquires its equipment, in particular its weapons, lawfully;
  - uses equipment, in particular weapons, that is not prohibited by international law;
  - has complied with contractual provisions concerning return and/or disposition of weapons and ammunition (para. 11);
- whether the PSC's internal regulations include policies on the use of force and firearms (para. 12).

Contracting states should also include in contracts with PSCs:

- a clause confirming the PSC's lawful acquisition of equipment, in particular weapons (para. 14);
- a requirement that the PSC respect relevant national regulations and rules of conduct, including rules on the use of force and firearms, such as using force and firearms only when necessary in self-defence or defence of third persons, and immediate reporting to and cooperation with competent authorities in the case of use of force and firearms (para. 18).

States where PSCs are operating should, in addition to incorporating the above provisions into their licensing laws, establish appropriate rules on the possession of weapons by PSCs and their personnel, such as:

- limiting the types and quantity of weapons and ammunition that a PSC may import, possess, or acquire;
- requiring the registration of weapons, including their serial number and calibre, and ammunition, with a competent authority;
- requiring PSC personnel to obtain an authorization to carry weapons that is to be shown upon demand;
- limiting the number of employees allowed to carry weapons in a specific context or area;
- requiring the storage of weapons and ammunition in a secure and safe facility when personnel are off duty;
- requiring that PSC personnel carry authorized weapons only while on duty;
- controlling the further possession and use of weapons and ammunition after an assignment is completed, including return to point of origin or other proper disposition of weapons and ammunition (para. 44).

Sources: FDFA and ICRC (2009); Parker (2009, pp. 10–11)

will certify PSCs and monitor their compliance, although the parameters of such mechanisms remain to be negoti-ated in 2011 (FDFA, 2010, p. 6).

The ICoC contains several clauses relating to arms management and use; these are largely derived from those contained in the Montreux Document. As such, the ICoC has the potential to address some of the regulatory gaps highlighted above, if implemented. Firearms-related provisions remain vague when it comes to establishing specific standards for the acquisition of firearms, the use of force, accounting and record-keeping of weapons, and training requirements, however. A significant challenge for future oversight and governance mechanisms involves developing more detailed operational guidelines to facilitate the implementation of firearms-related provisions, including techni-cal standards and training modules. As highlighted throughout this chapter, increased industry transparency on arms holdings, use, and regulations, as well as systematic data collection on incidents of weapons discharges, would facilitate monitoring of compliance with the code. Furthermore, although human rights aspects of the ICoC apply to all situations, the key audience of the initiative remains large international PSCs operating in conflict environments, which, as illustrated by this chapter, represent only a fraction (yet one that is well armed) of PSC personnel worldwide.

Other initiatives include proposed negotiations for a new international convention on PSCs, on the basis of draft text prepared by the independent experts of the UN Working Group on the Use of Mercenaries.[77] This legal instru-ment would apply to all situations, armed conflict or not. Mandated by the Human Rights Council and the General Assembly of the United Nations, the draft text would require states to develop national regimes for the licensing, regulation, and oversight of PSC activities and calls for the establishment of an international register of PSCs (Gómez del Prado, 2010). While the proposed convention has the potential to improve the regulation of PSC activities, it is only at the expert consultation stage. It thus remains unclear how much political support it will receive from con-cerned governments.

CONCLUSION

The private security industry has grown to a significant size across the globe, employing more personnel than the police in many countries. PSCs include small local outfits as well as large multinational firms that carry out contracts for diverse clients such as governments, international corporations, local businesses, and private households. While they operate overwhelmingly in countries considered at peace, they are often more conspicuous in conflict contexts, where their actions can raise concerns.

While debates on the legitimacy and inequality of the industry continue, identifiable trends in PSC personnel employment, industry forecasts, and government contracting suggest that the industry will keep expanding into the foreseeable future. As the industry develops, the controls designed to regulate it are not keeping pace. States are generally lagging behind in developing effective oversight mechanisms of PSCs, and they appear to take necessary measures only to respond to, as opposed to prevent, violations.

This chapter reveals that the level of regulatory control exercised over the firearms held by PSCs is no exception to this rule. Little is reported or known about the actual quantities and types of firearms held by PSCs. In many countries, official standards for the management and safeguarding of PSC weapons, as well as for the training of PSC personnel, are non-existent. More worrying, the monitoring of PSCs' firearm holdings and use has progressed only in isolated cases and in response to highly publicized abuses. Lack of effective regulation has meant that the industry has to a great extent developed its own firearm-related standards, which only the largest companies are able and willing to implement. Confidentiality of internal PSC regulations has meant that these standards have not been disseminated widely or shared within the industry, resulting in different PSCs abiding by different rules.

The ongoing effort to regulate the private security industry at the international, national, and industry levels following adoption of the Montreux Document has potential due to the buy-in of both industry and concerned states as well as the intent to create independent oversight mechanisms. Assessing its effectiveness will require increased transparency and information sharing on PSC personnel qualifications, levels of training, and incidence of abuses. Similarly, more information is required to assess whether controls of PSC firearms are actually being implemented and enforced.

Requiring greater transparency from PSCs with respect to their firearm holdings and discharges would significantly enhance the ability to measure progress and hold the industry to international standards. For the industry the stakes are potentially high: failing to provide evidence of compliance with acceptable standards would expose them to public criticism, lost business, and, ultimately, drastic government response, such as occurred in Afghanistan. ◾

ABBREVIATIONS

ICC	International Criminal Court
ICoC	International Code of Conduct
IHL	International humanitarian law
ISOA	International Stability Operations Association
PMC	Private military company
PSC	Private security company
RPG	Rocket-propelled grenade (launchers)
SOP	Standard operating procedure

ANNEXE

Online annexe at <http://www.smallarmssurvey.org/publications/by-type/yearbook/small-arms-survey-2011.html>

Annexe 4.1. Private security personnel in 70 countries

In addition to reproducing the figures shown in Table 4.1, this table provides a comprehensive list of sources.

ENDNOTES

1   Definition adapted from FDFA and ICRC (2009, p. 9).

2   See Abrahamsen and Williams (2009).

3   Some analysts have even proposed typologies to distinguish between different types of PMCs. Singer, for instance, proposes a typology based on a company's proximity to the frontline, classifying PMCs as military provider firms, military support firms, and military consultant firms (Singer, 2003, pp. 91–93).

4   See, for example, Abrahamsen and Williams (2009, p. 1); Holmqvist (2005, p. 7); Rosemann (2008, p. 9); and Singer (2004, p. 524).

5   The sources, a comprehensive list of which appears in Annexe 4.1, include Arias (2009); CoESS (2008); CoESS and APEG (2010); Page et al. (2005).

6   The 70 countries listed in Table 4.1 represent a total population of 4.5 billion. The median rate of PSC personnel for these countries is 298 per 100,000 people. Assuming that these 70 countries are documented because the scale of their PSC industry is significant, it is highly unlikely that the overall PSC personnel rate in the rest of the world will exceed this 298 per 100,000 median. Based on available world population figures (UN, 2008), countries for which there is no PSC personnel data available represent a population of two billion. Applying the median rate of PSC personnel from documented countries to this 'undocumented' population would mean that there could be a maximum of 2 billion x 298 / 100,000 = 6 million PSC personnel in undocumented countries, producing an upper-end estimate of 25.5 million PSC personnel.

7   Based on data for 20 countries, van Dijk suggests a global PSC personnel rate of 348 per 100,000 compared with 318 police officers per 100,000 (van Dijk, 2008, pp. 215, 368–69). Van Dijk's data does not cover China or India, however.

8   Author correspondence with Scott Horton, contributing editor, *Harper's* magazine, 31 October 2010.

9   Private contractors hired by the US government perform a variety of non-security related tasks, such as medical and laundry services and transportation. As of 30 September 2010, for instance, only 13,101 of 88,448 contractors (15 per cent) employed by the US Department of Defense, US Department of State, and US Agency for International Development in Iraq were classified as PSC personnel (SIGIR, 2010, p. 55).

10   This section draws partly from Richard (2010).

11   See, for example, Singer (2004, p. 521); Walker and Whyte (2005, pp. 651–87).

12   See, for example, Gillard (2006, pp. 527–28); Sossai (2009, p. 1); Bailes and Holmqvist (2007, p. 7).

13   Prosecution by the International Criminal Court requires that an individual's actions meet criteria for a crime under the ICC Statute. The ICC has jurisdiction over individuals only, not corporations (Schabas, 2007, p. 211). This means that the Court has jurisdiction over the managers of PSCs for negligence in the prevention of the commission of crimes by their employees.

14   Companies, as private entities, have no legal status under international humanitarian law.

15   Regardless of their legal categorization, the reality on the ground is that contractors in Afghanistan and Iraq are regularly subjected to attacks, with contractor casualties even exceeding military deaths for the period January–June 2010 (Isenberg, 2010a; Miller, 2010).

16   See, for example, Schmitt (2005, pp. 538–39); Doswald-Beck (2007, p. 129).

17   See, for example, Dinstein (2004, p. 27); Sossai (2009, p. 14).

18   The International Court of Justice, among other international bodies, has addressed the applicability of international human rights law during armed conflicts—both international and non-international. The Court first affirmed the applicability in its 1996 *Advisory Opinion on the Legality of the Threat or Use of Nuclear Weapons* (ICJ, 1996, para. 25). This was then confirmed in the *Advisory Opinion on the Legal Consequences of the Construction of a Wall in the Occupied Palestinian Territory* (ICJ, 2004, paras. 106–13) and subsequently in the binding judgement, *Armed Activities on the Territory of the Congo* (ICJ, 2005, para. 216). See also IACHR (2000, para. 20); UNHRC (2004, para. 11).

19   Author correspondence with Kennedy Mkutu, Dar es Salaam Business School, 11 August 2010.

20   See HRF (2008).

21   In this section, analysis of national legislation is derived from da Silva (2010).

22   Author correspondence with William Godnick, UN Regional Centre for Peace, Disarmament and Development in Latin America and the Caribbean (UN-LiREC), 21 October 2010.

23   CoESS (2008); da Silva (2010, p. 2); van Steden and Huberts (2006, p. 23); Yoshida and Leishman (2006, p. 228).

24   Author interview with private security representative 1, Geneva, 19 August 2010.

25   Author interviews with private security representatives 4 and 5, London, 14 July 2010.

26   Author correspondence with private security representative 2, 26 August 2010.

27   Author interview with private security representative 1, Geneva, 19 August 2010.

28   Author interview with private security representative 3, Geneva, 2 October 2010.

29   Author interview with private security representative 1, Geneva, 19 August 2010.

30   Author correspondence with Aaron Karp, 13 October 2010.

31   This ratio was confirmed during an interview by Sonal Marwah with Kunwar Vikram Singh, chairman, Central Association of Private Security Industry–India, Delhi, 20 October 2010.

32   Correspondence with Bryan de Caires, chief executive officer, Australian Security Industry Association Limited, 3 December 2010.

33   Author interviews with private security representatives 4 and 5, London, 14 July 2010.

34   In this section, analysis of national legislation is derived from da Silva (2010).

35   Author interview with private security representative 1, Geneva, 19 August 2010.

36   See Arias (2009, p. 79).

37   See South Africa (2000, ch. 2, sec. 4.1).

38   See photos of sub-machine guns on the website of the Russian company Alfa-Inform (n.d.).

39   Author interview with Christopher Beese, private security industry commentator, London, 14 July 2010.

40   Greystone Limited, a Blackwater subsidiary, reportedly asked prospective employees to check off their qualifications regarding the use of a variety of weapons, including the AK-47, GLOCK 19, M16, M4, machine guns, mortars, and shoulder-fired weapons such as RPGs and light anti-armour weapons (Scahill, 2007, p. 59).

41   Author interview with private security representative 5, London, 14 July 2010.

42   Industry and government representatives made this assertion consistently during author interviews and research for this study.

43   Author interview with private security representative 3, Geneva, 2 October 2010.

44   Author interview with Christopher Beese, private security industry commentator, London, 14 July 2010.

45   Author interview with Christopher Beese, private security industry commentator, London, 14 July 2010.

46   Author correspondence with private security representative 2, 26 August 2010.

47   Author interview with Christopher Beese, private security industry commentator, London, 14 July 2010.

48   Author correspondence with private security representative 2, 26 August 2010.

49   Author interview with Christopher Beese, private security industry commentator, London, 14 July 2010.

50   Author correspondence with former private security representative 6, 6 August 2010.

51   Author correspondence with former private security representative 6, 6 August 2010.

52   Examples of PSC arms misuse in Iraq are drawn primarily from Isenberg (2010b).

53   Author interview with private security representative 1, Geneva, 19 August 2010.

54   Author interview with private security representative 5, London, 14 July 2010.

55   Author correspondence with Kennedy Mkutu, Dar es Salaam Business School, 11 August 2010.

56   Author interview with private security representatives 4 and 5, London, 14 July 2010.

57   Author interview with Doug Brooks, president, International Stability Operations Association (ISOA), Geneva, 2 October 2010.

58   Author interview with private security representative 5, London, 14 July 2010.

59   Author interview with Christopher Beese, private security industry commentator, London, 14 July 2010; author correspondence with private security representative 2, 26 August 2010, and with former private security representative 6, 6 August 2010.

60   Author interview with private security representative 5, London, 14 July 2010.

61   Author interview with private security representative 5, London, 14 July 2010.

62   Author interview with Christopher Beese, private security industry commentator, London, 14 July 2010.

63   Author interview with private security representative 5, London, 14 July 2010. In addition, Blackwater head Erik Prince testified before the US Congress that his company's weapons were discharged in less than one per cent of 6,500 diplomatic escorts in 2006, and less than three per cent of 1,873 diplomatic escorts from January to October 2007 (Prince, 2007, p. 4).

64   Author interview with Christopher Beese, private security industry commentator, London, 14 July 2010.

65   Author interview with private security representative 1, Geneva, 19 August 2010.

66   Author correspondence with former private security representative 6, 6 August 2010.

67   Author correspondence with former private security representative 6, 6 August 2010.

68   Such policies are consistent with the UN Basic Principles for the Use of Force and Firearms by Law Enforcement Officials, for instance. These state that: 'Law enforcement officials shall not use firearms against persons except in self-defence or defence of others against the imminent threat of death or serious injury' (UN, 1990, para. 9).

69   Author interview with private security representative 5, London, 14 July 2010.

70   Author correspondence with private security representative 5, 14 July 2010.

71   Author interview with Christopher Beese, private security industry commentator, London, 14 July 2010.

72   Author correspondence with former private security representative 6, 6 August 2010.

73   Parts of this section draw from Richard (2010).

74   For details, see FDFA and ICRC (2009); FDFA (2009).

75   It should be noted that the industry began developing standards in the early 2000s, if not before. The US-based ISOA worked with human rights lawyers and NGOs to develop a code of conduct as early as 2001, and has revised it 12 times since. Version 12 contains three paragraphs on arms control, committing member companies to undertake responsible accounting, control, and disposal of weapons; to refrain from using unauthorized weapons; and to acquire weapons exclusively through legal channels (ISOA, 2009, paras. 9.4.1–9.4.3). The ISOA code also calls on companies to develop rules on the use of force that are in compliance with international humanitarian and human rights law (para. 9.2.2). The company has received a total of about 20 complaints since its code of conduct was established. In cases of credible allegations, the ISOA's

standards committee—composed of industry representatives—has required violators of the code to take measures to redress wrongdoing (author interview with Doug Brooks, president, ISOA, Geneva, 1 October 2010). The significance and effectiveness of such measures cannot be assessed, however, since the outcome of investigations is kept confidential. The only exception is the ISOA's initiation of an independent review to determine whether Blackwater—an ISOA member at the time—had violated the ISOA code of conduct during the 2007 Nisoor Square shootings. Blackwater withdrew its membership from ISOA a few days after the inquiry began (Fontaine and Nagl, 2010, p. 28; Rosemann, 2008, p. 35).

76   Author interviews with private security representatives 3 and 7, who were involved in the drafting of the ICoC, Geneva, 1–2 October 2010.

77   At the Human Rights Council's 15th session in September 2010, states voted to establish an open-ended intergovernmental working group to consider the possibility of elaborating an international regulatory framework on PSCs, including the option of a legally binding instrument (UNHRC, 2010).

BIBLIOGRAPHY

Abrahamsen, Rita and Michael Williams. 2005a. *The Globalization of Private Security: Country Report—Nigeria*. Aberystwyth: Department of International Politics, University of Wales. <http://users.aber.ac.uk/rbh/privatesecurity/country%20report-nigeria.pdf>

—. 2005b. *The Globalization of Private Security: Country Report—Sierra Leone*. Aberystwyth: Department of International Politics, University of Wales. <http://users.aber.ac.uk/rbh/privatesecurity/country%20report-sierra%20leone.pdf>

—. 2009. 'Security Beyond the State: Global Security Assemblages in International Politics.' *International Political Sociology,* Vol. 3, pp. 1–17. <http://onlinelibrary.wiley.com/doi/10.1111/j.1749-5687.2008.00060.x/pdf>

Afghanistan. 2010. 'Afghan Gov't Stands Firm to Disband Private Security Companies.' Press release. 24 October. Kabul: Office of the President, Islamic Republic of Afghanistan. <http://www.president.gov.af/Contents/88/Documents/2300/release_eng.html>

al-Ansary, Khalid. 2010. 'Iraq Confiscates Arms in Private Security Crackdown.' Reuters. 9 January.

Alfa-Inform. n.d. Company wesbite. <http://www.alfa-inform.ru/services/uslugi_po_obespecheniyu_bezopasnosti/sluzhebnoe_oruzhie/>

Arias, Patricia. 2009. *Seguridad Privada en América Latina: el lucro y los dilemas de una regulación deficitaria*. Santiago de Chile: FLACSO Chile. July.

Baker, Deane-Peter and James Pattison. 2010. *The Principled Case for Employing Private Military and Security Companies in Humanitarian Interventions and Peacekeeping*. Human Rights and Human Welfare Working Paper No. 56. Denver: Denver University.

Bailes, Alyson and Caroline Holmqvist. 2007. *The Increasing Role of Private Military and Security Companies*. Brussels: European Parliament. October.

Berg, Julie. 2007. *The Accountability of South Africa's Private Security Industry*. Newlands: Open Society Foundation for South Africa.

Bergner, Daniel. 2005. 'The Other Army.' *The New York Times*. 14 August.

Bushnell, Alexis. 2010. *Private Security Contractors: Legal Debates under International Humanitarian Law*. Unpublished background paper. Geneva: Small Arms Survey.

Carballido Gómez, Armando. 2008. *Seguridad pública y privada en América Latina*. Presentation prepared for the Organization of American States Working Group charged with preparing the first OAS ministerial meeting on public security in the Americas. <http://scm.oas.org/pdfs/2008/CP20722T01.ppt>

Cockayne, James and Emily Speers Mears. 2009. *Private Military and Security Companies: A Framework for Regulation*. New York: International Peace Institute.

CoESS (Confederation of European Security Services). 2008. *Private Security in Europe: CoESS Facts & Figures 2008*. Brussels: CoESS.

— and APEG (Association Professionnelle des Entreprises de Gardiennage). 2010. *Private Security in Belgium: An Inspiration for Europe?* Third White Paper. Brussels: CoESS and APEG. <http://www.coess.org/brussels_summit/CoESS-APEG-BVBO_White_Book_(EN).pdf>

Colombia. 1994. Decree 356 of 11 February 1994. *Diario Oficial,* No. 41.220. Bogotá: Ministerio de Defensa Nacional. <http://www.presidencia.gov.co/prensa_new/decretoslinea/1994/febrero/11/dec0356111994.doc>

CONOC (Contractors Operation Center). 2010. 'CONOC Website, Other Downloads.' <http://www.rocops.com/roc2/OtherDocuments.aspx>

Cotton, Sarah, et al. 2010. *Hired Guns: Views about Armed Contractors in Operation Iraqi Freedom*. Santa Monica: RAND Corporation.

CPDE (Center for Peace and Disarmament Education) and Saferworld. 2005. *Turning the Page: Small Arms and Light Weapons in Albania*. London and Tirana: Saferworld.

Cusumano, Eugenio. 2009. *Regulating Private Military and Security Companies: A Multifaceted and Multilayered Approach*. EUI Working Paper. AEL 2009/11. San Domenico di Fiesole: Academy of European Law.

da Silva, Clare. 2010. *Analysis of Selected Countries' National Legislation on Private Military and Security Companies (PSCs) and Firearms and Multinational Companies' Use of PSCs*. Unpublished background paper. Geneva: Small Arms Survey.

Das, Madhuparna. 2010. 'Illegal Arms Reach Naxals via Private Security Agencies.' *Express India*. 15 January.

de Goede, Meike. 2008. 'Private and Public Security in Post-war Democratic Republic of Congo.' In Sabelo Gumedze, ed. *The Private Security Sector in Africa*. ISS Monograph Series No. 146. July, pp. 35–68.

Dinstein, Yoram. 2004. *The Conduct of Hostilities under the Law of International Armed Conflict*. Cambridge: Cambridge University Press.

Doswald-Beck, Louise. 2007. 'Private Military Companies under International Humanitarian Law.' In Simon Chesterman and Chia Lehnardt, eds. *From Mercenaries to Market: The Rise and Regulation of Private Military Companies*. Oxford: Oxford University Press, pp. 115–38.

Dreyfus, Pablo, et al. 2010. *Small Arms in Brazil: Production, Trade, and Holdings*. Special Report No. 11. Geneva: Small Arms Survey.

Falalyev, Mikhail. 2010. 'Disarmament of PSCs.' *Rossiyskaya Gazeta* (Russian Gazette), Federal Iss. No. 5099(20). 2 February.
<http://www.rg.ru/2010/02/02/ohrana.html>

FDFA (Swiss Federal Department of Foreign Affairs). 2009. 'The Montreux Document on Private Military and Security Companies.' Berne: FDFA. 8 October.
<http://www.eda.admin.ch/psc>

—. 2010. International Code of Conduct for Private Security Providers. <http://www.news.admin.ch/NSBSubscriber/message/attachments/21143.pdf>

— and ICRC (International Committee of the Red Cross). 2009. *The Montreux Document on Pertinent International Legal Obligations and Good Practices for States Related to Operations of Private Military and Security Companies during Armed Conflict*.
<http://www.icrc.org/eng/assets/files/other/icrc_002_0996.pdf>

Fontaine, Richard and John Nagl. 2010. *Contracting in Conflicts: The Path to Reform*. Washington, DC: Center for a New American Security. June.

FPB (Federal Police of Brazil). 2009. 'Audiência Pública Câmara Federal.' PowerPoint presentation by the general coordinator for the control of private security.

Gee, Steve and Gemma Jones. 2007. 'Gun Theft Gangs Strike Again.' *Daily Telegraph* (Australia). 10 July.
<http://www.dailytelegraph.com.au/news/nsw-act/gun-theft-gangs-strike-again/story-e6freuzi-1111113921839>

Gillard, Emanuela-Chiara. 2006. 'Business Goes to War: Private Military/Security Companies and International Humanitarian Law.' *International Review of the Red Cross,* Vol. 88, No. 863. September, pp. 525–72.

Glanz, James and Andrew Lehren. 2010. 'Use of Contractors Added to War's Chaos in Iraq.' *The New York Times*. 23 October.

Godnick, William. 2009. *Private Security: A Preliminary Report for the Second Meeting of the Organized Crime Observatory for Latin America and the Caribbean, San Jose, Costa Rica, 8–9 September 2009*. Lima: United Nations Regional Centre for Peace, Disarmament and Development in Latin America and the Caribbean. Revised 15 October.

Gómez del Prado, José. 2010. *Why Private Military and Security Companies Should Be Regulated*. 3 September.
<http://www.reports-and-materials.org/Gomez-del-Prado-article-on-regulation-of-private-and-military-firms-3-Sep-2010.pdf>

Gould, Chandré and Guy Lamb. 2004. *Hide and Seek: Taking Account of Small Arms in Southern Africa*. Pretoria: Institute for Security Studies.

Heaton, J. Ricou. 2005. 'Civilians at War: Reexamining the Status of Civilians accompanying the Armed Forces.' *Air Force Law Review*, Vol. 57, pp. 155–208.

Holmqvist, Caroline. 2005. *Private Security Companies: The Case for Regulation*. SIPRI Policy Paper No. 9. Stockholm: Stockholm International Peace Research Institute. January. <http://books.sipri.org/files/PP/SIPRIPP09.pdf>

—. 2007. *The Private Security Industry, States, and the Lack of an International Response*. Prepared for the seminar on Transnational and Non-State Armed Groups convened by the Program on Humanitarian Policy and Conflict Research, Harvard University, Cambridge, 9–10 March.

Hope, Bradley. 2011. 'Firearms: An Odd Casualty of Piracy.' *National* (Abu Dhabi). 6 February.
<http://www.thenational.ae/featured-content/channel-page/business/middle-article/firearms-an-odd-casualty-of-piracy>

HRF (Human Rights First). 2008. *Private Security Contractors at War: Ending the Culture of Impunity*. New York and Washington, DC: HRF.

IACHR (Inter-American Court of Human Rights). 2000. *Bamaca Velásquez* v. *Guatemala*. Judgement. 25 November.

ICJ (International Court of Justice). 1996. *Advisory Opinion on the Legality of the Threat or Use of Nuclear Weapons*. 1996 I.C.J. 240 of 8 July.

—. 2004. *Advisory Opinion on the Legal Consequences of the Construction of a Wall in the Occupied Palestinian Territory*. ICJ Reports. 9 July.

—. 2005. *Armed Activities on the Territory of the Congo (Democratic Republic of the Congo* v. *Uganda)*. Judgement. ICJ Reports. 19 December.

Isenberg, David. 2009. *Shadow Force: Private Security Contractors in Iraq*. Westport, Connecticut: Praeger.

—. 2010a. *Private Military and Security Contractor Deaths in Iraq and Afghanistan*. Unpublished background paper. Geneva: Small Arms Survey.

—. 2010b. *Regulation, Oversight, Storage, Acquisition, and Use of Small Arms by PSCs in Iraq*. Unpublished background paper. Geneva: Small Arms Survey.

ISOA (International Stability Operations Association). 2009. 'Code of Conduct.' Version 12.
<http://ipoaworld.org/eng/codeofconduct/87-codecodeofconductv12enghtml.html>

JASG (Joint Area Support Group). 2008. *Approval of Weapons Authorization for Private Security Company*. Memorandum. Baghdad: JASG. 1 July.

Jones, Trevor and Tim Newburn, eds. 2006. *Plural Policing: A Comparative Perspective*. London: Routledge.

Joras, Ulrike and Adrian Schuster, eds. 2008. *Private Security Companies and Local Populations: An Exploratory Study of Afghanistan and Angola*. Working Paper. Berne: SwissPeace. <http://www.swisspeace.ch/typo3/fileadmin/user_upload/pdf/Working_Paper/WP_1_2008.pdf>

Karimova, Takhmina. 2010a. *Private Military Companies/Private Security Companies and Arms in Afghanistan*. Unpublished background paper. Geneva: Small Arms Survey.

—. 2010b. *Private Security Companies in Russia*. Unpublished background paper. Geneva: Small Arms Survey.

Karp, Aaron. 2008. *Highest Contemporary and Current Uniform-wearing Military Personnel, Active and Reserve*. Unpublished background paper. Geneva: Small Arms Survey.

—. 2010a. *America's Other Private Security Company*. Unpublished background paper. Geneva: Small Arms Survey.

—. 2010b. *Eccentricity and Illegality in Indian Armed Private Security*. Unpublished background paper. Geneva: Small Arms Survey.

Kougniazondé, Christophe. 2010. *L'état des lieux de la privatisation de la sécurité en Afrique Francophone: une revue de littérature*. Working Paper Series No. 1. Santiago de Chile: The Global Consortium on Security Transformation. <http://www.securitytransformation.org/images/publicaciones/ 158_Working_Paper_1_-_L_etat_de_la_privatisation_de_la_securite_en_Afrique_francophone.pdf>

Krzalic, Armin. 2009. *Private Security in Bosnia and Herzegovina*. Sarajevo: Center for Security Studies.

Marlowe, Lara. 2010. 'Notorious Blackwater Firm Awarded Lucrative Afghan Contracts.' *Irish Times*. 26 June.

Mbogo, Steve. 2010. 'Kenya: Dilemma over Arming Private Security Guards.' *Business Daily* (Nairobi). 13 September.

Melzer, Nils. 2009. *Interpretive Guidance on the Notion of Direct Participation in Hostilities under International Humanitarian Law*. Geneva: International Committee of the Red Cross.

Miller, T. Christian. 2010. 'This Year, Contractor Deaths Exceed Military Ones in Iraq and Afghanistan.' *ProPublica*. 23 September.

— and Aram Roston. 2009. 'Former Iraq Security Contractors Say Firm Bought Black Market Weapons, Swapped Booze for Rockets.' *ProPublica*. 18 September.

MNF–I (Multi-National Force–Iraq). 2009. 'Random PSC Compliance Inspections.' Baghdad: Armed Contractor Oversight Division. 5 February.

Modestov, Nicolai. 2009. 'PSCs Will be Armless.' *Evening Moscow*, No. 2 (25022). 13 January. <http://www.vmdaily.ru/article/70148.html>

Ocqueteau, Frédéric. 2006. 'France.' In Jones and Newburn, pp. 55–76.

Page, Michael, et al. 2005. *SALW and Private Security Companies in South Eastern Europe: A Cause or Effect of Insecurity?* Belgrade and London: South Eastern Europe Clearinghouse for the Control of Small Arms and Light Weapons, International Alert, and Saferworld.

Parker, Sarah. 2009. *Handle with Care: Private Security Companies in Timor-Leste*. Johannesburg and Geneva: ActionAid and Small Arms Survey.

PNG (Papua New Guinea) and Sandline. 1997. Agreement for the Provision of Military Assistance Dated This 31 Day of January of 1997 between the Independent State of Papua New Guinea and Sandline International. <http://www.privatemilitary.de/pool_doc/Documents_Sandline-PNG.pdf>

Prenzler, Tim. 2005. 'Mapping the Australian Security Industry.' *Security Journal*, Vol. 18, No. 4, pp. 51–64.

Prince, Erik. 2007. 'Statement of Erik D. Prince, Chairman and CEO, Blackwater, for the House Committee on Oversight and Government Reform.' Washington, DC: Committee on Oversight and Government Reform. 2 October.
<http://i.a.cnn.net/cnn/2007/images/10/02/statement.of.erik.d.prince.pdf>

PSIRA (Private Security Industry Regulatory Authority). 2010. *Annual Report 2009/2010*. Pretoria: PSIRA.

Renou, Xavier. 2005. *La privatisation de la violence: mercenaires et sociétés militaires privées au service du marché*. Marseille: Agone.

Richard, Emilia. 2010. *Private Military and Security Companies and Small Arms: Literature Review*. Unpublished background paper. Geneva: Small Arms Survey.

Richards, Anna and Henry Smith. 2007. 'Addressing the Role of Private Security Companies within Security Sector Reform Programmes.' *Journal of Security Sector Management*, Vol. 5, No. 1. May.
<http://www.ssronline.org/jofssm/issues/jofssm_0501_richards&smith.pdf?CFID=1714330&CFTOKEN=67526974>

RoA (Republic of Angola). 1992. Law of 31 July 1992 on Private Security Companies.

RoP (Republic of the Philippines). 2005. *Rules and Regulations on the Implementation of the 1969 Act on Private Security*.
<http://www.privatesecurityregulation.net/files/Philippines_2005_RulesFor1969_ActPS.pdf>

Rosemann, Nils. 2008. *Code of Conduct: Tool for Self-Regulation for Private Military and Security Companies*. Occasional Paper No. 15. Geneva: Geneva Centre for the Democratic Control of Armed Forces.

Rubin, Alissa. 2010. 'Karzai Delays Order to Ban Private Security Companies.' *The New York Times*. 27 October.

Scahill, Jeremy. 2007. *Blackwater: The Rise of the World's Most Powerful Mercenary Army*. New York: Nation Books.

Schabas, William A. 2007. *An Introduction to the International Criminal Court*, 3rd edn. Cambridge: Cambridge University Press.

Schmitt, Michael. 2005. 'Humanitarian Law and Direct Participation in Hostilities by Private Contractors or Civilian Employees.' *Chicago Journal of International Law*, Vol. 5, No. 2.

Schwartz, Moshe. 2010. *Department of Defense Contractors in Iraq and Afghanistan: Background and Analysis*. Report to Congress. Washington, DC: Congressional Research Service. 2 July.

Securitas. 2009. *Annual Report*. <http://www.e-magin.se/paper/kssqtdvk/paper/28>

—. n.d. Company website. Accessed February 2010. <http://www.securitas.com>

SIGIR (Special Inspector General for Iraq Reconstruction). 2010. *Quarterly Report to the United States Congress.* Arlington: SIGIR. 30 October.
    <http://www.sigir.mil/files/quarterlyreports/October2010/Report_-_October_2010.pdf#view=fit>

Singer, Peter Warren. 2003. *Corporate Warriors: The Rise of the Privatized Military Industry.* Ithaca: Cornell University Press.

—. 2004. 'War, Profits and the Vacuum of Law: Privatized Military Firms and International Law.' *Columbia Journal of Transnational Law,* No. 42.

Small Arms Survey. 2010. *Small Arms Survey 2010: Gangs, Groups, and Guns.* Cambridge: Cambridge University Press.

Sossai, Mirko. 2009. 'Status of PMSC Personnel in the Laws of War: The Question of Direct Participation in Hostilities.' *EUI Working Papers.* AEL 2009/6.
    San Domenico di Fiesole: Academy of European Law.

South Africa. 2000. Firearms Control Act 2000. <http://www.acts.co.za/firearms/whnjs.htm>

Thottam, Jyoti and Nilanjana Bhowmick. 2010. 'For the Arms Industry, India Is a Hot Market.' *Time Magazine.* 19 February.

Trevaskes, Susan. 2008. 'The Private/Public Security Nexus in China.' *Social Justice,* Vol. 34, Nos. 3–4 (2007–08).

UN (United Nations). 1990. Basic Principles on the Use of Force and Firearms by Law Enforcement Officials. Havana: UN.
    <http://www2.ohchr.org/english/law/firearms.htm>"

—. 2008. 'World Population Prospects: The 2008 Revision Population Database—2005 Medium Variant.' <http://esa.un.org/unpp/index.asp>

UNHRC (United Nations Human Rights Committee). 2004. *Nature of the General Legal Obligation Imposed on States Parties to the Covenant: General Comment No. 31.* CCPR/C/21/Rev.1/Add.13. 26 May.

—. 2010. 'Human Rights Council establishes Working Group on activities of Private Security Companies, renews mandates on Sudan and Somalia.' Press Release. 1 October.
    <http://www.reliefweb.int/rw/RWFiles2010.nsf/FilesByRWDocUnidFilename/EGUA-89TT38-full_report.pdf/$File/full_report.pdf>

UNODC (United Nations Office on Drugs and Crime). 2006. *Violencia, Crimen y Tráfico Ilegal de Armas en Colombia.* Bogotá: UNODC.

USASC (United States Armed Services Committee). 2010. 'Opening Statement of Senator Carl Levin, Senate Armed Services Committee Hearing on Contracting in a Counterinsurgency: An Examination of the Blackwater–Paravant Contract and the Need for Oversight.' Washington, DC: United States Senate. 24 February.

USDOD (United States Department of Defense). 2001. *Quadrennial Defense Review Report.* 30 September.

USDOS (United States Department of State). 2010. *Charging Letter.* Letter addressed to Xe Services regarding violations of the Arms Export Control Act and the International Traffic in Arms Regulations. 13 August.
    <http://www.pmddtc.state.gov/compliance/consent_agreements/pdf/Xe_PCL.pdf>

USHR (United States House of Representatives). 2007. *Private Military Contractors in Iraq: An Examination of Blackwater's Actions in Fallujah.* Washington, DC: Committee on Oversight and Government Reform. September.

—. 2010. *Warlord Inc.* Washington, DC: Subcommittee on National Security and Foreign Affairs Committee on Oversight and Government Reform. June.

van Dijk, Jan. 2008. *The World of Crime: Breaking the Silence on Problems of Security, Justice, and Development Across the World.* London: Sage.

van Steden, Ronald and Leo Huberts. 2006. 'The Netherlands.' In Jones and Newburn, pp. 12–33.

Volkov, Vadim. 2002. *Violent Entrepreneurs: The Use of Force in the Making of Russian Capitalism.* Ithaca: Cornell University Press.

Walker, Clive and Dave Whyte. 2005. 'Contracting out War? Private Military Companies, Law and Regulation in the United Kingdom.' *International and Comparative Law Quarterly,* No. 54, pp. 651–87.

Wood, Jennifer and Nancy Cardia. 2006. 'Brazil'. In Jones and Newburn, pp. 139–68.

Yoshida, Naoko and Frank Leishman. 2006. 'Japan.' In Jones and Newburn, pp. 222–38.

ACKNOWLEDGEMENTS

Principal author

Nicolas Florquin

Contributors

Emilia Richard, David Isenberg, Clare da Silva, Takhmina Karimova, Aaron Karp, and Alexis Bushnell

A security guard stands near an ExxonMobil rig in Kome, southern Chad.
© Tom Stoddart/Getty Images

Protected but Exposed

MULTINATIONALS AND PRIVATE SECURITY

5

INTRODUCTION

In November 2010, close to 60 private security providers signed an international code of conduct that committed them to 'respecting human rights and humanitarian law in their operations' (FDFA, 2010). The International Code of Conduct for Private Security Providers (ICoC) sets out standards in areas such as the use of force, vetting of private security personnel, and reporting of incidents. It emerged in response to alleged human rights abuses by private security contractors in conflict zones, and its creation reflects the growing scrutiny of private security companies (PSCs), particularly those employed by governments.

Another expressed aim of the ICoC is for PSC clients—'be they states, extractive industries or humanitarian organizations'—to embed the code in their contracts (FDFA, 2010). Ideally, this goal should apply to multinational corporations (MNCs) in general, extractive and otherwise, as they are major consumers of private security, particularly in countries where the rule of law is weak and state-provided security is inadequate or non-existent. Even where the rule of law is well established, MNCs often employ PSCs to protect assets, personnel, and property.

And yet, despite the heightened attention to PSCs, and the frequency with which MNCs turn to private security, research on MNC use of PSCs is scarce. While MNCs clearly rely on private security providers to play an important role in protecting their operations, this relationship is not always straightforward. In conflict or post-conflict areas, MNCs may face difficulty in finding disciplined, well-trained private security personnel who have not been linked to hostilities. In other cases, the lack of a distinction between public and private security forces can affect MNC control over their security operations and hamper efforts to establish liability with respect to the misuse of force. In some cases, particularly in the extractive industries, companies' PSC personnel are alleged to have killed, injured, or intimidated local community members, protesters, and others through excessive or improper use of force or arms.[1] The absence of global data on armed violence involving PSCs makes it difficult to assess the incidence of such abuses.

No international legal framework governs PSCs or MNCs, and national regulation of private security companies is weak or non-existent in many countries.[2] It is difficult to hold MNCs accountable in their home states for incidents of weapons misuse associated with their operations abroad (including abuses committed by their private security providers). Research for this study indicates that host states generally have limited legislation regulating MNC use of private security. The consequence is often a lack of accountability for MNC use of private security, particularly overseas and in countries with weak governance.

This chapter focuses on some of the problems surrounding MNC use of private security and associated misuse of force or arms. In so doing, it focuses on the extractive industries and on selected key issues:

- Under what conditions do MNCs use private security and what are some of the variations in these arrangements?
- How do governmental restrictions or local conditions affect private security arrangements of MNCs?

- Under what conditions do PSCs misuse force or firearms while in the employ of MNCs?
- What mechanisms exist, both legal and otherwise, for holding MNCs accountable regarding their use of PSCs?

Key findings include:

- Contrary to what may be expected, risks associated with in-house security point to the need for companies that opt for such an arrangement to engage in a high level of due diligence.
- The progressive blurring of distinctions between private and public security forces challenges the assumption that MNCs can turn to PSCs to bypass public security forces with a poor human rights record.
- Weaknesses in the regulation of PSCs and MNCs at the domestic and international levels, as well as gaps in oversight at the company level, may create conditions for violence, including excessive use of armed force, by private security contractors working for MNCs.
- While legal and non-binding mechanisms exist to hold MNCs accountable for their use of private security, significant obstacles to using them remain in place.
- Standards of good practice regarding MNC use of private security have begun to emerge, primarily through the Voluntary Principles on Security and Human Rights (VPs). No systematic research has been done on their implementation, however, and signatories face few consequences for failing to uphold agreed principles.

Weaknesses in regulation and gaps in oversight may create conditions for violence.

The chapter begins with a brief look at the kinds of private security MNCs typically use, placing them within the context of the international debate on the responsibility of companies to respect human rights. It then examines a broad set of issues that may increase the likelihood that PSCs will misuse armed force—including the degree of control of MNCs over private security providers, vetting and training, and the use of *public* security personnel for private security. The final section focuses on MNC accountability for private security use, identifying both the gaps and potential mechanisms for closing them. The chapter is based on desk research and interviews with representatives of both the private security and extractive industries, non-governmental organizations, academia, and others.[3] It also draws on expert contributions commissioned by the Small Arms Survey.

SETTING THE SCENE: MNC SECURITY NEEDS

This chapter focuses on MNCs that hire PSCs that can or do use preventive or defensive force (including in self-defence) to protect people or assets. It does not review private security companies that provide offensive or military services, which are sometimes referred to as private military companies (PMCs),[4] although some MNCs do use the services of such companies to protect their operations and staff in volatile areas.[5] PMCs account for a relatively small percentage of the security industry, if one defines private security broadly to include, for example, guards at shopping malls.[6] Some scholars use the collective term 'private military and security companies' (PMSCs); others dispute the division of companies into PSCs and PMCs, arguing that these categories can start to merge in conflict situations and that distinctions must therefore be made on a case by case basis.[7]

Nearly all multinational corporations use private security in some form.[8] They often use it in countries with weak state institutions or in volatile, conflict-prone areas.[9] They also use private security in developed countries, with the United States being the largest market for privatized security in the world (Abrahamsen and Williams, 2009, p. 2). They do so for a number of reasons, not least because public security forces cannot fill the growing demand for security and because these forces do not or cannot necessarily provide the services MNCs may require, such as guarding private property (PRIVATE SECURITY COMPANIES).

Corporations turn to PSCs for a variety of services, including unarmed security (patrols or static security such as staffing checkpoints, monitoring security cameras, and providing theft control), as well as advisory functions (such as technical advice, risk analysis, monitoring, and consultancy). Services can also entail armed security, such as convoy escorts, close protection of VIPs, bank transfers, intervention or alarm response at industrial sites, and the guarding of factories, gold refineries, fuel facilities, commercial centres, banks, and delivery trucks.

The chapter focuses on MNC use of PSCs for several reasons. First, MNCs are among the top consumers of private security and thus help drive PSC industry expansion.[10] Second, MNCs are at the heart of the current debate about corporate social responsibility. Third, they are particularly sensitive to controversies involving their private security contractors because of heightened scrutiny by civil society organizations.

A French businessman, the general director of Technip, prepares to enter an armoured vehicle during his Baghdad visit with the French industry minister in February 2010. Corporations sometimes use private security for close protection of VIPs. © Eric Gaillard/Reuters

Security personnel guard the chemical plant of the BASF-SINOPEC joint venture in Nanjing, China, September 2005.
© Eugene Hoshiko/AP Photo

The chapter narrows the focus further to extractive multinationals —specifically oil, gas, and mining enterprises—because they face the toughest security challenges. They often operate in conflict-prone or volatile regions that raise particular concerns regarding the use of force and firearms. Specifically, some extractive MNCs have been associated with high-profile controversies involving serious human rights abuses, often perpetrated by third-party security forces hired to protect personnel and assets.[11] While the examples in this chapter are primarily from the extractive industries, other multinationals, notably those active in agribusiness and food commodities, have also been implicated in human rights controversies involving security forces in their employ. Future research on corporations' use of private security might thus usefully examine a broad cross-section of multinationals in order to test some of the ideas discussed in this chapter and generate others.

A lack of data and the sensitivity of the subject posed challenges for the research of this chapter. For example, it is not known how many injuries or deaths per year are due to violence (particularly armed violence) perpetrated by PSCs guarding MNCs. Nor is it known how many accusations of arms-related abuses are lodged each year against these PSCs. While companies themselves may collect such data, they are not available at a global level. This lack of information makes it difficult to know whether MNC use of private security is more problematic, in human security terms, than the use of public security. Due in part to liability concerns, MNCs have little incentive to share this information publicly. As discussed below, certain international initiatives may lead to improved incident reporting, which could facilitate comparisons of different levels and types of armed violence by PSCs to those of public security forces.

Two recent international frameworks have brought increased attention to corporate responsibility in relation to human rights. They have also raised questions about the responsibility corporations have in relation to the human rights impacts of their business partners, such as private security contractors. In 2005, UN Secretary-General Kofi Annan appointed his Special Representative (SRSG) on Business and Human Rights, John Ruggie, to 'identify and clarify standards of corporate responsibility and accountability' for MNCs and other businesses with regard to human rights. In 2008 Ruggie published a framework based on three pillars: state duty to protect human rights, corporate responsibility to respect human rights, and access to remedies for victims of human rights abuses (Ruggie, 2008a).

Corporate-related human rights abuses are more frequent in countries with weak rule of law, and 'legitimate and well-recognized firms also may become implicated in human rights abuses, typically committed by others, *for example, security forces protecting company installations and personnel*' (Ruggie, 2009, p. 3, emphasis added). In such cases, the security provider is the perpetrator, while the MNC, as contractor, may be complicit (see below). In this context, Ruggie has stated that the exercising of 'human rights due diligence'—through policies, assessments, integration, and tracking and reporting—helps companies identify and manage human rights-related risks and 'address their responsibilities to individuals and communities that they impact' (Ruggie, 2010b, pp. 16–17). A key part of due diligence, according to the SRSG's framework, is for a company to identify and manage risks posed by its relationships to third parties and to assess the context in which it operates.

Ruggie's guidance captures the growing consensus that a company should conduct due diligence on relationships with third parties—such as PSCs—to ensure 'it is not implicated in third-party harm to rights' through these links (Ruggie, 2008b, p. 7). Companies' relationships with business or joint venture partners such as state and private security forces are already under increasing scrutiny. They are likely to be an integral part of the standards expected to emerge when Ruggie presents his recommendations to the UN Human Rights Council in 2011. It is noteworthy that the ICoC pledges signatory companies to endorse the Ruggie framework (FDFA, 2010, p. 3). Further, the framework's focus on the importance of human

---

Box 5.1 Less-lethal weapons and munitions

Given the range of MNC private security arrangements, it is difficult to generalize about the types of weapon their private security providers use. A meaningful discussion of weapons used would require a more comprehensive survey than was conducted for this chapter. Yet despite its limited scope, this study reveals that some MNCs use weapons and munitions often categorized as 'less-lethal'. These include TASERs and stun guns, tear gas, and paint ball rounds.

One security studies expert points out that firearms constitute a liability for private security companies. In some countries, such as South Africa, armed guards may be attacked for their weapons. MNCs also face liability risks if their armed guards misuse their weapons; in most cases, both MNCs and PSCs thus 'prefer as little violent capacity as possible'.[12]

Unarmed security guards, however, may face attack from illegally armed groups or individuals precisely because they carry no weapons. A security executive at AngloGold Ashanti asserts that, because the company's PSC personnel in some countries must, by law, be unarmed, they have become targets of a 'criminal element' in those countries. In response to attacks that left PSC personnel seriously injured, in 2008 the company began to consider providing security personnel with less-lethal munitions so they could protect themselves until backup help arrived. The company approached the government of Ghana, where only individuals (not PSCs) may be licensed to carry arms, and 'agreed in principle' with the government to investigate the use of less-lethal munitions.[13] As of late 2010, this process was 'pending', with the company awaiting government buy-in on the idea. AngloGold Ashanti is also looking at using less-lethal munitions for its private security personnel in South Africa, where it uses primarily in-house security personnel, some of whom are armed,[14] as an alternative to the use of firearms. The decision of whether to equip personnel with firearms is based on threat and risk assessments.[15]

In another case, an extractive company has developed a standard set of less-lethal munitions: 12-gauge stabilized bean bags, 40 x 46 mm long-range and short-range tear gas ammunition, and Triple Chasers, hand-held CS gas canisters. It also uses paint balls in training and simulation. A security executive for the company reports, however, that because legislation has not kept up with the development of this technology, it is more difficult to obtain less-lethal munitions than lethal ones.[16] The former security manager of a mining company echoes this point, noting a lack of legislation to license the use of these munitions.[17]

The extent of the use of less-lethal weapons and munitions among MNCs is not known.[18] Research is needed to understand the conditions under which PSCs guarding MNCs are using less-lethal materiel at their sites, the weapons' impact on death and injury rates, and the legal impediments to MNC acquisition and use of less-lethal munitions. MNCs may favour switching to less-lethal equipment primarily in countries where their PSCs are not allowed to carry firearms (and the company perceives a need to arm them for self-defence purposes). Or they might use less-lethal weapons and munitions, more broadly, as a way to cause less harm to civilians–it is too early to tell.

It is also worth remembering that less-lethal weapons, and the use of weapons other than firearms, can cause serious injury and death if used improperly (EMERGING TECHNOLOGY).

rights-related reporting and internal grievance mechanisms is likely to be particularly important in the context of MNC use of private security.

A second relevant framework is that of the Voluntary Principles on Security and Human Rights, a multi-stakeholder initiative to specifically address MNC use of security. The VPs, designed for the extractive industries and launched in 2000, cover MNC use of both public and private security. They are not legally binding but, as discussed below, have nonetheless led certain companies to introduce new standards aimed at improving respect for human rights.

KEY CHALLENGES IN MNC USE OF PRIVATE SECURITY

The dearth of research on MNC use of private security prevents comprehensive analysis across any particular industry, setting, or region. Yet interviews conducted for this study shed light on key characteristics of this type of security, as well as on the factors that enable the misuse of armed force by PSCs. This section presents these findings through an examination of in-house security; international versus local private security; private and public security; and the use of hybrid security.

> Security is a specialized and complex service, and not an MNC's core business.

In-house security

One of the decisions MNCs must make regarding the protection of their operations is whether to use their own staff for security provision (also called 'in-sourcing') or to outsource this function. According to numerous sources, MNCs generally do not use their own staff to provide security. If they do have security personnel, these are often managers who oversee outsourced security.

Several factors explain this tendency to outsource. Security is a specialized and complex service, and not an MNC's core business. In-house security can be more expensive than outsourcing, as it requires the corporation to take on the responsibility of training (and retraining) staff, and may also require registering as a PSC.[19] In contrast, outsourcing private security affords an MNC the capacity to terminate a contract if something goes wrong, granting it some distance from the PSC and possibly allowing it to avoid liability or reputational risks. It is harder for an MNC to fire its own staff if, for example, a security officer is involved in the improper use of force. If an MNC relied on staff, any incident of misconduct would be directly subject to board supervision and corporate whistleblower procedures, which can be avoided when contractors are used.[20] Further, external private security personnel are more removed from staff and therefore arguably more objective in monitoring MNC staff to prevent theft (including of sensitive information) or other misbehaviour, in addition to protecting them.[21]

Some MNCs do use their own security personnel. In one case, a multinational extractive company recruits, trains, and uses its own security personnel in some parts of the world, sometimes in combination with contracted private security. One security executive notes that using in-house security is more complicated but points out that the company does so because it has found private security providers in developing countries to be lacking in capacity and proper training.[22] In another case, the former security manager of a multinational mining company advocates a three-layered strategy, with the 'inner ring' composed of in-house 'trusted staff' who guard valuable assets and may be armed (depending on local law).[23] The second ring can be outsourced security (public or private) and the third ring represents the local community (Faessler, 2010, p. 18). This third ring serves as a 'strategic' one, in that the local community can provide (or deny) a company the 'license to operate' in the area.[24] Frictions between the community and the company can lead to conflict and increased security risks. If there is a good relationship between the two, however, the community 'often becomes a key source of information', providing early warnings to the company on potential security threats.[25]

Using in-house security has the potential to give an MNC better oversight and tighter control over the activities of its security personnel. But the degree of control, and its effect on weapons use and misuse, seems to depend on numerous variables. These include the local context and prevailing security situation; the profile of public security forces; the population from which an MNC recruits its private security personnel; and the MNC's internal procedures for handling infractions by such personnel. Use of in-house security can impede a company's ability to respond appropriately in the event that its own security personnel are involved in the improper use of force. Companies

---

Box 5.2 The Porgera joint venture mine

Security personnel at a gold mine in Papua New Guinea (PNG) have been accused of serious human rights abuses. Local community members have claimed that guards employed by the Canadian mining company Barrick Gold at the Porgera joint venture (PJV) gold mine[26] have used 'excessive or abusive force' in their work, killed individuals at the mine and outside its confines, and 'engaged in physical abuse and rape' (IHRC and CHRGJ, 2009, p. 1). In addition, armed PJV security personnel allegedly raped women at gunpoint and were accused of being involved in the shooting deaths or injuries of several individuals (pp. 13-17, 19-20).

In 2009, a team of researchers from Harvard Law School and the New York University School of Law testified before the Standing Committee on Foreign Affairs and International Development of the Canadian House of Commons about these alleged human rights abuses. The research team reported it had found 'a close relationship between PJV security personnel and PNG police', with the police also implicated in allegations of violence (IHRC and CHRGJ, 2009, p. 1). It also reported that armed members of PJV's private security force carried out many of the alleged abuses. While the research team acknowledged 'some of the actions of PJV security personnel may have been justifiable on the grounds that force was lawfully used to protect property or life', it urged independent investigation of 'all incidents of violence and death' (IHRC and CHRGJ, 2009, p. 1).

The case underscores some of the complexities of situations in which MNCs use both private and public security. The close relationship between PJV and the government raises questions about the latter's 'ability to independently investigate' claims of human rights abuses (IHRC and CHRGJ, 2009, p. 9). The case also highlights an important potential problem with the use of in-house security. As the research team argued, despite PJV's efforts to investigate violent incidents at the mine, an 'inherent conflict of interest' exists in its doing so, 'as it is PJV personnel who are alleged to have committed these abuses' (p. 2). In addition, the company's recruitment of security personnel among former police officers has raised concerns given the history of institutionalized police violence. These concerns call for a higher

level of vigilance regarding possible abuses by private security personnel than might be called for in other contexts.[27]

In early 2011, Human Rights Watch published a report on Porgera's impact on human rights, further examining the alleged abuses committed by members of the mine's private security force in 2009 and 2010. The report finds that mine security personnel were 'generally well disciplined' in responding to 'violent nighttime raids by illegal miners on the central areas of the mine'.[28] Human Rights Watch concludes, however, that in more remote parts of the mining area, 'some security personnel have committed violent abuses against men and women, many of them illegal miners engaged in nonviolent scavenging for scraps of rock'; the violence occurred 'on or near the sprawling waste dumps around the mine' (HRW, 2011, p. 9). The report identifies one of Barrick's 'most glaring failures' at the mine as 'its inadequate effort to monitor the conduct of mine security personnel working in the field'; Human Rights Watch also argues that the company had failed to create 'a safe and accessible channel' for individuals to register alleged abuses by security guards or other company employees (HRW, 2011, p. 14).

According to the report, Barrick has taken steps to address these and other problems, including commissioning a 'former police commissioner and ombudsman to investigate the allegations of abuse by PJV security personnel' (HRW, 2011, p. 10). In this context, it is notable that Barrick joined the VPs in November 2010 (Barrick Gold Corporation, 2010). Human Rights Watch also points to the government's failure to carry out 'meaningful day-to-day oversight' of the mine's private security force and urges the government to create a mechanism to 'oversee the conduct of all private security actors' in the country (HRW, 2011, p. 24).

In February 2011, Barrick issued a response to the report, thanking Human Rights Watch for providing it with information on the alleged abuses. The company writes that, among other measures, it has conducted an extensive internal investigation, has terminated employees who violated its code of conduct, and is introducing monitoring systems for its security personnel (Barrick Gold Corporation, 2011).

opting for internal security therefore need effective procedures to address these factors. Box 5.2 highlights one example in which a company's in-house security guards have been implicated in allegations of serious human rights abuses stemming from the misuse of force, as well as some of the questions that arise in such cases.

International versus domestic private security

MNCs often use domestic PSCs.[29] If they use international private security companies,[30] these are generally staffed with local employees, though upper management positions tend to be filled by expatriates. In some countries, local law stipulates that PSCs must be owned by nationals (as is the case in Nigeria);[31] in others, they must be staffed by nationals (as in Angola).[32] Yet one source points out that foreign PSCs operate in Nigeria by forming management agreements with Nigerian companies, highlighting 'the extent to which regulation of the private security sector is often circumvented' (da Silva, 2010, p. 9). Some countries (such as Colombia and Sierra Leone) place no restrictions or preferences on the use of foreign versus domestic PSCs (p. 9).

Various factors explain an MNC's decision to contract international PSCs, or at least their local branches. Sources close to the industry assert one reason is that in some regions local private security personnel lack training and are not familiar with international standards, such as the VPs or the UN Basic Principles on the Use of Force and Firearms by Law Enforcement Officials (UN, 1990). In their view, using such firms has legal and reputational risks for MNCs. The former security manager of a mining MNC notes that his company used the local registered office of a global private security company in Africa for a number of reasons. Despite the higher cost, the mining company was able to hire security personnel with better training. It also had the option of adding penalty clauses to contracts, and the security firm offered higher standards of licensing and bonding than did local PSCs.[33] MNCs also turn to international PSCs (or their local branches) because they possess highly sophisticated technological and knowledge-based security solutions and systems.[34]

On the other hand, local PSCs hold certain advantages over international ones. Several sources point out that cost is a key reason MNCs use the former.[35] In addition to being more affordable, local security personnel do not require housing or moving costs. They know the area and local conditions and speak the local language. In the words of one former private security representative, local PSCs 'tend to blend in better' and 'can often resolve issues before [they] escalate to violence'.[36] Further, certain countries (such as South Africa) grant firearms licences only to citizens or permanent residents.[37] In such cases, if an MNC plans to use armed PSC personnel, it must use local staff. If, however, potential security staff from the area around the MNC facilities are incompatible with the local population due to their military background or other personal characteristics, an MNC might not want or be able to use local personnel.[38]

Local market conditions can also help inform an MNC's choice between international and local private security. For example, as the market for private security is regulated in Colombia, there is a wide choice of PSCs.[39] MNCs in the oil and mining sectors in Colombia tend to hire domestic PSCs, but a variety of international and local PSCs operate in the country. Some of these provide banking, commercial, or residential security, as in other countries in Latin America, while other PSCs in Colombia include US government contractors involved in the 'war on drugs'.[40]

Private and public security

It is not possible—nor useful—to generalize about whether using private security in any particular situation poses greater risks of weapons misuse than using public security forces. The risk depends on the context, including the nature of both the public sector forces and those providing private security. The absence of comprehensive data on

Local PSCs hold certain advantages over international ones.

incident reporting further clouds the picture. This section looks at some of the factors that appear to lead MNCs to use private security, while acknowledging that it is not a simple question of choosing private over public security. In fact, one school of thought in security studies argues that categorizing security provision into private versus public ignores the major shift taking place in many countries towards 'global security assemblages'; in this view, public and private security increasingly work together and the lines between the two are becoming blurred (Abrahamsen and Williams, 2006, pp. 7–8; 2009, pp. 3, 6). This issue is explored further in the section on 'hybrid security', below.

The wide variety of security arrangements MNCs use around the world may be illustrated as straddling a spectrum. In certain countries, corporations are required to use public security in some form, perhaps because the host government requires public security at certain facilities (such as oil installations) if they are deemed sensitive or 'of national interest'.[41] Alternatively, a government may prohibit PSCs from carrying arms, obliging an MNC to call on the state when it needs or wants armed protection. Even where MNCs are not legally obligated to use public forces, they might choose to use them as a supplement to private security, as in areas of conflict in countries where the armed forces are considered capable and a functioning state exists. Elsewhere, a host government may require companies to

An armed security guard watches over an official diamond mine in Lunda Norte, Angola, 1992. The Brazilian company Odebrecht operates the mine on behalf of the Angolan government. © Paul Lowe/Panos Pictures

provide their own security. In Angola, for example, MNCs entering certain domestic market sectors are responsible for the provision of security. Specifically, the Angolan Diamond Law 16/94 calls for regulated diamond concessionaires to 'provide for their own security' in 'areas defined as "restricted" and "protection zones"' (da Silva, 2010, p. 8; Joras and Schuster, 2008, p. 41).

The reasons MNCs might choose private security include:

Cost-effectiveness and flexibility. PSCs can be cost-effective, supplying 'short-term and contract-bound services' and a 'level of flexibility that state security forces cannot provide'.[42] Some states that contract PSCs claim that this flexibility translates into cheaper costs compared to permanent in-house security (Schwartz, 2010, p. 2; PRIVATE SECURITY COMPANIES).

Concerns with public security. MNCs might turn to private security to avoid using public forces in countries where the police and army are unreliable, weak, or have a record of human rights violations. Public forces are sometimes poorly trained and disciplined, and several sources raise the issue of corruption among public security forces as a factor leading MNCs to avoid them if at all possible. One source asserts:

PSCs can offer MNCs a buffer where local communities have conflict with the state, where the state of course uses public security as its enforcement arm. In these situations, hiring PSCs can help MNCs distance themselves from abusive public security forces.[43]

Availability. Even where they are reliable, state forces might not be available to fill the demand for private security.[44] Abrahamsen and Williams point to the 'declining ability and/or willingness of the state to provide adequate protection of life and property' as a major factor in the growth of the private security sector in sub-Saharan Africa (Abrahamsen and Williams, 2006, p. 5). MNCs may also use private security because the services they commonly require (such as the guarding of private property or the protection of individuals) are not those that public security forces normally provide. PSCs commonly perform static surveillance; public security forces are needed for other functions.[45]

Enhanced control (real or perceived) over security provision. One of the most common reasons cited for using private security is the possibility of exerting more control over security providers through economic or contractual means. One source has observed a trend of companies trying to change from using public to private security because of the 'control factor'.[46] Moreover, MNCs can often pay private security personnel more than these individuals would be paid to serve in their country's public security forces. While an MNC has 'relatively limited leverage with state security forces', it can simply terminate a contract with a PSC or fire a private security guard in the event of improper use of weapons or other violations. When it engages a private security firm, it thus 'has a contract and additional leverage over its service delivery and performance' (International Alert, 2008, p. 8). In this way, contracts can increase the degree of oversight and control that an MNC has over its security personnel.

Another aspect of control is training. MNCs can train private security personnel, or demand a certain level of training and competence as part of their contracts. Indeed, several sources cite training and competence of PSCs as a factor in their favour over public security forces. Yet, as in other areas, there are regional and national variations. While police and military in Latin America have 'relatively sophisticated and standardized training for their personnel, private security entities exhibit little consistency in staff training throughout the region'—even among 'the more respected transnational private security companies with operations in Latin America' (Godnick, 2009, p. 11). A review of 19 countries' legislative frameworks for PSCs concludes that most of the countries have 'some loose legal requirements' on the 'vetting and training of PSC sector workers'. Nevertheless, 'they often contain merely a vague specification that the PSC is responsible for ensuring that its employees are properly trained' (da Silva, 2010, p. 5). The degree of control an MNC has over private security, in contrast to public security, is thus not a given, but can vary significantly depending on the situation.

Hybrid security: the blurring of public and private

The previous discussion touches on why and where MNCs might choose to use private security. The lines between public and private are not always clear, however. This section discusses the challenges this poses for MNCs in terms of control, accountability, and the impact of security operations on local communities. As elsewhere in the chapter, examples are drawn primarily from extractive MNC practice.

In numerous developing countries, the distinction between public and private security can be blurred by the fact that retired state security force personnel often own PSCs, and in many places ex-military or ex-combatants staff PSCs.[47] Where military forces have poor human rights records, there is an obvious risk of hiring private security personnel who have been involved in rights abuses. In addition, 'porousness' may exist between public and private security, meaning that private security personnel are active members of the armed forces. Although a number of informed sources assert they have not come across examples of such porousness, Dammert lists eight Latin American countries where police officers are permitted to work in private security in their off-hours (Dammert, 2008, p. 32).

Differing viewpoints exist regarding the implications of off-duty personnel providing private security. One source posits there could be some advantage to off-duty police officers working in PSCs, as the officers could receive training and bring those skills back to public security forces where these are lacking. Yet the same source warns that in some countries, such situations could be dangerous because the personnel who are supposed to be protecting people switch to 'protecting property from people',[48] raising potential conflicts of interest (Godnick, 2009, p. 9). Accountability is a further concern: 'How can one expect the police to investigate human rights abuses by PMSCs if members of the police work for these companies?' (Lazala, 2008, p. 9).

An important variation of security is 'hybrid policing', where private and public forces work together to provide security (Abrahamsen and Williams, 2006, p. 8). In some countries it is common for MNCs to use both PSCs and police or the army to fill their security needs. The decision to use both depends on factors such as whether the MNC is operating in an area of armed conflict or violence, whether the armed forces are considered competent, and whether local legislation prohibits PSCs from being armed.[49] In the latter case, MNCs that want armed security are dependent on the state security forces (da Silva, 2010, p. 7).

In a number of regions, MNCs pay into public funds intended to support state security forces and receive security services in return. These arrangements vary considerably. In Colombia, for example, the use of public security is optional, but extractive MNCs typically supplement private with state security. They do so by signing agreements with the Ministry of Defence for the provision of public security in exchange for their paying funds for the 'well-being' of the armed forces in the area where the MNC operates. This can include paying for flights home, food, and sometimes infrastructure. The use of the funds is restricted: according to sources in Colombia, they are not supposed to be used for salaries, weapons, or additional troops.[50] Arrangements in other countries differ, with MNCs paying supplementary wages or bonuses to armed forces personnel (see below).

MNC agreements with host governments sometimes provide for the use of private security firms to protect MNC facilities ('static security') while the army protects the company against external threats, such as armed groups. One industry source notes this is common among extractive firms in Colombia, but not in other sectors, as the former are more likely to operate in rural conflict zones that have illegal armed groups, and have less scope to leave those zones than do other MNCs.[51]

In Colombia, where large numbers of extractive MNCs have signed such agreements, they tend to hire local PSCs, many of whom are unarmed and more 'communications-oriented'. Since police and military provide the MNCs, and the geographical regions in which they operate, with armed protection, there is less of a need to arm PSCs.[52]

Faessler has argued that such hybrid arrangements work well in Colombia due to soldiers' training, their ability to deter 'militia threats', the army's improving 'human rights strategy', and the fact that the government regulates the private security sector. He contrasts this with the Democratic Republic of the Congo, where, in his experience, police 'lacked training, particularly in the fundamentals of respect for human rights' (Faessler, 2010, p. 20). Others have expressed concern over the human rights record of Colombian armed forces and gaps in regulation of PSCs in the country.[53] Box 5.3 provides an example of hybrid security in the Niger Delta, a region where security arrangements are so integrated that 'it is often difficult to determine where public force ends and private security begins' (Abrahamsen and Williams, 2009, p. 10). Box 5.4 presents a case of armed violence allegedly perpetrated by an MNC's private security guards and police in Peru. It highlights questions about state and company oversight of, and accountability for, armed security given that distinctions between public and private security are not always clear.

Box 5.3　Oil MNCs in the Niger Delta

The complex situation of oil exploitation in the Niger Delta provides an example of hybrid security provision. Multinational oil companies operate in, and have contributed to, 'an environment of hostility and violence', marked, among other things, by environmental damage from oil extraction, aggrieved (economically marginalized) local communities, and the emergence of armed militias (Abrahamsen and Williams, 2009, p. 10).

The security arrangements oil companies use in the Delta are a complicated mix of public and private. For example, Chevron Nigeria Ltd. has hired a G4S subsidiary, Outsourcing Services Ltd., to provide it with 1,200 security personnel, who by law are unarmed. Because of the violence in the region, G4S offers its client armed protection by cooperating with the Nigerian paramilitary Mobile Police, who often carry automatic weapons and are 'more or less permanently seconded to PSCs and integrated into their everyday operations' (p. 10). Although they are under police authority, these officers receive 'supplementary wages' from the MNCs. Outsourcing Services personnel also cooperate with Nigeria's Supernumerary Police, who are trained by the state and then assigned to the oil MNCs as unarmed police. They are paid by the MNCs, and have 'police powers only on company property' (p. 11). The oil MNCs also have

Box 5.4　The Minera Yanacocha case

In 2007, the UN Working Group on the Use of Mercenaries[55] undertook a mission to Peru to investigate the alleged involvement of a mining MNC and its private security personnel in human rights abuses against local community members.[56] The mission report places the incidents against the backdrop of extensive privatization of security in Peru due to inadequate numbers of police. But as the mining company case involved both police and the MNC-contracted PSCs, it also sheds light on some of the problems with hybrid policing.

The case in question involves the Yanacocha gold mining operation, co-owned by the Newmont Mining Corporation, the Peruvian company Buenaventura Group, and the World Bank's International Finance Corporation. According to a 2009 independent review (described below), Minera Yanacocha has in-house security managers who oversee security, but it uses a PSC, Forza, to guard its mining facilities.[57] It employs two other private companies 'for the supply of information' and has a 'cooperation agreement' with the Peruvian National Police (PNP) to protect its property, personnel, and executives' residences. A team of police officers is stationed at its mines and the MNC has a 'small mobile police station'. For these services, Yanacocha pays the police a 'special bonus' and makes a 'contribution' to the PNP (Costa, 2009, p. 11).

On its mission the Working Group investigated allegations from 2006 that PSC personnel and police officers providing private security for the MNC were 'intimidating the population of Cajamarca' (where Yanacocha has its facilities), especially environmental rights defenders (UN Working Group, 2008, p. 15). Community members from one village had clashed with Yanacocha that year over environmental pollution from one of the mines, and a farmer was shot to death. According to the Working Group report, 'three police officers working as private security guards at Yanacocha were identified as suspects by investigators' (p. 16). It was unclear if the mining company had hired them to provide security or if Forza, Yanacocha's private security contractor, had hired them (pp. 15–16).

increasingly used the services of military Government Security Forces and the navy, given the continued instability in the region.

Abrahamsen and Williams argue that private security, in this case, has 'a fundamental impact on the security situation in the Niger Delta, providing technology, expertise and expatriate personnel that substantially influence the practices of public security forces' (p. 12). In their view, oil MNCs might be trying to 'distance themselves from the coercion of public security services' due to previous accusations of human rights violations in connection with the protection of oil operations (p. 12).[54]

While admitting that 'the depth of this commitment remains questionable', Abrahamsen and Williams posit that international oil companies, through this very particular integration of private and public security, may be able to influence public forces (p. 12). Importantly, though, they conclude that although private security actors may be 'integrated within state structures', local communities still tend to perceive them as 'aggressive, disempowering, and exclusionary' (p. 15).

Source: Abrahamsen and Williams (2009, pp. 9-15)

A police officer and a man wearing an oil company uniform patrol the area near Port Harcourt, Nigeria, October 2004. Oil MNCs in the Niger Delta often use a mix of public and private security. © Jacob Silberberg/Getty Images

The Working Group also received allegations that Yanacocha was spying on and intimidating environmental leaders of a local group, GRUFIDES (UN Working Group, 2008, pp. 16-18).[58] The Working Group asserts these events are 'not isolated cases but repeated occurrences' in Peru, that PSCs appear to be buying information on environmental leaders from the government and selling it to mining companies, and that this is part of a larger campaign in Peru to discredit those who oppose mining projects (p. 19). It expresses concern over PSC hiring of off-duty security forces personnel, who use 'State property such as uniforms, weapons and ammunition' to guard mines (p. 21). And despite legal restrictions, 'it seems that private security companies can purchase unlimited quantities of arms and ammunition' (p. 21).

Oxfam America has also raised concerns about Yanacocha's security practices in Cajamarca. As a result, in 2007 it entered into mediated dialogue with Newmont under the VP framework, in which both participate. The company agreed to an independent review of its security policies and procedures, which two Peruvian consultants produced in 2009 (Costa, 2009).

To address the communities' 'deterioration of trust in the company', the consultants recommend that Newmont take several steps (Costa, 2009, p. 12). One is to reinforce Yanacocha's capacity to investigate and sanction alleged human rights abuses, for example, by keeping a 'record on use of force and alleged human rights violations' and publishing reports on results of investigations (p. 14). They recommend the company ensure 'stationed police forces and hired private security personnel perform all tasks with professionalism' and 'the strictest respect for human rights' (p. 12). Newmont should require its PSCs to submit to it background check information on all personnel before they are employed at the mine, and consider terminating its relationship with the three PSCs, given their 'background and track record' on human rights and the damage they have caused Minera Yanacocha's 'image and reputation' (pp. 15-16).

The consultants conclude that 'although police cooperation is beneficial to Minera Yanacocha, such collaboration affects the neutrality of the police force in the eyes of the population' (p. 14). They therefore recommend the company make public its cooperation agreements with the PNP; ensure police officers at the mine receive anti-riot equipment but not carry firearms; clearly define PSC tasks versus those of the PNP; and forbid PSCs to work with active-duty personnel (pp. 14-15). These recommendations are fundamental in clarifying potential liability for the misuse of force, as the blurring between public and private security provision points to significant ambiguities around who is responsible for alleged human rights violations.[59]

This is the only case to have gone all the way through the mediation process of the VPs,[60] and in which a public, independent review has resulted from incidents involving MNC use of private security (and the police, in this case). It serves as an important pilot of what the VPs process might achieve in addressing claims of human rights violations in the context of signatory operations, as well as growing concerns about the ambiguous line between private and public security. Key questions concern the extent to which Yanacocha will take up the steps—including the preventive ones—outlined in the review,[61] and the degree to which other signatories will take up this kind of recommendation as good practice.

A panoramic view of Yanacocha, Latin America's largest gold mine, by the Andean city of Cajamarca, north of Lima, Peru, November 2006. © Pilar Olivares/Reuters

The combining of public and private security for the protection of MNCs is a complex issue, as the examples in this section illustrate. Closer examination is needed of factors that could influence the misuse of force and firearms in such situations, including the training and vetting of private security personnel, the use of active-duty personnel to guard private property, and access of such personnel to firearms. Overall, the discussion of security challenges has highlighted several problems that can result from weak oversight and regulation of private security, such as the recruitment of private security from public forces with poor human rights records, and the way the blurring of private and public security can impede the investigation and punishment of the improper use of armed force. These factors can foster impunity among PSC personnel, as well as an erosion of community trust in MNCs and their security providers. The chapter now turns to a discussion of measures designed to address these regulatory and accountability gaps.

REGULATORY FRAMEWORKS AND OTHER APPROACHES TO ACCOUNTABILITY

The accountability of PSCs under international humanitarian law (IHL) and international human rights law is the subject of vigorous international debate,[62] as is the question of MNC accountability for the behaviour of joint venture or business partners, such as private security providers. Legal regulation of PSCs, generally weak at the national level, is non-existent at the international level, a situation that has led to the creation of international initiatives to address private security contractors' behaviour and clarify their responsibilities under international law. There have also been calls for the creation of international legal standards to govern MNC behaviour as well as for home-country regulation of MNC activities overseas. This section highlights relevant developments and prospective approaches, both legal and otherwise, to holding MNCs accountable for the actions of their private security providers.

National regulation of PSCs

In general, PSCs are unregulated or poorly regulated in many countries.[63] In Sierra Leone, for example, the private security sector is 'largely unregulated'; in Kenya, the vast majority of security companies are unregistered (Abrahamsen and Williams, 2006, pp. 12, 15). In Angola, implementation of the law on PSCs appears to have been 'limited and selective' (Joras and Schuster, 2008, p. 42). In Central America, many private security companies and their staff reportedly

---

Box 5.5 Private security and agro-industry in Brazil

In October 2008, the agrochemical MNC Syngenta turned over an experimental seed farm to the Paraná state government in Brazil. The farm, in Santa Tereza do Oeste, had been the site of a land dispute between landless workers' movements and the company. Amnesty International reports that in October 2007, 40 armed guards from a Syngenta-contracted PSC, NF Segurança, carried out an 'illegal and violent eviction' of landless workers at the farm, which resulted in the deaths of one of the movement's leaders and a security guard.

The following year, Amnesty International found that investigations into NF Segurança's conduct, including the death of the landless workers' leader, led to the security company losing its license. The human rights organization called on the Brazilian government to investigate all entities—including MNCs—using private security firms that commit human rights abuses, and to hold accountable those who fail 'to adequately vet or oversee their security company' (AI, 2008). Amnesty also urged the government to 'control the flood of irregular and/or illicit security companies, many of which are effectively acting as illegal militias in the service of landowners or agro-industry' (AI, 2008).

According to one source, at the time of the shooting, no one at NF Segurança had a licence to use firearms. As of 2010, the company's owner and several guards were on trial in connection with the events of 2007, yet the company continued to work in private security in Brazil, notwithstanding the loss of its operating licence.[64]

operate 'outside established legal parameters', with a 'lack of adequate control' over firearms, even among legal PSCs (Godnick, 2009, p. 7). Central and South America have large numbers of illegal or unregistered security companies and guards (Godnick, 2009, p. 8). Box 5.5 presents one example from the region.

In Colombia, the situation is particularly complex. The government regulates domestic PSCs, but one source asserts that 'the absence of certain laws' has exacerbated the growth of private security in that country 'without sufficient oversight and control by the state', including over aspects of possession and use of arms as well as the hiring of personnel (Cabrera and Perret, 2009, pp. 5, 7–8). In addition, under the bilateral military pact, Plan Colombia, US personnel of private security companies contracted by the United States as part of its 'war on drugs' are exempt from Colombian criminal jurisdiction (p. 10).[65]

Certain countries (such as South Africa and the Philippines) do have detailed regulations on PSCs (da Silva, 2010, pp. 13, 21), and others have moved towards legislation to regulate their private security sectors.[66] Some observers argue, however, that national regulation of private security is insufficient given the global nature of the security industry. Their recommendations include a 'global standards implementation and enforcement framework' (Cockayne et al., p. 1), or even a legally binding international treaty to regulate private military and security companies.[67]

The ICoC aims to set standards to which industry would commit.

International initiatives addressing private security

Two major international initiatives, both spearheaded by the Swiss government, have sought to address the lack of regulation of private security companies and prevent improper use of force and human rights violations (PRIVATE SECURITY COMPANIES).

The first is the Montreux Document, drafted by states, NGOs, and industry, and unifying existing legal obligations of states under IHL and international human rights law in relation to contracting and regulating PSCs (Cockayne et al., 2009, p. 10). Although the framework deals only with PSCs in armed conflict, it provides, on one account, 'the most coherent, precise and consensually developed statement of "good practice" ' to date (Cockayne et al., 2009, p. 53). The second is the ICoC, which aims to set standards to which industry would commit. An enforcement and accountability mechanism for these standards is under discussion.

Neither of these initiatives, which are aimed primarily at states and PSCs, is legally binding. Nevertheless, the Swiss government calls on clients of private security—including private enterprise—to consider making 'formal approval of the Code by service providers . . . [a] *precondition for future contracts*'[68] (see Box 5.6). Of particular relevance to MNCs, as major clients of PSCs, is the fact that the ICoC calls on signatory companies to prepare an incident report documenting the use of any weapon, to conduct an inquiry into the incident, and to provide the report to the client (FDFA, 2010, pp. 15–16).

MNC accountability for private security

The next sections focus on MNCs. Box 5.6 summarizes key points regarding MNC accountability for private security contractors under IHL and international human rights law, underscoring the obstacles to applying these norms to MNCs. The remainder of this section explores some of these points in more detail, looking at legal accountability at both the domestic and the international levels.

There are no international legal standards on human rights specific to MNCs and their business or joint venture partners, although human rights groups have continued to call for such standards.[69] In his framework on corporate responsibility to respect human rights, the SRSG on Business and Human Rights emphasizes that he uses the term 'responsibility'—not 'duty'—to respect:

Box 5.6 IHL and international human rights law[70]

It is difficult to hold *companies* liable under international law for the misuse of force or firearms by employees. International humanitarian law and international criminal law—applicable in situations of armed conflict—govern the conduct of *individuals*, including PSC employees (PRIVATE SECURITY COMPANIES). IHL and international human rights law apply to *states*, including the states that hire PSCs (contracting states), those where they operate, and those where they are incorporated.

Domestic (national) law offers more possibilities in relation to corporations. Domestic criminal law cases can be brought against private security contractors that have committed human rights abuses while under contract to a corporation.[71] There are, however, important practical obstacles to such prosecutions. States often lack the necessary laws and resources to prosecute multinational corporations and their security contractors. The expense of bringing a suit against an MNC can be a major barrier for individual plaintiffs, especially in developing countries. Jurisdictional problems may also arise if the accused PSC employee is a citizen of a third country (Cockayne and Speers Mears, 2009, p. 3).

Despite such constraints, systems of accountability do exist at the national level. A notable example is the US Alien Tort Claims Act (ATCA), which allows civil suits to be filed in US courts against non-US citizens

to indicate that respecting rights is not an obligation current international human rights law generally imposes directly on companies, although elements may be reflected in domestic laws. At the international level it is a standard of expected conduct acknowledged in virtually every voluntary and soft-law instrument related to corporate responsibility (Ruggie, 2010a, pp. 2–3).

This position is not without controversy, with several major human rights NGOs questioning the notion that companies do not have direct obligations under international human rights law.[76]

In a handful of countries (such as Canada, the Netherlands, and the UK), discussions have taken place at the parliamentary level about holding MNCs domiciled in those countries legally accountable in their home states in relation to human rights abuses associated with their activities overseas. Such legislation is unlikely in the near future and, for the moment, there is 'no serious regulation of MNCs' human rights conduct overseas by their home governments'.[77] The SRSG is studying the question of extraterritorial jurisdiction in relation to business and human rights as part of his final recommendations to the UN Human Rights Council.

In general, few countries seem to have specific legislation on MNC use of private security; where it does exist, this legislation does not appear to target MNCs specifically (da Silva, 2010, p. 6).[78] For example, the UK government regulates activities of PSCs operating in the UK and prohibits them from using or carrying firearms. It does not, however, regulate the overseas activities of British PSCs or the use of British PSCs abroad by foreign companies or British MNCs (da Silva, 2010, p. 19). This loophole could contribute to the lack of oversight of PSCs in connection with their activities overseas, particularly in countries with weak governance.[79]

Significant scholarly work has been done recently on the concept of corporate complicity, which entails corporations 'knowingly providing practical assistance or encouragement that has a substantial effect on the commission of a crime' (Ruggie, 2008a, p. 20). For example, in 2008 the International Commission of Jurists published a three-

for violations of customary international law, including fundamental human rights norms (Martin-Ortega, 2008, p. 8). Several MNCs have been sued under ATCA for alleged complicity in human rights violations committed by their security providers (primarily public security forces).[72] As of late 2010, however, the applicability of ATCA to corporations was being challenged.[73]

Truth and reconciliation commissions, such as the one in Liberia, have included non-state actors under their statute for economic crimes and could be adapted to cover corporate accountability in the future (Ramasastry, 2010, p. 2). Scholars are also exploring the concepts of aiding and abetting, as well as pillage, as international crimes that may potentially be applicable to multinational corporations and security contractors.[74] Codes of conduct, such as the ICoC, may provide further means of accountability. The ICoC requires signatory firms to adhere to the human rights and humanitarian law obligations expressed in its articles. In future, multinational corporations may be able to adhere to the code as well, binding themselves to employ only those security firms that are signatories.[75]

Author: Alexis Bushnell

A foreign security contractor guards a drilling site of the Norwegian oil company DNO in northern Iraq, November 2005. © Safin Hamed/AFP Photo

volume study on corporate complicity, which examines the conditions under which corporations could be liable under criminal or civil law for involvement in gross human rights abuses committed by other actors (ICJ, 2008, p. 3).

Referring to companies using private or state security forces, the study points out that contracting companies and security forces often have a close relationship ('proximity'), largely through information sharing, payment of fees, or security providers' presence on company property. Therefore, if the security forces (public or private) commit human rights abuses, 'criminal or civil courts may hold that a company knew of the risk that the abuses would occur', especially if the security forces 'have a record of gross human rights abuses' (ICJ, 2008, p. 29).[80]

Litigation offers a possible avenue for holding MNCs accountable for the misuse of force by their PSCs,[81] although challenges to bringing such claims remain in place, as discussed in Box 5.6. One recent example of litigation involves allegations of complicity against a mining MNC in connection with alleged human rights abuses by the Peruvian police and the corporation's contracted PSC. In 2009, a group of Peruvian villagers filed a lawsuit in the UK against a London-registered multinational mining company, Monterrico Metals plc, and its Peruvian subsidiary, Rio Blanco Copper SA, over events that took place at Monterrico's Rio Blanco copper mine in Peru in 2005. The claimants allege they were 'tortured by the Peruvian police assisted by mine employees and mine security guards, following their environmental protest' at the mine (Leigh Day & Co., 2009).[82] One claimant is the widow of a protester who bled to death after being shot by police. Monterrico has denied 'its officers or employees had any involvement with the alleged abuses' (Leigh Day & Co., 2009). The British lawyer who brought the villagers' claim for damages, however, has argued 'it is inconceivable that the company did not know of the protestors' harsh treatment' (Leigh Day & Co., 2011, p. 22). According to the law firm representing the claimants, a trial is due to begin in June 2011 (p. 22).

The preceding sections analyse national and international mechanisms, both regulatory and, in the case of the ICoC, self-regulatory, which could address PSC activities and MNC accountability for these. The next section discusses the VPs, the only dedicated, multi-stakeholder initiative on MNCs and security provision.

The Voluntary Principles on Security and Human Rights

The most important soft law initiative addressing MNC use of private security is the VPs.[83] Launched by the UK and US governments, the VPs emerged in 2000 in response to allegations of human rights abuses by security forces contracted by extractive MNCs. Their primary and specific aim is to end abuses committed by security forces protecting company facilities (International Alert, 2008, p. 17). While they are not legally binding, the principles may serve to raise standards among companies involved, as signatories are expected to incorporate them into their operations.

The principles

The VPs provide companies with specific guidance on how to maintain the security of their operations while still respecting human rights. They cover risk assessment on security and human rights and company use of both public and private security, including general principles on the use of force and firearms (see Box 5.7). As of May 2010 adherents of the VPs included seven states, nine NGOs, and 17 companies, with three organizations having observer status (BHRRC, 2010). One country, Colombia, has a formal 'in-country' process to integrate the VPs into national practice.[84] Participation criteria include a commitment to public reporting by signatories (VPSHR, 2009, p. 1). The principles were designed for the extractive sector because of the particular challenges it faces in relation to private security. There have been attempts, however, to broaden the VPs to other industries.[85]

The VPs are not meant to be rules of engagement per se.[86] Rather, they call on companies to promote and observe existing international guidelines on the use of force, including the UN Code of Conduct for Law Enforcement Officials and the UN Basic Principles on the Use of Force and Firearms (UNGA, 1979; UN, 1990).[87] While the VPs have no legal status, they call for companies, 'where appropriate', to include the Voluntary Principles in (legally binding) contracts with private security companies. The International Peace Operations Association claims that, as of 2010, all companies belonging to the initiative were including the VPs 'in at least some of their contracts, particularly with private security' (IPOA, 2010). It is not clear how this is checked. According to one source involved in the process, most large signatory companies (and those with the greatest reputational risks) include either clauses on the Voluntary Principles in contracts with PSCs, or clauses on human rights more generally, as local PSCs are often unaware of the VPs.[88]

Implementation

In 2008, International Alert produced indicators designed to guide MNCs in implementing the VPs and measure company adherence to the principles. The expectation is that companies will test the indicators and eventually an 'industry standard' will emerge from the process (International Alert, 2008, p. 1). Indicators include:

Box 5.7 Private security use of force and firearms

According to the section of the VPs on interactions between companies and private security:

- Private security 'may have to coordinate with state forces (law enforcement, in particular) to carry weapons and to consider the defensive local use of force'.
- Private security 'should maintain high levels of technical and professional proficiency, particularly with regard to the local use of force and firearms'.
- Private security should have policies- or rules of engagement-on the local use of force.
- Private security should 'provide only preventative and defensive services and should not engage in activities exclusively the responsibility of state military or law enforcement authorities'.
- Private security should 'use force only when strictly necessary and to an extent proportional to the threat'.
- Where physical force is used, 'private security should properly investigate and report the incident to the Company'.
- 'To the extent practicable, agreements between Companies and private security should require investigation of unlawful or abusive behavior by private security personnel.'

Source: VPSHR (n.d.a)

A security officer stands guard over the In Salah Gas Krechba project, run by Sonatrach, BP, and StatoilHydro, in Algeria's Sahara Desert, December 2008.
© Adam Berry/Bloomberg/Getty Images

- Evidence of mainstreaming the principles in relations with security forces.
- Evidence of staff training on the VPs and human rights and of training for public security forces and private security contractors.
- Company scrutiny of the human rights record of public and private security providers.
- Company oversight of equipment transfers (International Alert, 2008, pp. 8–16).

The VPs call for companies to monitor the conduct of security providers. For this purpose, International Alert's guidelines provide a table of recommended data companies should seek from public and private security forces, such as the number of staff, the number of lethal weapons, and the number of small arms. While the authors acknowledge that obtaining some of this information from public security may be difficult 'for reasons of national security', they note that the company 'has the right to demand such information from its private security contractors' (International Alert, 2008, p. 15). The indicators also provide a model log for companies to report incidents. It is not clear to what extent companies are collecting this information, though this data is crucial for gauging the use of armed force by MNCs' private security (see Box 5.8).

Under the VPs, companies are asked, in relation to public security forces, to 'use their influence' to keep individuals with records of human rights abuses from providing security services. They are also urged to review the backgrounds of private security companies, in a way that includes 'an assessment of previous services provided to the host government and whether these services raise concern about the private security firm's dual role as a private security provider and government contractor' (VPSHR, n.d.a). Sources familiar with the VPs, including from civil society and industry, indicate this particular clause has not been a focus of discussions within the initiative. Nevertheless, it speaks to some of the concerns about hybrid security and porousness that the Yanacocha case clearly illustrates (see Box 5.4).

Although company adoption, incorporation, and implementation of the principles is increasingly seen as good practice in relation to corporations' use of security, the VPs remain, in essence, a declaration of good intentions. To the extent companies incorporate them into contracts, they could eventually help form a basis for holding private security providers legally accountable for human rights abuses committed in relation to the protection of private sector operations. But while the VPs have begun to clarify industry standards around MNC use of private (and public) security, they are not a replacement for state enforcement of IHL and international human rights law.[89]

Gaps and weaknesses

Monitoring of, and compliance with, the Voluntary Principles is up to individual MNCs, a point both participants and independent observers have acknowledged as a weakness. Various sources describe examples of signatory company implementation of the principles.[90] Non-signatory companies also report that they are implementing the provisions of the VPs in some form.[91] And in addition to the independent review of Minera Yanacocha's security arrangements described above, at least one external evaluation of the implementation of the VPs is available in the public domain.[92] The monitoring and evaluation of these implementation efforts, however, remain piecemeal.

Box 5.8   The leading edge: company incident reporting

Some signatory companies incorporate reporting on security issues in their public sustainability reports.[93] In its 2009 sustainability review, for example, the mining company AngloGold Ashanti includes a discussion of armed security at its operations, shooting incidents, and the company's response (although no details are provided on the type of firearms used). The company reports that the need for armed security guards has risen at its sites in recent years due to the increase in attacks on its security staff by 'armed criminals' (AngloGold Ashanti, 2009a, p. 82).

As part of its implementation of the VPs, the company states that in 2010 it 'aims to ensure that all potential violations of the principles are reported and investigated and that no violations of the Voluntary Principles occur' (AngloGold Ashanti, 2009b, p. 34). Both AngloGold Ashanti's annual report on the VPs, which the company posts on its website, and its sustainability review provide data on the number of injuries and fatalities of third parties involved in illegal activities. The sustainability review also indicates the number of major security interventions, including those involving the discharge of firearms, and the resulting deaths and injuries among community members and company personnel (AngloGold Ashanti, 2009b, p. 35).[94]

A former AngloGold security manager has suggested that what companies need, in line with the VPs, is a register to catalogue grievances, including around the use of weapons.[95] It could include evidence of the company's investigations and information about how a company communicated with local communities about issues of security. This register could be used to improve transparency, with stakeholders as well as internal or external auditors.[96] The independent review of security at Minera Yanacocha also calls for such recording and reporting (see Box 5.4).

This type of disclosure in public reporting is relatively rare among MNCs, whether as part of their work in implementing the Voluntary Principles or otherwise. A notable exception, besides that of AngloGold Ashanti, is Newmont's *Community Relationships Review* (Newmont Mining Corporation, 2009), which the company published in response to a shareholder resolution asking it to carry out a global review of its policies and practices in relation to local communities worldwide. The review discusses issues integral to the VPs, such as community perception of security provision (it also discusses Yanacocha). Systematic evaluation of the implementation of the VPs, ideally by an independent third party, and public disclosure of these assessments, would provide a more comprehensive picture of the degree to which MNCs are applying these principles.

Further, the initiative makes no provision for penalties in case of non-compliance, other than the possibility of being expelled from the VPs. Critics have expressed concerns that 'free-riding' companies appear to endorse the principles without actually implementing them, and Human Rights Watch has pointed out that the initiative's grievance mechanism can address only specific violations as they emerge, as opposed to systemic problems in a company (Cockayne et al., 2009, p. 156). Signatories seem aware of these weaknesses and gaps. In an undated report on company efforts to implement the principles, the Voluntary Principles Information Working Group notes that the VPs are 'difficult to monitor and audit' and that 'some form of independent verification is needed' to ensure implementation (IWG, n.d., p. 4). Public reporting, which the Working Group also calls for, is not mandatory for signatories.

The question of what impact the VPs have had, including on the human security of communities surrounding company operations, calls for further research. According to one source, the impact is being measured 'anecdotally'.[97] In relation to private security and local communities, it is considered increasingly important for MNCs to undertake formal consultations with communities regarding their concerns about the use of PSCs and to address grievances as they arise.[98] While the principles themselves merely call on companies to consult civil society 'regarding experiences with private security', International Alert's first two implementation indicators for the VPs explicitly address assessment of the impact of company operations on human rights and stakeholder consultation (International Alert, 2008, pp. 2–6).

Despite their weaknesses, the VPs are considered an important initiative and are likely to provide a 'key forum' for discussions on implementation and enforcement of improved standards in the global security industry (Cockayne et al., 2009, p. 144). Along with the ICoC and the Montreux Document, the VPs constitute part of an emerging set of standards that address expected behaviour of both PSCs and their employers.

> The ICoC, the Montreux Document, and the VPs could be mutually reinforcing.

Converging initiatives?

As yet, the ICoC and the Montreux Document are not formally linked to the VPs. Each one targets a different audience: the Montreux Document is aimed primarily at states; the ICoC at PSCs; and the VPs at extractive companies. In a number of ways, however, these initiatives are mutually reinforcing—at least in theory. First, the Swiss government has been the driving force behind the Montreux Document and the ICoC. It has also recently joined the VPs, and has used the plenary of the VPs as a forum to promote the Montreux Document and the ICoC to signatories of the VPs. Second, because signatories are major users of private security, it is in their interest to demand that PSCs abide by the ICoC, to limit the possibility of misconduct. Third, with time, proponents of the code expect governments to encourage all those contracting private security, including signatories of the VPs, to include signature of, and abidance by, the ICoC as an integral part of their security provision contracts. In this sense, members of the VPs may eventually be expected to reinforce the ICoC.[99]

Some of the same stakeholders, such as Switzerland, are already involved in both the ICoC and the VPs, or have begun to explore possible linkages. For example, companies in the Colombia Voluntary Principles process are assessing their private security as it compares to international principles, such as those in the ICoC.[100] Several leading extractive companies, governments, and NGOs that are members of the VPs have also been involved in the ICoC process.[101] But the need to explore the connection between these initiatives and the Ruggie framework is also evident. As Cockayne et al. (2009, p. 26) point out, 'discussions of improved regulation of the global security industry remain notably disconnected from this broader discussion of business and human rights', despite Ruggie's efforts to underscore these connections.

In the end, these initiatives cannot replace international or national law. The effectiveness of the ICoC will depend in part on the ability of its accountability mechanism (which has yet to be created) to monitor and build industry capacity to implement standards (PRIVATE SECURITY COMPANIES). Likewise, the legitimacy of the VPs will depend in large part on increased uptake of its standards and a greater capacity to monitor compliance and to sanction non-compliance.

CONCLUSION

Multinational corporations are among the most important users of private security services. And yet there has been little research on their use of private security in comparison to that of other contractors, such as governments or humanitarian organizations. Private security forces employed by MNCs have been involved in incidents of alleged human rights abuses and armed violence, though a lack of data makes it difficult to gauge the incidence of such violence.

Multinationals face a complex set of challenges related to their use of security. Their control over private security personnel varies significantly depending on the context. The use of public and private security together—whether 'hybridized' or otherwise intimately related to one another—calls for particularly close attention to the factors that influence the misuse of force and firearms. These include training and vetting of personnel, the use of active-duty security personnel to guard private property, and the access of these personnel to firearms. The blurring of private and public can also impede investigation and punishment of the improper use of armed force.

Weak oversight and regulation of private security forces create accountability gaps and potential conflicts of interest. These weaknesses have allowed MNCs to recruit security personnel with poor human rights records; in some cases, they have led to an erosion of public trust in MNCs and their private security providers. It remains difficult to hold MNCs accountable for the misuse of force by their private security providers, though domestic law offers some possible avenues. Yet an international consensus is developing that companies have a responsibility to ensure they are not complicit in abusing human rights, including through third-party relationships with partners such as private security providers.

Specific international initiatives have also emerged to address the lack of regulation of PSCs and the prevention of the improper use of force and human rights abuses. All three pillars of the Ruggie framework—state duty to protect, corporate responsibility to respect, and access to remedies—are relevant in addressing the problems and challenges associated with MNC use of private security. Standards for MNC use of public and private security are emerging through the VPs; despite the initiative's weaknesses, it serves as an important forum for addressing human rights protection while still ensuring the security of MNC operations. Implementation of these standards remains very limited, as do public reporting and the independent monitoring of compliance with the principles. Yet it is in the interest of MNCs to work towards the success of these initiatives and the strengthening of the standards they promote. Otherwise, they will continue to contribute—and be exposed—to the risks currently entailed in the reliance on private security. ■

LIST OF ABBREVIATIONS

ATCA	Alien Tort Claims Act
ICoC	International Code of Conduct for Private Security Providers
ICRC	International Committee of the Red Cross
IHL	International humanitarian law
MNC	Multinational corporation
PJV	Porgera joint venture
PMC	Private military company
PMSC	Private military and security company
PNG	Papua New Guinea
PNP	Peruvian National Police
PSC	Private security company
SRSG	Special Representative of the Secretary-General of the United Nations
VPs	Voluntary Principles on Security and Human Rights

ENDNOTES

1   See, for example, IHRC and CHRGJ (2009); Leigh Day & Co. (2009); Oxfam America (2007).

2   See, for example, Abrahamsen and Williams (2006, pp. 12, 15) on Sierra Leone and Kenya, and Joras and Schuster (2008, p. 42) on Angola.

3   As part of this study, 24 author interviews were conducted between June and November 2010 with representatives of PSCs, extractive sector MNCs, industry groups, academic institutions, the legal profession, and NGOs. These were complemented with interviews conducted by Nicolas Florquin for the chapter on private security companies in this volume, as well as written responses to a questionnaire sent to PSCs covering various aspects of their work. The PSC and MNC interviewees represent companies with operations all over the world.

4   Abrahamsen and Williams (2009, p. 3) distinguish between 'commercial private security' and 'privatized military'.

5   Scahill (2008, pp. 236–43) documents, for example, the Pentagon's contracting of Blackwater in the mid-2000s to help protect the operations of Western oil and gas MNCs in the Caspian Sea region. Joras and Schuster (2008, p. 50) report that in 2003 Chevron–Texaco contracted an Israeli PMC, Aeronautics Defense Systems, Ltd., for 'unmanned air surveillance' in Angola. According to a number of sources interviewed, PSCs in Iraq provide armed services not only to government clients but also to MNCs.

6   Author correspondence with Scott Horton, contributing editor, *Harper's* magazine, and lecturer in law, Columbia Law School, 20 September 2010.

7   See, for example, Joras and Schuster (2008, pp. 5–6); Gaston (2008, pp. 227–28); Perret (2008), cited in Godnick (2009, p. 2).

8   Despite the availability of country-level data on the number of registered PSCs, there is apparently no disaggregation of this data at the global level, which would show the distribution of PSC use across different clients, such as MNCs, local businesses, NGOs, and international organizations.

9   Avant (2007, p. 153) writes of the 'phenomenal growth in the private security industry providing services to transnational corporations working in risky environments'.

10   According to a number of PSC representatives interviewed, MNCs and large companies rank second in importance as clients—in terms of volume—after governments; according to one source, oil and gas companies tied with governments for first place. Another private security representative notes that extractive MNCs in particular are important clients of British PSCs, as the British government is not a major consumer of private security services (correspondence between Nicolas Florquin and Andy Bearpark, director general, British Association of Private Security Companies, 3 September 2010).

11   The SRSG on Business and Human Rights notes that, in a survey of 65 alleged corporate human rights abuses, the extractive sector 'utterly dominates this sample of reported abuses, with two-thirds of the total' (Ruggie, 2006, p. 8). A more recent survey of 320 cases of alleged corporate-related human rights abuses finds the extractive sector accounted for the single largest proportion (28 per cent) of these allegations (Wright, 2008, p. 7).

12   Author interview with Rita Abrahamsen, associate professor, University of Ottawa, 8 September 2010.

13   Author interview with Brian Gonsalves, vice president, Global Security, AngloGold Ashanti, 8 September 2010.

14   South Africa's 2000 Firearms Control Act permits private security to carry firearms (South Africa, 2000, art. 20(2); da Silva, 2010, pp. 14–15; correspondence with Clare da Silva, consultant in international law, 6 February 2011).

15   Author interview with Brian Gonsalves, vice president, Global Security, AngloGold Ashanti, 8 September 2010.

16   Author interview with a security executive of an extractive company, 1 September 2010.

17   Author interview with Mike Faessler, president, Oversight Risk Consulting, 24 August 2010.

18   Anecdotal evidence suggests there is more of a shift towards using less-lethal weapons and munitions within public security forces than in the private sector. However, discussion is ongoing in India, for example, about whether to allow MNC security guards to use stun guns and rubber bullets. Interview by Sonal Marwah with Kunwar Vikram Singh, chairman, Central Association of Private Security Industry, Delhi, 20 October 2010.

19   Author interview with private security representative 1, Geneva, 19 August 2010.

20   Author interview with Scott Horton, contributing editor, *Harper's* magazine and lecturer in law, Columbia Law School, 24 August 2010; author correspondence with Horton, 7 December 2010.

21   Author interview with Scott Horton, contributing editor, *Harper's* magazine and lecturer in law, Columbia Law School, 24 August 2010.

22   Author interview with a security executive of an extractive company, 1 September 2010.

23   Author interview with Mike Faessler, president, Oversight Risk Consulting, 24 August 2010.

24   Author correspondence with Mike Faessler, president, Oversight Risk Consulting, 5 December 2010.

25   Author correspondence with Mike Faessler, president, Oversight Risk Consulting, 5 December 2010.

26   Since 2006, the mine has been operated by Barrick and is co-owned by Barrick, the PNG government, and local landowners.

27   Author interview with Chris Albin-Lackey, senior researcher, Human Rights Watch, 7 September 2010; author correspondence with Albin-Lackey, 20 September 2010 and 5 December 2010.

28   Illegal mining is not uncommon in gold-mining areas.

29   Note that local and international PSCs are not always distinct. As Cockayne et al. write, 'In many cases, small local contractors and large multi-national [security] companies are connected, through subcontracting arrangements, joint ventures, personnel movements and subsidiary structures' (Cockayne et al., 2009, p. 16).

30   Examples of 'transnational PSCs of truly global scale' (in terms of their international presence) include G4S (formerly Group 4 Securicor), Securitas, and Prosegur (Abrahamsen and Williams, 2009, pp. 2–3).

31   The relevant legislation in Nigeria is part I, article 1(1)(c), of the Private Guards Companies Act (da Silva, 2010, p. 9; author correspondence with da Silva, 6 February 2011).

32   The relevant legislation in Angola is the Law on Private Security Companies (19/92), article 10 (Joras and Schuster, 2008, p. 40).

33   Author interview with Mike Faessler, president, Oversight Risk Consulting, 24 August 2010.

34   Author interview with Rita Abrahamsen, associate professor, University of Ottawa, 8 September 2010; author correspondence with Abrahamsen, 5 December 2010.

35   One of these sources states that cost 'is always the first concern'. Correspondence between Nicolas Florquin and former private security representative 6, 6 August 2010.

36   Correspondence between Nicolas Florquin and former private security representative 6, 6 August 2010.

37   See South Africa's Firearms Control Act 2000 (South Africa, 2000, art. 9 (2)(b)); author correspondence with Clare da Silva, consultant on international law, 22 September 2010 and 6 February 2011.

38   Author interview with Yadaira Orsini, International Alert, 26 August 2010.

39   Author interview with the executive officer of a mining company, 19 July 2010. See below for more on the regulation of PSCs in Colombia.

40   Correspondence with William Godnick, coordinator, Public Security Programme, UN Centre for Peace, Disarmament and Development in Latin America and the Caribbean (UN-LiREC), 24 September 2010. Godnick notes that the US contractors include companies such as DynCorp, which could be categorized as 'hybrid intelligence and surveillance contractors'.

41   This is the case in Nigeria, for example.

42   Interview by Nicolas Florquin with private security representative 5, London, 14 July 2010.

43   Correspondence with Mark Taylor, deputy managing director, Fafo, 20 September, 2010.

44   Author interview with private security representative 1, Geneva, 19 August 2010. Table 4.1 in this volume provides data showing that PSC employees outnumber police in a number of countries; Florquin argues that at the global level, PSC personnel exceed police due to the ratios in large countries such as China, India, and the United States (PRIVATE SECURITY COMPANIES).

45   Author interview with Doug Brooks, president, International Peace Operations Association, 9 July 2010; author interview with the executive officer of a mining company, 19 July 2010.

46   Author interview with Luc Zandvliet, director, Triple R Alliance, 9 July 2010.

47   For example, in Colombia 'demobilized paramilitaries have on occasion ended up working in the private military and security sector' (Lazala, 2008, p. 1). In Sierra Leone, many ex-combatants have gone to work for PSCs (Abrahamsen and Williams, 2006, p. 10). On Africa, see Abrahamsen and Williams (2006, p. 7) and on Latin America, see Godnick (2009, p. 9).

48   Author interview with Krista Hendry, executive director, Fund for Peace, 12 August 2010.

49   This is the case, for example, in the UK and several African countries, such as the Democratic Republic of the Congo, Kenya, Nigeria, and Sierra Leone. Although Sierra Leone's National Security and Intelligence Act of 2002 'does in principle allow PSCs to hold arms', the UN arms embargo of 1997, which prohibits the sale of arms to non-state actors, has overruled this provision. There is one exception in Sierra Leone, where PSCs guarding the Sierra Rutile mine are allowed to be armed (da Silva, 2010, pp. 11–12).

50   Author interviews with Yadaira Orsini, International Alert, 26 August 2010, and an executive officer of a mining company, 19 July 2010.

51   Author interview with an executive officer of a mining company, 19 July 2010.

52   Correspondence with William Godnick, Public Security Programme Coordinator, UN Centre for Peace, Disarmament and Development in Latin America and the Caribbean (UN-LiREC), 20 July 2010.

53   Cabrera and Perret (2009, p. 5) have called into question the level of regulation of private security in Colombia. For more details on regulatory frameworks and accountability, see below.

54   In two cases, oil MNCs were sued in the 1990s for alleged complicity in human rights violations committed by Nigerian security forces. Both lawsuits were brought in the United States under the Alien Tort Claims Act (ATCA; see Box 5.6). Chevron was cleared of charges in 2008, and Royal Dutch/Shell settled with plaintiffs in 2009. For one account of the allegations of human rights violations by Nigerian security forces, and the role and responsibility of oil MNCs, see AI (2005).

55   Its full title is the Working Group on the Use of Mercenaries as a Means of Violating Human Rights and Impeding the Exercise of the Right of Peoples to Self-determination.

56   On the Working Group and the tensions between it and the global security industry as well as certain states, see Cockayne et al. (2009, pp. 51–53).

57   Securitas AB acquired Forza, a Peruvian firm, in 2007.

58   GRUFIDES stands for Grupo de Formación e Intervención para el Desarrollo Sostenible.

59   Author interview with Keith Slack, program manager, Extractive Industries, Oxfam America, 14 September 2010.

60   Author interview with Keith Slack, program manager, Extractive Industries, Oxfam America, 14 September 2010.

61   As of 2010 it was not clear whether Newmont had acted on the recommendations, though Oxfam America was working to encourage the company to implement them. Author interview with Keith Slack, program manager, Extractive Industries, Oxfam America, 14 September 2010.

62   See, for example, the bibliography of Cockayne et al. (2009), which cites a number of key works from this debate.

63   See, for example, Renouf (2007, p. 5). See also Cockayne et al. (2009, pp. 18–19; ch. 2) on criticisms of national regulation of the global security industry; the authors note that the United States and South Africa are the two states that have made the greatest efforts to strengthen regulation of PMSCs (p. 39).

64   Author correspondence with Terra dos Direitos, 20 September 2010.

65   Cabrera and Perret distinguish between PSCs, which provide security, and PMCs, which are 'geared up to facilitating warfare'. They report that 'the use of PSCs is more common in Colombia' and companies contracted in Plan Colombia would be categorized as PMCs (Cabrera and Perret, 2009, p. 4). Outside of mentioning this distinction, however, they refer simply to PMSCs.

66   See, for example, Cockayne et al. (2009, ch. 2) and Abrahamsen and Williams (2006, p. 17). Note that, in addition to having detailed regulations on PSCs, South Africa is one of the few countries to regulate its PMCs as well (author correspondence with Clare da Silva, consultant in international law, 22 September 2010).

67   This proposal comes from the Working Group on the Use of Mercenaries (OHCHR, 2010).

68   Ambassador Claude Wild, head, Political Division IV–Human Security, cover letter to International Code of Conduct for Private Security Service Providers, Berne, 27 August 2010. Emphasis added.

69   See, for example, 151 Organizations (2007), which calls for 'global intergovernmental standards' in this area.

70   Box 5.2 draws on Bushnell (2010).

71   A 2006 study finds that as countries ratify the Rome Statute and incorporate IHL and international criminal law into their domestic law, the potential liability for legal persons (such as corporations), as well as for natural persons, might increase (Ramasastry and Thompson, 2006, p. 27).

72   For example, Unocal (now part of Chevron) was sued under ATCA in 1996 for alleged complicity in human rights abuses committed by the Myanmar military during the building of a gas pipeline in Myanmar (the parties settled out of court in 2005). For a list of documents on the case, see BHRRC (n.d.). See also endnote 54 of this chapter for two other examples of ATCA cases.

73   See *Kiobel* v. *Royal Dutch Petroleum* (2010). For commentary on ATCA in the federal courts, see Altschuller (2010).

74   This section draws on panel discussions held at the conference 'Corporations in Armed Conflict: The Role of International Law' at the National University of Ireland, Galway, 9 April 2010.

75   Correspondence between Alexis Bushnell and Doug Brooks, president, International Peace Operations Association, 12 September 2010.

76   See, for example, HRW and ESCR-Net (2010); other NGOs have made similar statements. In a related vein, while a number of human rights groups have welcomed the SRSG's framework, some have called for more emphasis on what is required—as opposed to what is encouraged—of companies in relation to human rights. See, for example, AI et al. (2011, pp. 1–2).

77   Author interview with Chris Albin-Lackey, senior researcher, Human Rights Watch, 7 September 2010. In September 2010, European NGOs launched a campaign to press the European Union to hold EU-based multinationals legally accountable for social and environmental harm they or their subsidiaries cause, including overseas (ECCJ, n.d.).

78   Similarly, Cockayne et al. (2009, pp. 39–40) point out that most national regulation of private security contractors in home and contracting states focuses on government contracting, ignoring international PMSCs in the employ of private clients overseas.

79   As noted above, extractive companies are major clients of British PSCs (Cockayne et al., 2009, p. 17).

80   One source reports that in Colombia PMSCs can be held liable under civil law for their employees' activities, though not under criminal law (as criminal liability applies only to individuals, not companies). It notes that civil liability could possibly 'extend to the particular or public contractor', with the former presumably including MNC clients (Cabrera and Perret, 2009, p. 9).

81   See, for example, Lam (2009). Focusing on Iraq, Lam makes the case for using the Alien Tort Claims Act to hold private military contractors accountable for torts.

82   The private security firm involved was Forza, the same firm implicated in the Yanacocha case.

83   Senden (2005) defines soft law as '[r]ules of conduct that are laid down in instruments which have not been attributed legally binding force as such, but nevertheless may have certain—indirect—legal effects, and that are aimed at and may produce practical effects'.

84   See, for example, VPSHR (n.d.b) and Cockayne et al. (2009, pp. 154–55).

85   In 2005, Fundación Ideas para la Paz and the International Business Leaders Forum spearheaded efforts to bring the kind of guidance provided in the VPs to other companies. As part of this initiative, in 2006 guidelines based on the VPs were published that aimed at non-extractive companies, such as food and agriculture, with a focus on Colombia. See Guaqueta (2006). A discussion paper prepared in 2006 for the UN SRSG on Business and Human Rights calls specifically for the VPs to be expanded to include 'state-owned enterprises and smaller and/or non-Western companies' (OHCHR, 2006, p. 3).

86   Correspondence with Krista Hendry, executive director, Fund for Peace, 25 September 2010.

87   For more information on these instruments, see Small Arms Survey (2004, ch. 7).

88   Author interview with Yadaira Orsini, International Alert, 26 August 2010.

89   Author interview with Scott Horton, contributing editor, *Harper's* magazine and lecturer in law, Columbia Law School, 24 August 2010.

90   See, for example, ICMM (2009, p. 19). For a detailed description of BP's implementation of the VPs at its Tangguh liquefied natural gas project in West Papua, Indonesia, as well as the Baku–Tbilisi–Ceyhan pipeline (which BP operates and co-owns), see Cockayne et al. (2009, pp. 152–54).

91   See, for example, Eni (2009).

92   See On Common Ground (2010, pp. 168–79). The review is a human rights assessment of GoldCorp's Marlin Mine in Guatemala. GoldCorp is not a formal member of the VPs.

93   Sustainability reports are those in which companies report on their social and environmental policies, programmes, and, sometimes, impact. These are generally voluntary reports, in contrast to legally mandated financial reporting.

94   Note that AngloGold Ashanti itself has faced accusations of complicity in alleged security-related human rights abuses in Ghana. See FIAN (2008) and CHRAJ (2008). On the company's response, see AngloGold Ashanti (2008).

95   For example, were weapons used? Should they have been? Were people's rights violated?

96   Author interview with Mike Faessler, president, Oversight Risk Consulting, 24 August 2010.

97   Author interview with Krista Hendry, executive director, Fund for Peace, 12 August 2010.

98   See, for example, On Common Ground (2010, p. 207) and Joras and Schuster (2008, p. 64).

99   This paragraph draws on an author interview with Claude Voillat, economic adviser, International Committee of the Red Cross (ICRC), 17 November 2010. The ICRC was a co-sponsor of the Montreux Document, was regularly consulted in the drafting of the ICoC, and is an observer to the VPs.

100   Author interview with Yadaira Orsini, International Alert, 26 August 2010.

101   Author correspondence with participants involved in public, multi-stakeholder workshops to draft the ICoC, 24 September 2010.

BIBLIOGRAPHY

151 Organizations. 2007. 'Open Letter to John Ruggie.' 10 October. <http://www.fidh.org/IMG/pdf/OpenLetter_Ruggie_FINAL_wOct10Endorsements.pdf>

Abrahamsen, Rita and Michael Williams. 2006. 'Security Sector Reform: Bringing the Private In.' *Conflict, Security & Development,* Vol. 6, No. 1. April, pp. 1–23.

—. 2009. 'Security Beyond the State: Global Security Assemblages in International Politics.' *International Political Sociology,* Vol. 3, No. 1, pp. 1–17.

AI (Amnesty International). 2005. *Nigeria Ten Years On: Injustice and Violence Haunt the Oil Delta.* AI Index AFR 44/022/2005. London: AI. 3 November.

__. 2008. 'Contested Land in Brazil Handed to State.' London: AI. 22 October.

<http://www.amnesty.org/en/news-and-updates/good-news/contested-land-brazil-handed-state-20081022>

— et al. 2011. 'Joint Civil Society Statement on the Draft Guiding Principles on Business and Human Rights.' 31 January.

Altschuller, Sarah. 2010. 'The Federal Courts and Corporate Liability under the Alien Tort Statute.' Corporate Social Responsibility and the Law (blog). Washington, DC: FoleyHoag LLP. 27 September.

<http://www.csrandthelaw.com/2010/09/articles/litigation/alien-tort-statute/the-federal-courts-and-corporate-liability-under-the-alien-tort-statute/>

AngloGold Ashanti. 2008. 'AngloGold Ashanti Response to the FIAN/WACAM Commentary on Mining in Ghana.' 2 June.

<http://www.reports-and-materials.org/AngloGold-Ashanti-response-FIAN-WACAM-2-Jun-2008.doc>

—. 2009a. 'Case Studies: Implementation of the Voluntary Principles on Security and Human Rights.' *AngloGold Ashanti Sustainability Review: Supplementary Information 2009.*

<http://www.anglogold.co.za/subwebs/informationforinvestors/reports09/SustainabilityReview09/f/AGA_SD09_20.pdf>

—. 2009b. 'Sustainability Review: Environment, Community and Human Rights.'

<http://www.anglogoldashanti.co.za/subwebs/informationforinvestors/reports09/SustainabilityReview09/f/AGA_SR09_12.pdf>

Avant, Deborah. 2007. 'NGOs, Corporations, and Security Transformation in Africa.' *International Relations,* Vol. 21, pp. 143–61.

Barrick Gold Corporation. 2010. 'Barrick Joins Voluntary Principles and Announces New Corporate Social Responsibility Initiatives.' Press release. 19 November. <http://www.barrick.com/News/PressReleases/PressReleaseDetails/2010/Barrick-Joins-Voluntary-Principles-and-Announces-New-Corporate-Social-Responsibility-Initiatives/default.aspx>

—. 2011. 'Statement in Response to Human Rights Watch Report.' 1 February.

<http://www.barrick.com/CorporateResponsibility/KeyTopics/PorgeraJV/Response-to-Human-Rights-Watch-Report/default.aspx>

BHRRC (Business & Human Rights Resource Centre). 2010. 'Voluntary Principles on Security and Human Rights.'

<http://www.business-humanrights.org/ConflictPeacePortal/Specialinitiatives/VoluntaryPrinciples>

—. n.d. 'Case Profile: Unocal Lawsuit (re Burma).'

<http://www.business-humanrights.org/Categories/Lawlawsuits/Lawsuitsregulatoryaction/LawsuitsSelectedcases/UnocallawsuitreBurma>

Bushnell, Alexis. 2010. *Private Security Contractors: Debates under International Humanitarian Law.* Unpublished background paper. Geneva: Small Arms Survey.

Cabrera, Irene and Antoine Perret. 2009. 'Colombia: Regulating Private Military and Security Companies in a "Territorial State."' *PRIV-WAR Report Colombia, National Reports Series 19/09.* 15 November. <http://priv-war.eu/wordpress/wp-content/uploads/2009/12/nr-19-09-col.pdf>

CHRAJ (Commission for Human Rights and Administrative Justice, Ghana). 2008. 'CHRAJ Indicts Mining Companies of Human Rights Abuses.' 27 June. <http://www.ghanaweb.com/public_agenda/article.php?ID=10591>

Cockayne, James and Emily Speers Mears. 2009. *Private Military and Security Companies: A Framework for Regulation.* New York: International Peace Institute.

Cockayne, James, et al. 2009. *Beyond Market Forces: Regulating the Global Security Industry.* New York: International Peace Institute.

Costa, Gino. 2009. *Comprehensive Review of Minera Yanacocha's Policies Based on the Voluntary Principles on Security and Human Rights.* Lima: Newmont Mining and Minera Yanacocha. 12 May. <http://www.business-humanrights.org/Links/Repository/321717/jump>

Dammert, Lucia. 2008. *Seguridad Pública y Privada en las Américas.* Washington, DC: Organization of American States. March.

<http://www.oas.org/dsp/documentos/Publicaciones/Seg%20Pub-%20LasAmericas.pdf>

da Silva, Clare. 2010. *Analysis of Selected Countries' National Legislation on Private Military and Security Companies (PMSCs) and Firearms and Multinational Companies' Use of PSCs.* Unpublished background paper. Geneva: Small Arms Survey.

ECCJ (European Coalition for Corporate Justice). n.d. 'Rights for People Rules for Business.' Campaign website. <http://www.rightsforpeople.org/>

Eni. 2009. *Sustainability Report 2009.* Rome: Eni. <http://www.eni.com/attachments/sostenibilita/sustainability-report-09-eng.pdf>

Faessler, Mike. 2010. 'Working with Local Security: A Case Study from the D.R. Congo.' *Journal of International Peace Operations,* Vol. 5, No. 5. March–April, pp. 18, 20.

FDFA (Swiss Federal Department of Foreign Affairs). 2010. International Code of Conduct for Private Security Providers. <http://www.news.admin.ch/NSBSubscriber/message/attachments/21143.pdf>

FIAN (FoodFirst Information and Action Network). 2008. 'Universal Periodic Review—Ghana (May 2008): Human Rights Violations in the Context of Large-scale Mining Operations.' May. <http://www.fian.org/news/resources/documents/others/mining-related-human-rights-violations-ghana/pdf>

Gaston, E.L. 2008. 'Mercenarism 2.0? The Rise of the Modern Private Security Industry and Its Implications for International Humanitarian Law Enforcement.' *Harvard International Law Journal*, Vol. 49, No. 1. Winter, pp. 221–48.

Godnick, William. 2009. *Private Security: A Preliminary Report for the Second Meeting of the Organized Crime Observatory for Latin America and the Caribbean, San José, Costa Rica, 8–9 September 2009*. Lima: United Nations Regional Centre for Peace, Disarmament and Development in Latin America and the Caribbean. Revised 15 October.

Guaqueta, Alejandra. 2006. 'Company Operations in Weak Governance Zones: A Practical Guide for Non-Extractive Industries.' Bogotá: Fundación Ideas para la Paz. November.

HRW (Human Rights Watch). 2011. *Gold's Costly Dividend: Human Rights Impacts of Papua New Guinea's Porgera Gold Mine*. New York: HRW. February.

— and ESCR-Net (International Network for Economic, Social & Cultural Rights). 2010. 'Statement to the UN Human Rights Council on Business and Human Rights: Human Rights Watch and ESCR-Net Address the Council's 14th Session.' Geneva. 4 June.

ICJ (International Commission of Jurists). 2008. *Corporate Complicity & Legal Accountability: Facing the Facts and Charting a Legal Path*. Vol. 1. Geneva: ICJ.

ICMM (International Council on Mining and Metals). 2009. *Human Rights in the Mining and Metals Industry: Overview, Management Approach and Issues*. London: ICMM. May. <http://www.icmm.com/page/14809/human-rights-in-the-mining-and-metals-industry-overview-management-approach-and-issues>

IHRC and CHRGJ (International Human Rights Clinic and Center for Human Rights and Global Justice). 2009. *Legal Brief before the Standing Committee on the Foreign Affairs and International Development (FAAE), House of Commons, Regarding Bill C-300*. Ottawa: IHRC, Harvard Law School, and CHRGJ, New York University School of Law. 16 November. <http://www.reports-and-materials.org/Harvard-testimony-re-Porgera-Main.pdf>

International Alert. 2008. *Voluntary Principles on Security and Human Rights: Performance Indicators*. London: International Alert. June.

IPOA (International Peace Operations Association). 2010. 'Principled Security: The Voluntary Principles on Security and Human Rights.' *Journal of International Peace Operations*, Vol. 5, No. 5. March–April, p. 10. <http://web.peaceops.com/pdf/journal_2010_0304_hires.pdf>

IWG (Information Working Group of the Voluntary Principles on Security and Human Rights). n.d. 'Overview of Company Efforts to Implement the Voluntary Principles.' <http://www.voluntaryprinciples.org/files/vp_company_efforts.pdf>

Joras, Ulrike and Adrian Schuster, eds. 2008. *Private Security Companies and Local Populations: An Exploratory Study of Afghanistan and Angola*. Working Paper. Berne: SwissPeace. <http://www.swisspeace.ch/typo3/fileadmin/user_upload/Media/Publications/WP1_2008.pdf>

Kiobel v. *Royal Dutch Petroleum*. 2010. Docket Nos. 06-4800-cv, 06-4876-cv. US Court of Appeals, 2nd Circuit. 17 September.

Lam, Jenny. 2009. 'Accountability for Private Military Contractors under the Alien Tort Statute.' *California Law Review*, Vol. 97. 10 October, pp. 1459–99.

Lazala, Mauricio. 2008. 'Private Military and Security Companies and their Impacts on Human Rights in Contexts Other than War.' London: Business and Human Rights Resource Centre. January.

Leigh Day & Co. 2009. 'Peruvian Torture Victims Obtain Worldwide Freezing Injunction over Mining Company Assets.' 19 October.

—. 2011. 'Trial of Peruvian Torture Victims' Claim against UK Miner Draws Closer.' In Leigh Day & Co. *2010 in Words and Pictures*, p. 22. <http://www.business-humanrights.org/Links/Repository/1004028/jump>

Martin-Ortega, Olga. 2008. 'Deadly Venture: Multinational Corporations and Paramilitaries in Colombia.' *Revista Electronica de Estudios Internacionales*, No. 16. <http://www.reei.org/reei%2016/doc/MARTINORTEGA_Olga.pdf>

Newmont Mining Corporation. 2009. 'Community Relationships Review.' <http://www.beyondthemine.com/2009/?pid=470>

OHCHR (Office of the High Commissioner for Human Rights). 2006. 'Security of People and Assets.' Discussion paper prepared for the UN Special Representative to the Secretary-General on Business and Human Rights. 21 July. <http://www.reports-and-materials.org/Discussion-paper-security-Jul-2006.pdf>

—. 2010. 'Mercenaries: UN Expert Panel Pushes for Stronger Regulation of Private Military and Security Companies.' Geneva. 23 July. <http://www.ohchr.org/en/NewsEvents/Pages/DisplayNews.aspx?NewsID=10227&LangID=E>

On Common Ground. 2010. *Human Rights Assessment of GoldCorp's Marlin Mine*. Vancouver: On Common Ground Consultants, Inc.

Oxfam America. 2007. 'Oxfam Calls on Newmont Mining Company to Publicly Renounce Human Rights Abuses at Peruvian Gold Mine.' 30 July.

Perret, Antoine. 2008. *El uso de contratistas en Colombia: ¿una política equivocada?* Bogotá: Universidad Externado de Colombia.

Ramasastry, Anita. 2010. 'The Role of Truth Commissions in Articulating Norms.' Presentation for the conference on 'Corporations in Armed Conflict: The Role of International Law,' National University of Ireland, Galway. 9 April.

— and Robert Thompson. 2006. *Commerce, Crime and Conflict: Legal Remedies for Private Sector Liability for Grave Breaches of International Law—A Survey of Sixteen Countries*. Fafo Report 536. Oslo: Fafo. September.

Renouf, Jean. 2007. 'Do Private Security Companies Have a Role in Ensuring the Security of Local Populations and Aid Workers?' September. <http://www.eisf.eu/resources/library/psmcs_local_pop_jsf.pdf>

Ruggie, John. 2006. *Promotion and Protection of Human Rights: Interim Report of the UN Special Representative of the Secretary-General on the Issue of Human Rights and Transnational Corporations and Other Business Enterprises*. E/CN.4/2006/97 of 22 February.

—. 2008a. *Protect, Respect and Remedy: A Framework for Business and Human Rights*. Report of the Special Representative of the Secretary-General on the Issue of Human Rights and Transnational Corporations and Other Business Enterprises. A/HRC/8/5 of 7 April.

—. 2008b. *Clarifying the Concepts of 'Sphere of Influence' and 'Complicity.'* Report of the Special Representative of the Secretary-General on the Issue of Human Rights and Transnational Corporations and Other Business Enterprises. A/HRC/8/16 of 15 May.

—. 2009. 'Consultation on Operationalizing the Framework for Business and Human Rights Presented by the Special Representative of the Secretary-General on the Issue of Human Rights and Transnational Corporations and Other Business Enterprises: Opening Remarks.' Geneva, 5–6 October. <http://www2.ohchr.org/english/issues/globalization/business/docs/OpeningSpeechJohnRuggie.pdf>

—. 2010a. 'Engaging Business: Addressing Respect for Human Rights.' Keynote address at the Coca-Cola Company, Atlanta. 25 February.

—. 2010b. *Business and Human Rights: Further Steps toward the Operationalization of the 'Protect, Respect and Remedy' Framework*. Report of the Special Representative of the Secretary-General on the Issue of Human Rights and Transnational Corporations and Other Business Enterprises. A/HRC/14/27 of 9 April.

Scahill, Jeremy. 2008. *Blackwater: The Rise of the World's Most Powerful Mercenary Army*. New York: Nation Books.

Schwartz, Moshe. 2010. *Department of Defense Contractors in Iraq and Afghanistan: Background and Analysis*. Report to Congress. Washington, DC: Congressional Research Service. 2 July.

Senden, Linda. 2005. 'Soft Law, Self-Regulation and Co-Regulation in European Law: Where Do They Meet?' *Electronic Journal of Comparative Law*, Vol. 9.1. January. <http://www.ejcl.org/91/art91-3.html#N_74_>

Small Arms Survey. 2004. *Small Arms Survey 2004: Rights at Risk*. Oxford: Oxford University Press.

South Africa. 2000. Firearms Control Act 2000. <http://www.acts.co.za/firearms/whnjs.htm>

UN (United Nations). 1990. Basic Principles on the Use of Force and Firearms by Law Enforcement Officials ('UN Basic Principles'). Havana: UN. <http://www2.ohchr.org/english/law/firearms.htm>

UNGA (United Nations General Assembly). 1979. Code of Conduct for Law Enforcement Officials. Resolution 34/169 of 17 December. <http://www2.ohchr.org/english/law/codeofconduct.htm>

UN Working Group. 2008. *Report of the Working Group on the Use of Mercenaries as a Means of Violating Human Rights and Impeding the Exercise of the Right of Peoples to Self-Determination: Mission to Peru (29 January to 2 February 2007)*. A/HRC/7/7/Add.2 of 4 February.

VPSHR (Voluntary Principles on Security and Human Rights). 2009. 'Amendments Approved at VPs 2009 Oslo Plenary.' May. <http://voluntaryprinciples.org/files/vp_amendments_200905.pdf>

—. n.d.a. 'The Principles: Interactions between Companies and Private Security.' <http://www.voluntaryprinciples.org/principles/private_security>

—. n.d.b. 'Voluntary Principles: Colombia Case Study.' <http://voluntaryprinciples.org/files/vp_columbia_case_study.pdf>

Wright, Michael. 2008. *Corporations and Human Rights: A Survey of the Scope and Patterns of Alleged Corporate-related Human Rights Abuse*. A study conducted for John G. Ruggie, United Nations Special Representative for Business and Human Rights. Cambridge: John F. Kennedy School of Government, Harvard University. April.

ACKNOWLEDGEMENTS

Principal author

Elizabeth Umlas

Contributors

Alexis Bushnell and Clare da Silva

A member of Madagascar's security forces takes up position as the army storms a barracks housing dissident officers aiming to overthrow Andry Rajoelina, 20 November 2010.
© Siphiwe Sibeko/Reuters

Ethos of Exploitation
INSECURITY AND PREDATION IN MADAGASCAR

6

INTRODUCTION

In 2001 Madagascar stepped back from the brink of probable civil war. The country's new leader, President Marc Ravalomanana, seemed poised to pave the way for long-term stability and economic prosperity. The international community provided widespread support as prospects appeared to improve, but Ravalomanana had built his power on fragile foundations. In the years that followed, opposition grew from key economic, military, and political stake-holders, and, by late 2008, Ravalomanana's power base began to unravel. A series of disturbing events ensued—including a mutiny at an army barracks and the massacre of civilians by presidential security forces on the main square in the capital, Antananarivo—precipitating a political crisis that continued through late 2010. By that time, international donors had largely suspended their non-humanitarian assistance.

Madagascar is a deeply impoverished country. In the absence of financial support from and oversight by the international community, the 'transition' government is signing unfavourable contracts, primarily with Chinese and other Asian investors, who are gaining unchecked access to the island's wealth of resources. The extraction and transportation of these resources—including timber, seafood, beef and rare animal species, coal, uranium, gold, diamonds, and other precious stones—allegedly relies on the tacit cooperation and collusion of members of the country's security forces. Moreover, rates of armed robbery, often committed with military weapons, appear to have risen sharply since 2008, as has the presence of international criminal networks. Research suggests that these networks are taking advantage of the political disarray to turn Madagascar into a hub for regional and global trafficking—predominantly of drugs.

Madagascar's security sector has always been weak, undermined by external influence and, since independence in 1960, instrumentalized by successive heads of state and their entourages. As a result, the military, gendarmerie, and police forces do not constitute effective units with a clear vocation; instead, the regular forces are severely underpaid while a hugely inflated number of high-ranking officers are pursuing their own political and economic agendas. Combined with Madagascar's strategic location, lack of basic infrastructure, difficult terrain, and porous borders (see Map 6.1)—which attract predatory actors who plunder the natural resources and engage in illegal trafficking—a dysfunctional security sector has generated the conditions for armed violence of worrisome proportions.

This chapter offers an analysis of Madagascar's insecurity, with a special focus on the role of security sector actors. Policy-makers and researchers have largely ignored the fragile relationships between the Malagasy state and its security forces and between political stability and economic predation. Information on these topics is extremely limited and fragmentary, often historically inaccurate, and frequently contradictory. To offer a critical and more coherent perspective on political violence on the island, the United Nations Office of the Resident Coordinator in Madagascar commissioned a Peace and Conflict Impact Assessment in early 2010; the study built on nearly 200 key informant interviews, background papers, and a preliminary survey of security perceptions in Antananarivo and environs (Jütersonke and Kartas, 2010). This chapter draws on all of the source material developed for that assessment as well as additional desk and in-country research.

This chapter's principal conclusions include the following:

- To a large extent, Madagascar's inability to develop effective state security forces can be attributed to its colonial heritage and strategic location. As a result, the main rationale for a career in the military or gendarmerie is the pursuit of personal gain.
- Since their politicization and instrumentalization in the 1970s, Madagascar's armed forces have constantly been embroiled in struggles over political power and economic access to the country's wealth of resources.
- Today, Madagascar's security sector is characterized by severely underpaid and ill-equipped regular forces, far too many high-ranking officers, and a mushrooming of special intervention units with questionable mandates.
- Collusion between elements of the country's security sector and both foreign and domestic business interests has sharply intensified since the political crisis of early 2009. In the resulting security vacuum, armed criminality is on the rise, rural banditry has expanded, and Madagascar is gaining in importance as an international trafficking hub.
- The state administration has encouraged the organization of neighbourhood watch initiatives and village self-defence groups; it has also turned a blind eye to the operations of highly aggressive indigenous private security companies that hunt down rural bandits.

The chapter has two main sections. The first section reviews the historical context and development of the security sector in Madagascar, noting the specific ways in which the armed forces have continuously been subject to strategies of dispersion and co-optation by colonial and post-colonial governments. Not only has this process led officers to abuse their positions in the pursuit of personal gain, but it has also transformed Madagascar's armed forces into pawns of predatory actors seeking to further their own political and economic agendas. The second section analyses

Rebel soldiers loyal to Andry Rajoelina take over one of the presidential offices in downtown Antananarivo, 16 March 2009.
© Jerome Delay/AP Photo.

the sources of current political instability and the dynamics of today's dysfunctional security sector. It considers three main types of insecurity: armed criminality, large-scale rural banditry, and international trafficking networks on the island. In so doing, it focuses on the role of state security actors either in failing to prevent insecurity or in perpetuating it.

FROM COLONY TO COUNTRY: A HISTORICAL PERSPECTIVE

Madagascar's political process and its administrative, societal, and economic institutions and practices were all shaped by the impact of colonial politics on traditional society. Yet while Madagascar is thus a classic 'imported state' (Badie, 2000), the island features a series of structural particularities. In fact, Madagascar's plight cannot be grasped without an understanding of the historical dynamics that have led to the blurring of national interest and private benefit—and, ultimately, to the country's dysfunctional security sector.

Map 6.1 **Madagascar**

The emergence and co-option of a politico-military elite

In the later decades of the 19th century, the Imerina Kingdom—and its people, the Merina—conquered roughly three-quarters of the Great Island and rose from a small 'principality' to become the isle's imperial master (Allen, 1987, p. 135). This rise to power was the result of an alliance between the royalty and resourceful *hova* family clans that had helped to finance the king's war. In return, successive kings and queens had provided the clans with forced labour and a regulatory system to build irrigated rice fields and marketplaces. Crucially, the military expansion of the Imerina Kingdom depended on trade: the export of rice and slaves and the import of textiles and firearms.[1]

France encountered Imerina dominance when it invaded the island and subsequently took control of Antananarivo in 1895. The *hova* plutocrats had effective control over a highly repressive kingdom that had 'colonized' the other people of the Great Island through merciless forced labour and the slave trade. Thus, the French colonizers sought to 'pacify' an island experiencing constant anti-*hova* insurgencies and civil strife on the one hand, and worked to contain the power of the plutocrats on the other. By co-opting the kingdom's administrative structure, France adopted a divide-and-rule strategy between the Merina, emblematic of *hova* domination, and the people from the coastal regions (the so-called *côtiers*), thereby emphasizing ethnic differences between them (Deschamps, 1972, pp. 244–47; Allen, 1987, p. 137; Rajoelina, 1988).

The Merina *hova,* who began organizing themselves politically to restore their dominance and oust the new colonial power, expressed their claims in nationalistic terms. In contrast, the emerging 'non-*hova*' middle class (*côtiers* and less privileged Merina and Betsileo) placed their support behind the French, whose rule was less repressive than the Imerina Kingdom's had been and provided some opportunity for social and economic advancement.

The 1947 insurgency provided France with the opportunity to weaken Malagasy aspirations for independence (see Box 6.1). The colonial administration supported pro-French political forces for a future transfer of power but relied on Merina and Betsileo officers for the build-up of indigenous forces (Razafindranaly, 2000, p. 253). From the French perspective, with the election of Philibert Tsiranana, a *côtier* teacher, as president of the First Republic and the parallel consolidation of the Parti Social Démocrate (Social Democratic Party, PSD), the necessary conditions for maintaining a neo-colonial regime after independence had been met.

Box 6.1 Major episodes of political violence in Madagascar

Late 18th century As the Imerina kings and queens unify the island with the help of the British, thousands of people lose their lives in a series of rebellions and resistance movements from rival kingdoms and chieftaincies (Razafindranaly, 2000; Rasamoelina, 2007; Stadelmann, 2009).

1890s Following the two Franco-Malagasy wars (1883–85 and 1895) and the establishment of a French protectorate, Gen. Joseph Gallieni is sent to Madagascar in 1896 to 'pacify' the island based on his experience in Indochina. The insurgency ends in 1898 with more than 50,000 dead from the fighting and related famine and disease (Covell, 1995, p. 141; Rabinow, 1995).

1940s Malagasy nationalism and calls for self-government surge again in the 1940s. The resulting crackdown by the colonial administration in March 1947–and the subsequent state of emergency upheld until 1956–lead to a death toll that continues to be contested to this day, with estimates ranging from 11,200 to 100,000 people (Deschamps, 1972, p. 269; Covell, 1995, p. 212).

1971 The gendarmerie, headed by Col. Richard Ratsimandrava, ruthlessly crushes a regional revolt over tax collection with the support of the prime minister's Israeli-trained Force Républicaine de Sécurité (Republican Security Force, FRS). Between 500 and 1,000 people are reportedly killed. Famously, the Malagasy armed forces decline to participate in the operations (Schraeder, 1994; Rakotomanga, 1998; Archer, 1976, p. 49).

1972 In response to student protests, the FRS raids the University of Madagascar, sparking the May Revolution. Once again, the FRS fires into the crowd at the Hôtel de Ville in Antananarivo, while the gendarmerie and army refuse to shoot at the demonstrators (Rakotomanga, 1998; Althabe, 1980).

A French military convoy prepares to leave for a reconnaissance mission during violent rebellions that led to innumerable casualties, 25 September 1947. © AFP Photo

Neo-colonial foundations of the security sector

Madagascar did not gain independence through armed struggle, and there was no liberation army to unite popular sentiments. Instead, the basic structure of the security forces of the First Republic were those inherited from the colonial system—in particular, the tripartite system of army, gendarmerie, and police (Rakotomanga, 1998, p. 12; see Box 6.2). Moreover, according to the cooperation agreement, France continued to guarantee Madagascar's external defence and internal security (Covell, 1995, p. 73). In the state administration, French technical assistants still played a key role; in the military, they continued to be the dominant force.

At the time of independence in 1960, Madagascar had only ten Malagasy officers, all of whom had previously been active in the French armed forces; two of them had trained at the famous St. Cyr academy. Practically all of them were drawn from Merina aristocrats or *hova* families.[2] Tsiranana kept the defence portfolio under the control of the presidential office and appointed the then highest ranking officer, Gabriel Ramanantsoa (a descendant of the royal family), as head of the joint staff, with the task of organizing the military (Rakotomanga, 1998, pp. 12, 32, 39). The first regiments were soon created, made up of indigenous soldiers equipped with antiquated MAS-39 and MAS-49 rifles.[3]

1985 The practice of martial arts is banned after members of the 'Kung Fu' movement attack a state-run youth association that was widely believed to have been involved in street crime and informing the security police about political dissidents. President Didier Ratsiraka's security forces launch an attack on the movement, killing the leader, Pierre-Michel Andrianarijoana, and several hundred members (Gow, 1997).

1991 After fraudulent elections in 1989 and 1991, students initiate a new protest movement. On 10 August 1991, more than 400,000 people participate in the 'Great March' towards Ratsiraka's presidential palace. When the crowd reaches the 'red zone' around the palace, members of the presidential guard opened fire, killing about 130 people (Covell, 1995, p. xliv).

2001 The presidential candidates–Ratsiraka and the mayor of Antananarivo, entrepreneur Marc Ravalomanana–accuse each other of election fraud. Ratsiraka responds by blockading the capital along the roads linking the highlands to the coastal regions. Both sides supply reservists and civilian supporters with weapons, and several hundred people are killed in episodic clashes along the blockade over a six-month period (Rakotomanga, 2004, pp. 66-67; Vivier, 2007, pp. 60-65).

2009 Antananarivo's young new mayor, Andry Rajoelina, a former disc jockey and the owner of an entertainment business, demands the resignation of President Ravalomanana. Supporters of Rajoelina storm and set fire to the national broadcasting company and loot the Magro shops of the President's Tiko Group. The riots spread to several other cities and claim more than 70 lives. Ravalomanana then removes Rajoelina from office, and, in a subsequent demonstration in the capital on 7 February 2009, presidential guards shoot into the crowd, killing at least 30 people and injuring more than 200 (ICG, 2010a, p. 5).

Fearful of the possibility of a military coup, Tsiranana sought to control and counterbalance the power of the military through two principal measures. First, he ordered the minister of the interior and the secretary-general of the PSD, André Resampa, to create the Force Républicaine de Sécurité (FRS), a paramilitary police force trained and armed by France and Israel and recruiting exclusively non-Merina men.[4] Second, Tsiranana placed the former French gendarmerie (Zandarmaria Nasionaly) directly under his authority. It was commanded by a separate, special joint staff headed by the French Col. Bocchino and placed hierarchically above the army's joint staff (Rakotomanga, 1998, pp. 34–37). At the time, the gendarmerie was a much-feared tool of French colonial oppression, serving as the main security and policing force with the broadest coverage of the territory (Razafindranaly, 2000, pp. 253–57). Its soldiers were thus hardly trained to perform regular policing duties, a fact that continues to haunt Madagascar's security system today (Milburn, 2004).

French technical assistants continued to be present at all levels, especially in the president's special staff. Two factors explain France's high interest in Madagascar during the cold war. First, the island's long coastline offered a privileged position for military reconnaissance, notably for submarines; Soviet bases on the island would have been a major threat to Western naval forces. Second, with the demise of Portuguese colonialism in 1974–76 and the simultaneous rise of socialist, nationalist regimes in Mozambique and Angola, the anti-communist buffer between Central and Southern Africa disappeared. The ports of Mozambique were a

---

Box 6.2 State security actors in Madagascar

Not only did Madagascar inherit the French political system of presidential republicanism, but it was also bestowed with the French security triad of the army, the gendarmerie, and the police, which remains in place today.

The **Malagasy Army** (Armée Malgache) churns out enough high-ranking officers per year to cater to an army of several hundred thousand soldiers (Rakotomanga, 2004, p. 71),[5] far more than its 12,500 troops actually need (IISS, 2009, p. 309). Although Madagascar is an island, it has no navy, no functioning air force, and its battalions are built around materiel (such as tanks and artillery) that the country happened to acquire, rather than on what it needs to fulfil its operational functions. In practice, a career in the military is seen as a path to social status, political influence, and personal self-enrichment.

As is the case in France, the **national gendarmerie** (Gendarmerie Nationale) in Madagascar constitutes a branch of the armed forces that is tasked not with the defence of the country against external attack but with the maintenance of law and order—predominantly in rural areas. Yet, unlike in France (or Italy), there is no effective system of civilian oversight or, consequently, public accountability. The gendarmerie is characterized by an inflated proportion of high-ranking officers, a meddling in domestic politics, and entrepreneurial enrichment—as is the army. The 8,100 gendarmes are ultimately an ineffective service of public order on the island's vast territory (IISS, 2009, p. 309).

Madagascar's **national police** (Police Nationale), by contrast, has generally tried to distance itself from the political scene. That aim, however, is hampered by the fact that its *directeur général* is, together with his counterparts from the army and the gendarmerie, subordinate to the Mixed National Operational Joint Staff (Etat-major mixte opérationnel national, EMMONAT). It is further undermined by the gendarmerie's insistence on having a substantial presence in the island's urban centres. The result is direct competition among security services, rather than effective collaboration.[6]

Since the events of 2009, the theoretical division of labour in the triad has been undermined further with the creation of the Special Intervention Forces (Forces d'intervention spéciales, FIS) by the current transitional head of state, Andry Rajoelina. Composed of elements from the military and the gendarmerie, the FIS was supposedly tasked with combating rural banditry (*Madagascar Tribune*, 2009). In reality, the FIS has become a prominent actor in urban centres, notably the capital, where it is perceived as a security arm of the Rajoelina regime. Crucially, the existence of the FIS has not led to a reduction in the multitude of special forces operating in Madagascar. Instead, it constitutes another instance of an unconstitutional arrangement set up to reward officers who were instrumental in Rajoelina's coming to power (Rajaofera, 2010a).

major access point for weapon supplies in the so-called Frontline States, which were supporting nationalist movements such as the South West Africa People's Organization in Namibia, the African National Congress in South Africa, and the two Zimbabwe African National Union factions (Verrier, 1986; Leys, 1995).

Recognizing Madagascar's strategic position, external and internal forces alike actively impeded the armed forces from evolving into an effective and independent entity (Allen, 1987, pp. 145, 152–53; Milburn, 2004). This containment resulted in one of the most striking features of Madagascar's armed forces: the complete lack of a navy or operational coast guard. Indeed, even today, Madagascar's armed forces are not equipped for (or even reflect strategically about) defence of the territory from external threats (Jütersonke and Kartas, 2010, p. 68).

In short, the neo-colonial character of Madagascar's security sector not only undermined the formation of accountable, disciplined, and independent forces, but also sowed the seeds of division between Merina notables and *côtiers* aspirants.[7] Since the 1960s, internal disunity and external influence have been the fundamental characteristics of Madagascar's armed forces, exposing the army and gendarmerie to instrumentalization and co-option and, ultimately, transforming them into pawns in a political power struggle (Rakotomanga, 2004, pp. 71–84; 1998, p. 116; Rabenirainy, 2002).[8]

The politicization of the military and its control through dispersion

The division and competition among and within the different security forces were further strengthened with the demise of the First Republic in the early 1970s. Student revolts in May 1972 forced Tsiranana to hand over power to Gen. Gabriel Ramanatsoa and a transitional military government. Although Ramanantsoa accepted political power only reluctantly, officers soon recognized the establishment of a military directorate as an opportunity to remove the French technical military assistants and create new positions for young Malagasies (Allen, 1987, p. 162). The armed forces became highly politicized as a result (Archer, 1976, pp. 58–60).

> Madagascar does not have a navy or operational coast guard.

The assassination of Col. Richard Ratsimandrava and the mutiny of the Groupe Mobile de Police (the successors of the FRS) under Col. Bréchard Rajaonarison in February 1975 once again highlighted the continual tensions and political bickering that were splitting the armed forces into rival factions. On 15 June 1975, the military directorate nominated Frigate Captain Didier Ratsiraka as the new head of state, with the additional title of head of the Supreme Revolutionary Council (*Conseil Suprême de la Révolution*) (Archer, 1976, p. 155). Yet, despite all his rhetoric about socialist revolution and civilian government, Ratsiraka, who had come to power thanks to the armed forces, feared that a military coup would be his ultimate downfall if the top brass were not pleased with the benefits they accrued from his rule. Indeed, an effective separation of the armed forces from the civilian state administration would have forced many officers to return to their normal military functions and (lower) salaries, a change they wanted to avoid at all costs (Rakotomanga, 1998, p. 90).[9]

As a result, Ratsiraka did not opt for civilian oversight of the armed forces but sought to control the military through dispersion. Proclaiming himself 'admiral' in 1983,[10] he followed a three-pronged strategy to rid his presidency of any contestation of his rule from the ranks of the military (Rabenirainy, 2002, pp. 90–92). The first consisted of a 'banalization' of the military establishment through a redefinition of the armed forces as 'militants in uniform' fighting for the socialist cause. The second entailed a major restructuring of the army into three new joint staffs (*état-majors*), all of which had a primarily domestic orientation.[11] A general joint staff within the presidential office superseded these joint staffs, and a defence ministry was created simultaneously—but without clear lines of command or accountability. Ratsiraka's third strategy involved caring for the long-term careers of his high-ranking officers, who received high posts in the military development committee, the national revolutionary council, the state administration, and the nationalized enterprises. Indeed, a National Military Office for Strategic Industries (Office militaire national pour les industries stratégiques, OMNIS) was set up for this purpose, and officers deemed potentially subversive were sent for training to the Soviet Union and North Korea, the new 'cooperation partners' (Galibert, 2009, p. 80; Rabenirainy, 2002, p. 91).[12]

This was the setting the International Monetary Fund and the World Bank encountered when Madagascar began negotiating structural adjustment loans in the 1980s. No attempt was made to restructure and reform what had, by this time, become a completely ineffective and fragmented security apparatus. In fact, a variety of special armed units mushroomed, such as the much-feared General Directorate of Information and Documentation (Direction générale de l'information et de la documentation, DGID), a secret political police force created in the early days of Ratsiraka's rule. Instead of inter-agency collaboration, the national police entered into increasing competition with the gendarmerie in policing the urban centres (Rakotomanga, 1998, p. 90; Gow, 1997, pp. 417, 425). In contrast, rural areas—characterized by numerous *zones rouges* (red zones) to which the security forces had no access due to a lack of transportation and road networks— remained severely under-patrolled. The paradox of rising insecurity in a bloated police state reflected the highly centralized administration that had lost its capacity to govern in the periphery.

A Third Republic—but still no security sector reform

The student and civil servant movement of 1990–92, a result of the failed economic policies of Ratsiraka, paved the way for a new, liberal Third Republic, accompanied by great hopes for a democratic awakening of Madagascar. Next to the Malagasy Council of Churches, the military and retired generals were instrumental in negotiating the transfer of power to the Forces vives of Albert Zafy, who was proclaimed president of the High State Authority (Haute Autorité de l'Etat) in August 1992, before winning the presidential elections against Ratsiraka in February 1993. This intervention of military actors only demonstrated again, despite all claims of impartiality, just how deeply they continued to pursue their own interests in the national political and economic spheres (Rakotomanga, 1998, p. 116).

The 1992 constitution of the Third Republic, which set itself the task of creating a parliamentary regime focusing on a devolution of centralized power, soon transformed into a semi-authoritarian pres-

With Madagascar on the brink of civil war, forces loyal to Marc Ravalomanana patrol the area around the town of Ambilobe, 2 July 2002. © Mike Hutchings/Reuters

idential system through constitutional reform projects in 1995 and 1998 (Jütersonke and Kartas, 2010, pp. 46–50; Marcus, 2010, pp. 123, 128). The reform of the security forces remained largely unaddressed. No prerogatives were introduced for civilian leadership of the military and gendarmerie, for the separation of external defence and domestic security, or for parliamentary control of the armed forces. Generals Désiré Ramakavelo and Marcel Ranjeva introduced a number of changes under the Zafy administration, among them a White Book, which affirmed the neutrality of the armed forces, avoided all references to 'the citizen in uniform', and defined new aims adapted to global threats. Yet all of this was lip service rather than the start of an actual process of restructuring (Rabenirainy, 2002, p. 92).

The dynamic businessman Marc Ravalomanana became president in 2002, vowing to reform and depoliticize the armed forces (Rakotomanga, 2004, pp. 46–47).[13] But it was largely business as usual in the security sector. Above all, the lead-up to his presidency had placed serious doubts on the credibility of any reform pledges. When Ravalomanana proclaimed himself victorious after a first round of elections in February 2002, Ratsiraka declared martial law, came to an agreement with five of the country's six provincial governors, and designated the port city of Toamasina provisional capital (BBC, 2002b; Vivier, 2007, pp. 48–53). He also set up a physical blockade of Antananarivo, in what amounted to an economic embargo of the highlands (Raison, 2002, pp. 124–28). Madagascar was on the brink of civil war.[14]

The armed forces were split down the middle into Ratsiraka's *loyalistes* and Ravalomanana's *légitimistes*. Faced with a bloated officer corps and a lack of trained soldiers, each side had to resort to reservists, militias, and civilians, who received weapons and fought by proxy along the barricades (Rakotomanga, 2004, pp. 65–68; Vivier, 2007, ch. 8). In fact, instead of constituting a usable fighting force, soldiers and gendarmes profited from the blockades and the resulting chaos by colluding with criminal gangs to loot and rob the island and residents (Rakotomanga, 2004; Raison, 2002, p. 123).

Foreign pressure eventually prevailed, and Ravalomanana was confirmed as president, but his rhetoric in favour of democratic control of the armed forces was incompatible with his immediate need to assuage the *légitimiste* camp of the military to prevent a reunited army from turning against him (Rakotomanga, 2004, pp. 46, 56, 129). Any attempts to reform the corrupt and inept security forces would have required fundamental changes, threatening the personal interests of its officers. Instead of actively pursuing security sector reform, therefore, he tried simply to ignore and trivialize the armed forces during his first term in office. High-ranking officers eligible for promotion waited in vain, while other, lower-ranking officers jumped ranks when it suited Ravalomanana.[15]

By the time Western consultants had begun elaborating a sensible security sector reform package in late 2007, the army and gendarmerie were already so resentful of Ravalomanana's top-down style that any notions of reform were met with downright animosity.[16] Indeed, it was finally the mutiny of officers at the CAPSAT[17] army barracks in Antananarivo that forced him to hand over power to a military directorate in March 2009; those same officers then coerced the directorate into nominating Andry Rajoelina as the president of the High Transitional Authority (Haute Autorité de la Transition, HAT) that has ruled the country into 2011 (Rajaofera, 2010b; ICG, 2010a, p. 6).

The 2009 coup

The process by which President Marc Ravalomanana lost power in March 2009 is illustrative of the ways in which the security sector has itself become part of Madagascar's security challenges. Technically, Ravalomanana was not overthrown by a violent military coup but simply lost control of the security apparatus. The protest movement led by Andry Rajoelina had been brutally repressed on 7 February 2009 and the military and gendarmerie, under the direction of the joint staff, EMMONAT, had successfully quelled the riots in the capital. Andry Rajoelina quickly

Two men take cover as police fire tear gas and warning shots to disperse anti-government crowds, 16 February 2009. © Walter Astrada/AFP Photo

It remains unclear exactly how many people died when security forces guarding the presidential offices of Marc Ravalomanana opened fire on protesters, 7 February 2009. © Walter Astrada/AFP Photo

lost momentum again, not least because his protesters did not emerge from any sort of popular movement, but were recruited from the poor neighbourhoods and rural suburbs of Antananarivo in exchange for small sums of cash. The 'regime change' from Ravalomanana to Rajoelina and the HAT was only made possible by a mutiny of non-commissioned officers of the CAPSAT barracks on 8 March, a mutiny brought about by those whose business interests had been made to suffer under Ravalomanana's rule (*Midi Madagasikara,* 2011).

A few days before the mutiny, CAPSAT soldiers had refused to leave the barracks to perform patrol duties, as they had not received their regular allowances, which were being cashed in by high-ranking officers seeking to profit from the political turmoil. As an International Crisis Group report and confidential interviews had already hinted at in early 2010 (ICG, 2010a, p. 5, n. 41),[18] and as was recently confirmed by two leading figures of the CAPSAT unit (Rajaofera, 2010b; Razafison, 2011), a number of commissioned and non-commissioned officers received substantial amounts of money from economically influential individuals close to Rajoelina. These officers set up barricades on the main road leading up to the barracks, claiming that the presidential guard was planning to storm their base. A paralysed military, already at loggerheads with Ravalomanana following a reshuffling of the highest ranks, failed to act, and the emboldened CAPSAT mutineers went out to occupy the (empty) presidential palace.

On 17 March 2009, Ravalomanana resigned and handed over power to a military directorate, led by Vice Admiral Hyppolite

Ramaroson, loyal to Ravalomanana. Yet the directorate was short-lived. On the very day the presidential decree granted power to Ramaroson, the mutinous group 'kidnapped' the members of the military directorate in front of the rolling media cameras, shoved them into a van, and drove them to the CAPSAT barracks, where they were ordered to transfer all powers to Rajoelina (*Courrier de Madagascar,* 2010; YouTube, 2009).

The 2009 coup shed light on how predatory economic actors exploit a security sector that does not, on the whole, identify with the state and society that it is meant to serve. The fact that the country's government is a highly centralized apparatus, without effective control of large parts of its territory, facilitates rent-seeking in a country abundant with natural resources. Indeed, foreign and domestic businesses have always relied on their privileged access to the presidency and its ineffective administration to protect their market interests. Since the demise of the First Republic and the politicization of the military, the state administration has also constituted a major resource for an officer's economic and social advancement. In 2009, with the collusion of 'rent-maximizing' economic and military actors, the state itself finally became the central agent in an ethos of exploitation.

Madagascar is at risk of becoming a major hub for international trafficking.

POLITICAL INSTABILITY AND RISING INSECURITY

As a result of the events of 2009, Madagascar's state apparatus has effectively been incapacitated. As a 2010 World Bank report starkly illustrates, public expenditures have shifted towards the HAT presidency, while all government ministries, with the notable exceptions of the armed forces and the ministry of internal security, have faced drastic budget cuts (World Bank, 2010). For the better part of two years, all democratic institutions, including the parliament, have been mothballed, and an already struggling media sector has been further curtailed. The international community has also suspended the vast majority of its funding and development programming (Jütersonke and Kartas, 2010, pp. 52–60; ICG, 2010b, p. 2). Meanwhile, the rent-maximizing continues unabated, to the detriment of the country's ecology and internal security.

This section explores the major repercussions of this dynamic—armed criminality, rural banditry, and international trafficking. These troubles are not new, but they have been exacerbated by the events of 2009. The triad of a dysfunctional security sector, a lame state apparatus, and dynamic predatory actors is at the heart of these inter-related phenomena.

Armed criminality

Comprehensive crime statistics do not exist for Madagascar. Interviewed officials hint that available data is regularly manipulated when it is deemed useful for relations with the international community or to secure institutional funding for training or equipment.[19] Nevertheless, media reports and preliminary surveys suggest that the use of firearms in violent crime is on the rise (see Box 6.3). Moreover, there appears to be a high degree of under-reporting, either because victims do not contact the security forces or because the security forces are themselves directly or indirectly involved in the criminal acts committed. Indeed, interviews conducted with *chefs de fokontany* (elected neighbourhood representatives) in the port city of Toamasina and elsewhere paint a disturbing picture of groups of bandits wreaking havoc at night, without police intervention. This trend has led to the formation of neighbourhood watch initiatives—*andrimasom-pokonolona,* or self-defence units—which have taken the provision of security into their own hands (Rasoanaivo, 2010; *Express de Madagascar,* 2010). The scope and scale of these initiatives, however, remain unclear.

Small arms are circulating freely in Madagascar, adding to the fragility of an already dire situation. The state's security forces undoubtedly represent one source of these weapons. A complete lack of small arms stockpile management is one of the most striking features of contemporary Madagascar's security sector. Assault rifles regularly disappear from arsenals during times of political crisis–never to be returned, as was the case when reservists and militia forces were armed in 1991 and 2002 and, again, according to confidential interviews, during the events of 2009 (Rakotomanga, 2004, p. 69; Hauge, 2005, p. 14; Rabako, 2010, p. 14). Interviews conducted with members of the military, gendarmerie, and police corroborate media reports that many officers rent out firearms while off duty to supplement low salaries. Indeed, a recent investigation undertaken by the gendarmerie reportedly found proof of such misconduct (*Les Nouvelles*, 2010a).

The incidence of armed robbery appears to be on the rise. Media reports compiled for this study reveal that, during the month of May 2010, attacks with firearms made up around two-thirds of all crimes reported, constituting an increase of more than 40 per cent compared to the previous year. Yet while automatic pistols and assault rifles are reportedly being used in criminal activities, the majority of arms being seized by custom officials at Toamasina, Madagascar's largest port, are hunting rifles (Rabako, 2010, p. 14). Given Antananarivo's thriving black market for small arms–where anything from a revolver to an AK-47 can be bought or rented– allegations over the widespread involvement of current and former members of the state's security sector appear quite credible (Rabako, 2010, p. 15; Andriamarohasina, 2010b; *Midi Madagasikara*, 2010). According to confidential sources, more than 1,500 permits for civilian firearms have been issued in the inner city of Antananarivo alone, with eligibility criteria becoming less, rather than more, stringent.

A full-scale survey on armed violence and security perceptions in Madagascar has yet to be designed and conducted. Only a small, preliminary (and unpublished) survey was commissioned in early 2010 as part of the UN's Peace and Conflict Impact Assessment. Questionnaire interviews (n = 80) and a number of informal focus groups were conducted in 15 *fokontany* of Antananarivo and in two *fokontany* of the small highland town Ambatolampy, apparently the island's main site of craft production of small arms. Of the respondents, 71 per cent said that they lived in insecurity and social disorder, and 67 per cent named political instability as the main source of this turmoil. Three-quarters of the respondents stated that the situation was deteriorating. While the sample is small and unrepresentative, it matches insight generated by a survey of Antananarivo that was also conducted in early 2010, in the context of the UN Multi-Cluster Rapid Assessment Mechanism. According to this survey, 50.7 per cent of households (n = 500) consider the lack of security a major concern (UN, 2010).

In this context, it is troubling that, as one high-ranking informant exclaimed during an interview in early 2010, 'Madagascar has no security but many securities!' Indeed, today's Malagasy security forces are characterized by a dizzying array of special units geared towards muscular interventions, while the day-to-day activities of enforcing law and order—and, above all, *preventing* crime—are largely neglected. In the capital, the national police created a rapid intervention group, an anti-gang service, a special intervention unit, and a special intervention brigade, each with its own mandate and territorial jurisdictions. Many such entities can also be found in France, where some members of these elite corps have been trained (*Midi Madagasikara*, 2005; Rakotomanga, 1998; 2004).[20] Regular forces, by contrast—be they police, gendarmerie, or army—are under-represented. As Organès Rakotomihantarizaka, minister of internal security, pointed out in September 2010, Madagascar has one police officer for every 3,000 citizens—while the international norm is three well-equipped police officers for every 1,000 citizens (*Les Nouvelles*, 2010b).

The rural quagmire: *zones rouges* and *dahalo*

Insecurity in Madagascar is far more than an urban phenomenon. Indeed, Madagascar's population is still predominantly rural, and, in the face of large-scale cattle rustling and with an increasing number of economic predators taking control of parts of the island's territory,

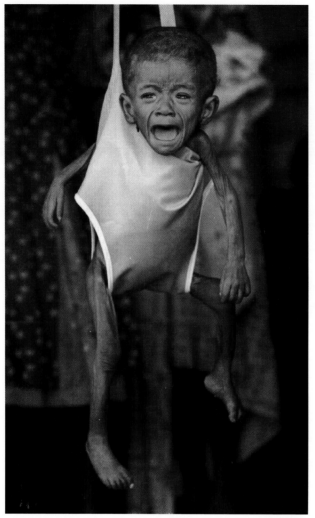

A one-year-old famine victim cries at a feeding centre in Amboasary, southern Madagascar, 1992. Famine continues to haunt parts of the island in 2011. © Alexander Joe/AFP Photo

rural communities are the ones suffering most from the current lawlessness. While the gendarmerie and specially designed military units are operating in rural areas, they may be poorly equipped, lack means of transportation, or even collude with attackers. Regularly undermined by spontaneous roadblocks, the *routes nationales* have become so unsafe that travel along certain sections is now only permitted in a convoy that is escorted by the gendarmerie; bus passengers are prohibited from using their mobile phones, in order to avoid tip-offs (Andriamarohasina, 2010a).

It is remarkable to what extent Madagascar continues to present an inhospitable terrain for any force trying to patrol and control it. These *zones rouges* encompass large areas of territory that are inaccessible to security personnel. Providing disaster relief is thus extremely difficult for the Malagasy state apparatus, with famines breaking out in the south of the island and cyclones hitting the eastern shoreline with devastating regularity.[21] But even on a day-to-day basis, Madagascar's topography, coupled with a fragmented road network of dirt tracks that is unusable during the rainy season, makes the provision of rural security a daunting affair. The most persistent source of insecurity continues to be the *dahalo,* the Malagasy word for bandit, typically used to designate the rustlers of zebu, the local hump-backed cattle.

While cattle theft is a crime punishable by severe penalties,[22] *dahalo* raids are a daily occurrence in many parts of the country. A 2001 survey on insecurity in rural areas estimated that 1.78 zebus were being stolen per week per commune; extrapolating that figure to Madagascar's 1,392 communes yields almost 130,000 stolen zebus per year (Programme ILO, 2003, p. 2). In 2006, 3,668 cases of cattle theft were brought before penal courts created specifically to deal with *dahalo* activities (Ignace, 2010). More recent media reports suggest raiding has increased further since the crisis of 2009. In the region of Mahajanga alone, media accounts have documented more than 160 *dahalo* attacks, with more than 3,000 zebus stolen between May and July 2010. It also appears increasingly likely that the *dahalo* phenomenon involves collusion with elements of the armed forces and international organized crime (*Madagascar Tribune,* 2010).[23]

Some have sought to downplay the importance of the *dahalo* by arguing that their activities constitute a ritualized form of cattle raiding that has been part and parcel of certain regions of Madagascar for centuries—similar to the practices observed with the Karimajong in northern Uganda. Indeed, studies have shown that for the Bara people in the south of the island, the stealing of zebu belonged to a complex system of reciprocal village rituals to channel aggression, secure the rite of passage from boy- to manhood, and fulfil the preconditions for marriage.[24] Moreover, the *dahalo* phenomenon must be understood in the larger historical context and the push and pull factors explaining the rise or decline of banditry in Madagascar. The *dahalo* are a modern type of criminal armed group that did not emerge out of traditional rites of passage (although young men might indeed find their path into banditry through these rites); specifically, they developed as a consequence of the oppression of the population under the Imerina Kingdom and its extensive reliance on forced labour and the slave trade to consolidate its rule (Rasamoelina, 2007, pp. 64–67, 94–98).

Today, the escalation of *dahalo* criminality and the resulting insecurity in rural areas is linked to the appropriation of cattle theft into a modern large-scale venture, with the capacity to launder the documentation of cattle, stock the cattle (sometimes for years) in the herds of influential and wealthy cattle owners (some have herds of more than 10,000 animals), and organize their transportation to the main harbours for trafficking (McNair, 2008, p. 16; Fauroux, 1989, p. 72).[25] In contemporary Madagascar, cattle raiders play an important role in trafficking dynamics and the circulation of weapons on the island. The largest and most powerful *dahalo* groups provide assault rifles (mainly AK-47s) and ammunition to all their members (*Madagascar Tribune,* 2010). They escort, in broad daylight, large herds of stolen cattle, carrying their guns openly over several hundred kilometres while crossing entire regions and provinces of the island. As many interviewees pointed out in 2010, these illiterate *dahalo* would not have the organizational, logistical, or financial capacity to conduct such operations without the support of influential networks reaching high into the echelons of the state administration. A new trend witnesses bandits operating in larger units of about 30 men, rather than the small groups of five to six that were common previously. They are now conducting veritable raids, taking women and children as hostages and burning down houses (*Madagascar Tribune,* 2010).

> Cattle raiders play an important role in the circulation of assault rifles.

The escalation of the *dahalo* phenomenon is emblematic of the fragmentation and co-option by economic interests of Madagascar's security sector. It seems beyond doubt that many gendarmes and soldiers accept bribes to turn a blind eye, rent out their firearms, and collaborate directly with the *dahalo*. Since the famous Keliberano case of 1989 (in which *dahalo* were sentenced to death for the killing of 11 people), it came to light that military figures were involved in cattle rustling operations and connected to arms trafficking. As noted in the *Country Reports on Human Rights Practices for 1989*:

> *The presiding judge [in the Keliberano case] called for an investigation and a hearing on these charges. On November 15 the court of Fianarantsoa rendered light sentences of 6 months to a year for those involved in supplying the weapons used in the Keliberano killings* (USDOS, 1990, p. 196).

As a result, security forces are not keen on having *dahalo* testify in court. Such verdicts may also explain why special gendarmerie and army units, when they are sent out to hunt down *dahalo,* tend to gun down their targets on the spot, rather than attempting to capture them alive.[26]

The main impediment to substantially reducing the recurrence of cattle raids continues to be the lack of policing, investigative capacities, and means of transportation of the security forces. None of the gendarmerie outposts have helicopters or functioning four-wheel-drive vehicles at their disposal. Catching a *dahalo* is thus a tricky endeavour, made even more ineffective by the fact that the detainee must then be marched on foot to the nearest courthouse, often several days away. There is thus ample opportunity for the detainee to escape or be freed by his group. Even if that does not happen, collusion with the local administration often results in the prisoners being acquitted in court or being allowed to escape from prison.

In an effort to counter its incapacity, the state administration has in the past decade not only encouraged the organization of village self-defence groups but also tolerated the operations of indigenous private security companies, called *zama,* hired by the villages to hunt down *dahalo* 'illegally' (Hogg, 2008). The backlash to this aggressive and highly punitive approach has been the *dahalo*'s radicalization and increasing reliance on firepower.

Illicit trafficking by sea

Despite more than 5,000 km of shoreline, the Malagasy state apparatus still has no navy or coast guard, nor functioning helicopters with which to patrol its borders. The remote beaches and small coastal hamlets and villages are littered with large quantities of foreign-looking waste—such as cans and plastic water bottles. The trash comes not from locals trudging through the bush and onto the beach, nor from the small trickle of eco-tourists who are still brave enough to visit. The litter comes from people arriving by sea.

Madagascar and, in particular, smaller adjacent islands, such as Ile St. Marie, have for centuries been a refuge for pirates and smugglers (Stadelmann, 2009). It has also, given its location, been a place of geo-strategic importance for numerous regional and global powers (Allen, 1987). Spotting a foreign ship in

Illegal drugs and counterfeit bills are often seized together, as here in Antananarivo, September 2010. © Moncef Kartas

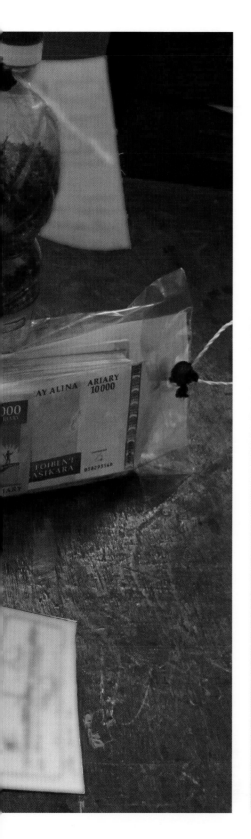

Box 6.4 Madagascar's 'rosewood massacre'

One of the few contemporary issues facing Madagascar that has caught the attention of the global public is the rampant illegal logging of rosewood, often in the country's nature reserves. The island is blessed (or cursed) with more species of rosewood than any other place in the world, with most of them unique to the island and dozens still unnamed and in need of urgent taxonomic work–the first step in efforts to save them from extinction.

The trafficking of precious woods is not a new phenomenon in Madagascar; but at no point has it reached the scale witnessed since the political crisis in early 2009. Commentators thus rightly speak of a rosewood 'massacre', with Global Witness estimating that the illegal Malagasy timber trade may be worth up to USD 460,000 per day (Global Witness, 2009; Schuurman and Lowry, 2009). According to the non-governmental organization Vakan'ala, 100,000 hectares of the island's forests–estimated at a total of 8.5 million hectares in 2005–are currently being lost per year (Vakan'ala, 2010; Muttenzer, 2010, p. 15). Of greatest concern is the disappearance of the eastern rainforests, home to most of the island's endemic species of flora and fauna. According to statistics of the Food and Agriculture Organization, the annual rate of primary forest loss in Madagascar is 0.65 per cent, three times higher than in Indonesia (EIA, 2010).

This massacre is organized, armed, and conducted with the direct and indirect collusion of elements of the current transitional regime. The trade is allegedly financed with advance payments from Chinese buyers and by loans from several international banks with branches in Antananarivo (Gerety, 2009). Heavily armed 'militias' are regularly reported to be operating in the rainforests, threatening park staff and villagers, and escorting the logs to makeshift ports on Madagascar's uncontrolled coastline, most notably at Vohémar (York, 2009). The identity of these militias remains a mystery, though there was wide-spread consensus among individuals interviewed that members of the country's armed forces are collaborating with the exporters. One undercover investigation quotes the commander at a military post at the entrance to the Masoala Nature Reserve as saying: 'I do not work for the government. I am an independent' (*GEO Magazin*, 2010). In September 2010, it was reported that a foreign journalist was arrested by the police and, in the presence of rosewood traffickers, forced to delete images of timber stockpiles from his camera (Wild Madagascar, 2010). And, in December 2010, a military truck carrying 50 rosewood logs was reportedly intercepted by police forces in the south of the island (Limbira, 2010).

In early 2010, the environment minister, Edelin Callixte Ramiandrisoa, was cited in the media as proudly proclaiming that, in the space of just two months, the state's coffers had received MGA 30 billion (USD 15 million) from the sale of 300 containers (roughly 30,000 logs) of rosewood (Rabary-Rakotondravony, 2010). These were supposedly legal exports derived from trees that had been brought down by cyclones or taken from old, existing stocks. But the 'cyclone' currently sweeping over the island is man-made. As long as the current political impasse is not overcome, ecological impunity will reign supreme, and armed actors willing to engage in coercion and violence to protect their economic interests will continue to fuel insecurity and hollow out the state's capacity to safeguard and promote the well-being of its people.

its territorial waters has always been common-place for the Malagasy living on the island's picturesque shores. Yet, in the political vacuum that has flourished since early 2009, the odd ship sighting has given way to a veritable invasion.

Apart from naval vessels, the 'illicit' ships that skirt the Malagasy coast can be placed into three rough categories: fishing trawlers, container ships, and regional oil tankers. To simplify somewhat, the first raid Madagascar's seafood stocks, the second cart away the island's natural resources, and the third, it is claimed, use their half-empty hulls to transport stolen cattle and other such commodities around the island.

What do ships bring with them when they arrive to cart away rosewood (see Box 6.4), lemurs, tortoises, and precious stones and metals? Speculations abound in Antananarivo, but one reoccurring rumour is that Madagascar is fast becoming a major hub for the global drug trade. Indeed, the allegation that the island is a transhipment point for heroin has been around for a while (CIA, 2010; Hübschle, 2008). 'Drug mules' are regularly caught by the authorities (*La Gazette de la Grande Ile*, 2010). While they have no proof, many locals suspect the government's involvement through collusion or tacit cooperation.

The political economist Ronen Palan has coined the term 'antisovereign spaces' to describe some of the conditions favouring international criminal networks (Palan, 2009, p. 36). Such operations are very much territorially oriented: they thrive in areas where central authority is weak but where the means of capital accumulation is still the traditional one of slowly amassing ill-gotten

The small port of Vohémar is alleged to be one of the main shipping points for rosewood trafficking. © Toby Smith/EIA/Getty Images

profits (as opposed to the focus of the modern economy on anticipated future earnings). As a result, their operations seek to prolong a fragile state's competitive advantage as a transhipment hub for illicit sectors by co-opting elements of the regime and the security forces. For every day that Madagascar's political crisis is not solved and that the dysfunctional security sector prevails, criminal actors will continue to consolidate their position on the island.

CONCLUSION

Since the island's independence in 1960, a host of factors have prevented the Malagasy armed forces from becoming a professional and disciplined security apparatus. Regional strategic considerations of the former colonial masters generally outweighed calls for the establishment of security forces that could meet the needs of an island defined by long coastlines and difficult terrain. As a result, the hypothetical division of labour and jurisdiction between the army, gendarmerie, and police were never upheld in practice, and, instead of precisely delineated activities, the security sector became increasingly embroiled in the political and economic spheres. Indeed, the military was never tasked or equipped to perform its constitutional mandate of defending the territory from external threats. It has been confined to a purely domestic role and, in the absence of a sense of duty, personal gain became the ultimate rationale for joining and pursuing a career in the security forces. Not surprisingly, such venality has attracted—and has relied on—political and economic predators who are eager to pursue their own interests.

A striking feature of post-independence Madagascar is that it has never been possible to address instability or public discontent through party politics. Instead, the armed forces were increasingly drawn into the void left by an ineffective polity; unprepared and internally divided, however, they lacked the discipline and unity to act as a stabilizing force. Since the politicization and instrumentalization of the armed forces, consecutive governments and the state administration have fallen prey to military careerists and the economic interests of both domestic and foreign actors. Yet the lack of state capacity to enforce the law, to protect and guard the island's coasts, and to put an end to the continuous outflow and misuse of small arms from its arsenals has not been a major impediment to doing businesses—quite the contrary. With the current transitional government not internationally recognized, the lines between legal extraction and illicit trafficking begin to blur; when faced with charges of buying and reselling raw materials that were not extracted according to international or national regulations, foreign companies react by showing officially stamped documentation legitimizing the transaction.

Madagascar's media reports appear to intimate that the government, the security sector, and major business interests have all merged into one. The distinctions between public and private and between legal and illicit seem to have evaporated. Armed robbery, often committed with military weapons, illicit trafficking with probable state assistance, and vigilantism are increasingly common (Andriamarohasina, 2010c; *Les Nouvelles,* 2010c). Meanwhile, the armed forces are prospering. To celebrate 50 years of independence in June 2010, Rajoelina raised the retirement age for the armed forces by a year, giving its top brass an additional bonus (Ranjalahy, 2010). In December 2010, a further 29 officers were promoted to the rank of general (Iloniaina, 2010).

As of this writing, with a few exceptions, the majority of bilateral and multilateral development programmes remain suspended (ICG, 2010b, p. 2, n. 8). Such work, it is said, requires a legitimate national partner. Just as in 2002, when the country was on the brink of civil war, the international community has adopted a wait-and-see strategy, sitting out the crisis and hoping for the return of an elected government so that its development cooperation can recommence. In the meantime, abuse of power, economic predation, ecological degradation, and a suffering population continue to be the key characteristics of the world's fourth-largest island. ◾

LIST OF ABBREVIATIONS

CAPSAT	Corps des personnels et des services administratifs et techniques (Army Corps of Personnel and Administrative and Technical Services)
EMMONAT	Etat-major mixte opérationnel national (Mixed National Operational Joint Staff)
FIS	Forces d'intervention spéciales (Special Intervention Forces)
FRS	Force Républicaine de Sécurité (Republican Security Force)
HAT	Haute Autorité de la Transition (High Transitional Authority)
PSD	Parti Social Démocrate (Social Democratic Party)

ENDNOTES

1   Stadelmann (2009); Covell (1995, pp. 225–26); Prou (1987, ch. 3); Allen (1987, pp. 134, 137).

2   Confidential author interviews, Antananarivo, 2010.

3   Confidential author interview, Antananarivo, 2010.

4   Confidential author interviews, Antananarivo, 2010; see also Archer (1976, p. 49); Rabenoro (1986); Rakotomanga (1998, p. 32).

5   Confidential author interviews, Antananarivo, 2010.

6   Confidential author interviews, Antananarivo and Toamasina, 2010.

7   While there is no real ethnic conflict between Merina and *côtiers,* the distinction has long been exploited for political or career ends; see Archer (1976, pp. 24–26).

8   Confidential author interviews, Antananarivo, 2010.

9   Confidential author interviews, Antananarivo, 2010.

10   In declaring himself admiral, Ratsiraka became the highest-ranking officer in the history of Madagascar (albeit with no navy to command).

11   These new joint staffs were the *armée de développement, forces aéronavales,* and *forces d'intervention.* Rakotomanga (1998, p. 89); Archer (1976, p. 160); Gow (1997); Rabenirainy (2002).

12   Confidential author interviews, Antananarivo, 2010.

13   Confidential author interviews, Antananarivo, 2010.

14   See also BBC (2002a); Randrianja (2005, p. 17).

15   Confidential author interviews, Antananarivo and Berlin, 2010.

16   Confidential author interviews, Antananarivo, Berlin, and Munich, 2010.

17   CAPSAT stands for Corps des personnels et des services administratifs et techniques (Army Corps of Personnel and Administrative and Technical Services).

18   Confidential author interviews, Antananarivo, 2010.

19   Confidential author interviews, Antananarivo, 2010.

20   Confidential author interviews, Antananarivo, 2010.

21   In December 2010, the US Agency for International Development gave USD 3 million in emergency food aid to southern Madagascar, where, according to UN predictions, 720,000 people will probably be affected by severe drought. Indeed, more than half of the island's children are stunted as a result of chronic malnutrition, a situation that is worse only in Afghanistan and Yemen (McNeish, 2010).

22   The law against cattle theft dates from 1960, the year of independence (Madagascar, 1960). Depending on the severity of the attack (as determined by factors such as the number of bandits, whether the raid was violent, whether it resulted in injury or death, whether it took place at night or during the day, and whether it involved false documents or the unauthorized use of uniforms), penalties range from five years to life-long forced labour or even the death penalty.

23   Confidential author interviews, Antananarivo, 2010.

24   Beaujard (1995, p. 572); Fauroux (1989, p. 67); Rasamoelina (2007, p. 223); McNair (2008).

25   Confidential author interview, Antananarivo, 2010. As early as 1985, the media reported that cattle were being smuggled to the Comores, Mauritius, and Reunion with the collusion of various levels of the 'civil and military hierarchies' (*Le Monde,* 1985; cited in Rasamoelina, 1993, pp. 29–30).

26   Confidential author interviews, Antananarivo, 2010.

BIBLIOGRAPHY

Allen, Philip. 1987. *Security and Nationalism in the Indian Ocean: Lessons from the Latin Quarter Islands.* Boulder: Westview Press.

Althabe, Gérard. 1980. 'Les luttes sociales à Tananarive en 1972.' *Cahiers d'études africaines,* pp. 407–47.

Andriamarohasina, Seth. 2010a. 'Madagascar: Opération coup-de-poing contre les bandits des routes nationales.' *Express de Madagascar.* 9 April. <http://fr.allafrica.com/stories/201004050860.html>

—. 2010b. 'Insécurité: Les bandits s'équipent en armes de guerre.' *Express de Madagascar.* 24 April.

—. 2010c. 'Ivato: Deux militaires lynchés.' *Express de Madagascar.* 29 October.

Archer, Robert. 1976. *Madagascar depuis 1972: la marche d'une révolution.* Paris: L'Harmattan.

Badie, Bertrand. 2000. *The Imported State: The Westernization of the Political Order.* Stanford: Stanford University Press.

BBC (British Broadcasting Corporation). 2002a. 'Largest Military Clash in Madagascar.' 4 June. <http://news.bbc.co.uk/2/hi/africa/2022811.stm>

—. 2002b. 'Madagascar Governors Stand Firm.' 23 April. <http://news.bbc.co.uk/2/hi/africa/1945443.stm>

Beaujard, Philippe. 1995. 'La violence dans les sociétés du sud-est de Madagascar.' *Cahiers d'études africaines,* pp. 563–98.

CIA (Central Intelligence Agency). 2010. *World Factbook.* <https://www.cia.gov/library/publications/the-world-factbook/fields/2086.html>

Courrier de Madagascar. 2010. 'Interprétation du 17 mars 2009: Un an déjà et toujours la discorde.' *Courrier de Madagascar.* 18 December.

Covell, Maureen. 1995. *Historical Dictionary of Madagascar.* Lanham, MD: Scarecrow Press.

Deschamps, Hubert. 1972. *Histoire de Madagascar.* Paris: Berger & Levrault.

EIA (Environmental Investigation Agency). 2010. 'Luxury Market Fuelling Destruction of Madagascar's Forests.' Press release. 26 October. <http://www.eia-international.org/cgi/news/news.cgi?t=template&a=616>

Express de Madagascar. 2010. 'Faits divers: Vol matinal.' 23 December.

Fauroux, Emmanuel. 1989. 'Bœufs et pouvoirs: Les éleveurs du sud-ouest et de l'ouest malgaches.' *Politique Africaine,* No. 34, pp. 63–73.

Galibert, Didier. 2009. *Les Gens du pouvoir à Madagascar: Etat postcolonial, légitimités et territoire (1956–2002)*. Paris: Karthala.

GEO Magazin. 2010. 'Illegaler Holzhandel: Raubbau am Regenwald.' No. 04/10.

Gerety, Rowan More. 2009. 'Major International Banks, Shipping Companies, and Consumers Play Role in Madagascar's Logging Crisis.' MongaBay.com News. 16 December. <http://news.mongabay.com/2009/1215-rowan_madagascar.html>

Global Witness. 2009. 'Illegal Malagasy Timber Trade Worth up to $460,000 a Day.' Press release. 2 December.
    <http://www.globalwitness.org/media_library_detail.php/890/en/illegal_malagasy_timber_trade_worth_up_to_460000_a>

Gow, Bonar. 1997. 'Admiral Didier Ratsiraka and the Malagasy Socialist Revolution.' *Journal of Modern African Studies*, Vol. 35, No. 3, pp. 409–39.

Hauge, Wenche. 2005. *Madagascar—Past and Present Political Crisis: Resilience of Pro-Peace Structures and Cultural Characteristics*. Report to donors. Oslo: International Peace Research Institute, Oslo. June.

Hogg, Jonny. 2008. 'Cattle "War Zone" in Madagascar.' BBC News. 21 June.
    <http://news.bbc.co.uk/2/hi/programmes/from_our_own_correspondent/7465726.stm>

Hübschle, Annette. 2008. 'Drug Mules: Pawns in the International Narcotics Trade.' Pretoria: Institute for Security Studies. 18 November.
    <http://www.iss.co.za/pgcontent.php?UID=13845>

ICG (International Crisis Group). 2010a. 'Madagascar: sortir du cycle de crises.' *Africa Report*, No. 156. Brussels: ICG. 18 March.

—. 2010b. 'Madagascar: la crise à un tournant critique?' *Africa Report*, No. 166. Brussels: ICG. 18 November.

Ignace, Rakato. 'L'insécurité rurale liée au vol de bœufs: quelques propositions de solution.' *TALOHA*, No. 19. 30 January.
    <http://www.taloha.info/document.php?id=906>

IISS (International Institute of Strategic Studies). 2009. *Military Balance 2009*. London: Routledge.

Iloniaina, Alain. 2010. 'Forces armées: Promotion massive de généraux.' *L'Express de Madagascar*. 16 December.
    <http://www.lexpressmada.com/forces-armees-madagascar/19182-promotion-massive-de-generaux.html>

Jütersonke, Oliver and Moncef Kartas. 2010. *Peace and Conflict Impact Assessment (PCIA) for Madagascar*. Geneva: Centre on Conflict, Development and Peacebuilding.

La Gazette de la Grande Ile. 2010. 'Trafic à Maurice: Une Malgache arrêtée avec 1,5 milliard Fmg de drogue.' 7 October.
    <http://www.lagazette-dgi.com/index.php?option=com_content&view=article&id=6796:trafic-a-maurice-une-malgache-arretee-avec-15-milliard-fmg-de-drogue&catid=41:politique&Itemid=55>

Le Monde. 1985. 'Les voléurs de zébus.' 30 March.

Les Nouvelles. 2010a. 'Banditisme: des contrats de location d'armes au menu.' 22 July. <http://www.les-nouvelles.com/spip.php?article2556>

—. 2010b. 'Banditisme: un réseau de malfaiteurs est démantelé par l'USI.' September. <http://www.les-nouvelles.com/spip.php?article4208#>

—. 2010c. 'Ankatso-I: Un voleur lynché par les étudiants.' 16 October.

Leys, Colin and John S. Saul, eds. 1995. *Namibia's Liberation Struggle: The Two-Edged Sword*. London and Athens, OH: James Currey and Ohio University Press.

Limbira, Mosa. 2010. 'Bois de rose . . . à Fort-Dauphin aussi!', *Moov*. 20 December. <http://www.moov.mg/actualiteNationale.php?articleId=648398>

Madagascar. 1960. Ordonnance No. 60-106. 27 September. *Journal Officiel*, No. 124. 1 October, p. 1949.

Madagascar Tribune. 2009. 'En attendant le FIS: Sécurité rurale à Ankazoabo Sud.' 1 July.

—. 2010. 'Contre les Dahalo, la gendarmerie préconise l'autodéfense villageoise.' Madagascar-Tribune.com. 4 September.
    <http://www.madagascar-tribune.com/Contre-les-Dahalo-la-gendarmerie,14648.html>

Marcus, Richard. 2010. 'Marc the Medici? The Failure of a New Form of Neopatrimonial Rule in Madagascar.' *Political Science Quarterly*, Vol. 125, No. 1, pp. 111–31.

McNair, John. 2008. 'Romancing *Dahalo*: The Social Environment of Cattle Theft in Ihorombe, Madagascar.' *ISP Collection*. 1 April.
    <http://digitalcollections.sit.edu/isp_collection/69>

McNeish, Hannah. 2010. 'US Gives $3 Million for Emergency Food Aid Program in Madagascar.' Voice of America. 13 December.
    <http://www.voanews.com/english/news/africa/east/-US-Gives-3-Million-for-Emergency-Food-Aid-Program-in-Madagascar-111803984.html>

Midi Madagasikara. 2005. 'Police: 3 corps d'élite opérationnels.' *Midi Madagasikara*. 29 December. <http://fr.allafrica.com/stories/200512300028.html>

—. 2010. 'Trafic d'armes: Un colonel et des "zazavao" dans le collimateur.' *Midi Madagasikara*. 4 May.

—. 2011. 'Révélations du Lt-col Charles : 2 milliards pour un double coup d'Etat le 17 mars 2009.' 6 January.
    <http://www.midi-madagasikara.mg/index.php?option=com_content&view=article&id=8280:revelations-du-lt-col-charles-2-milliards-pour-un-double-coup-detat-le-17-mars-2009&catid=12:a-la-une>

Milburn, Sarah. 2004. 'La Chasse Gardée: Post–World War II French West Africa, 1945–1970.' In Edward Rhodes et al., eds. *Presence, Prevention, and Persuasion: A Historical Analysis of Military Force and Political Influence*. Lanham, MD: Lexington Books, pp. 197–280.

Muttenzer, Frank. 2010. *Déforestation et droit coutumier à Madagascar: Les perceptions des acteurs de la gestion communautaire des forêts*. Paris: Karthala and the Graduate Institute of International and Development Studies.

Palan, Ronen. 2009. 'Crime, Sovereignty, and the Offshore World.' In H. Richard Friman, ed. *Crime and the Global Political Economy*. Boulder, CO, and London: Lynne Rienner, pp. 35–48.

Programme ILO. 2003. *The Security Situation Post-crisis: The Impact of the New Policies to Improve Security*. Post-Crisis Policy Brief No. 4. Ithaca, New York: Cornell University. January. <http://www.ilo.cornell.edu/polbrief/pbpc4.pdf>

Prou, Michel. 1987. *Malagasy: 'un pas de plus', vers l'histoire du royaume de Madagascar au XIXe siècle [1793–1894].* Paris: L'Harmattan.

Rabako, Aina. 2010. 'Armes à vendre.' *L'Hebdo de Madagascar*, No. 271. 23 April.

Rabary-Rakotondravony, Lova. 2010. 'Le grand défi des éco-citoyens pour 2010.' *L'Express de Madagascar.* 5 January.
<http://fr.allafrica.com/stories/201001041909.html>

Rabenirainy, Joana. 2002. 'Les forces armées et les crises politiques (1972–2002).' *Politique africaine*, No. 86, pp. 86–101.

Rabenoro, Césaire. 1986. *Les relations extérieures de Madagascar de 1960 à 1972.* Paris: L'Harmattan.

Rabinow, Paul. 1995. *French Modern: Norms and Forms of the Social Environment.* Chicago: University of Chicago Press.

Raison, Jean-Pierre. 2002. 'Economie politique et géopolitique des barrages routiers.' *Politique africaine*, No. 86. June, pp. 120–37.

Rajaofera, Eugène. 2010a. 'Madagascar: Col Charles Andrianasoavina—"Le FIS, sans bureau ni statut."' *Midi Madagasikara.* 9 March.

—. 2010b. 'Coup de force du 17 mars 2009: Révélations du Gal Noël Rakotonandrasana.' *Midi Madagasikara.* 7 November.
<http://www.midi-madagasikara.mg/index.php?option=com_content&view=article&id=6717:coup-de-force-du-17-mars-2009--revelations-du-gal-noel-rakotonandrasana&catid=12:a-la-une>

Rajoelina, Patrick. 1988. *Quarante années de la vie politique de Madagascar: 1947–1987.* Paris: L'Harmattan.

Rakotomanga, Mijoro. 1998. *Forces armées malgaches: entre devoir et pouvoir.* Paris: L'Harmattan.

—. 2004. *Forces armées malgaches face à la crise 2002.* Paris: L'Harmattan.

Randrianja, Solofo. 2001. *Société et luttes anticoloniales à Madagascar: de 1896 à 1946.* Paris: Karthala.

—. 2005. *Ravalomanana, 2002–2005: Des produits laitiers aux affaires nationales—FAST Country Risk Profile Madagascar.* Swisspeace Working Paper. Berne: Swisspeace.

— and Stephen Ellis. 2009. *Madagascar: A Short History.* Chicago: University of Chicago Press.

Ranjalahy, Sylvain. 2010. 'Retraite aux lambeaux.' *L'Express de Madagascar.* 28 June, p. 5.

Rasamoelina, Henri. 1993. 'Le vol de bœufs en pays Betsileo.' *Politique Africaine*, No. 52, pp. 22–30.

—. 2007. *Madagascar: état, communautés villageoises et banditisme rural—l'exemple du vol de zébus dans la Haute-Matsiatra.* Paris: L'Harmattan.

Rasoanaivo, Anjara. 2010. 'Tana ville : La population sur le qui-vive.' *Midi Madagasikara.* 18 November.

Razafindranaly, Jacques. 2000. *Les soldats de la Grande île: d'une guerre à l'autre, 1895–1918.* Paris: L'Harmattan.

Razafison, Rivonala. 2011. 'Soldier: I was paid $10,000 to topple Marc Ravalomanana.' AfricaReview.com. 6 January.
<http://www.africareview.com/News/-/979180/1084666/-/view/printVersion/-/11urxva/-/index.html>

Schraeder, Peter. 1994. 'Independence, the First Republic, and the Military Transition, 1960–75.' In Library of Congress. *A Country Study: Madagascar.* Washington, DC: Library of Congress. <http://memory.loc.gov/frd/cs/mgtoc.html>

Schuurman, Derek and Porter P. Lowry II. 2009. 'The Madagascar Rosewood Massacre.' *Madagascar Conservation & Development*, Vol. 4, Iss. 2. December, pp. 98–102.

Stadelmann, Franz. 2009. *Madagaskar: Das PRIORI-Buch.* Antananarivo: PRIORI.

UN (United Nations). 2010. *Situation socioéconomique des ménages de la ville d'Antananarivo et impact de la crise sociopolitique au niveau des ménages en mai 2010, Madagascar.* UN Multi-cluster Rapid Assessment Mechanism. Antananarivo: UN. June.

USDOS (United States Department of State). 1990. *Country Reports on Human Rights Practices for 1989*, Vol. 1989. Washington, DC: US Government Printing Office, p. 196. <http://www.archive.org/details/countryreportson1989unit>

Vakan'ala. 2010. Website. <http://vakanala.org/en>

Verrier, Anthony. 1986. *The Road to Zimbabwe 1890–1980.* London: Jonathan Cape.

Vivier, Jean-Louis. 2007. *Madagascar sous Ravalomanana: la vie politique malgache depuis 2001.* Paris: L'Harmattan.

Wild Madagascar. 2010. 'Police in Eastern Madagascar Arrest Foreign Journalist Investigating Illegal Timber Trafficking.' 17 September.
<http://news.mongabay.com/2010/0917-madagascar_arrest.html>

World Bank. 2010. 'Madagascar—Economic Update: A Closer Look at Three Strategic Areas.' Antananarivo: World Bank. October.
<http://siteresources.worldbank.org/INTMADAGASCAR/Resources/EcoUpdate_Oct.pdf>

York, Geoffrey. 2009. 'The Gangs of Madagascar.' *Globe and Mail.* 18 April.
<http://www.theglobeandmail.com/news/world/the-gangs-of-madagascar/article1138495/>

YouTube. 2009. 'Kidnapping à l'Episcopat antanimena.' 23 March. <http://www.youtube.com/watch?v=Rl0_MG1Yz_I>

ACKNOWLEDGEMENTS

Principal authors

Oliver Jütersonke and Moncef Kartas

Soldiers of the Forces Nouvelles wait for their chief in May 2009, prior to a ceremony in Bouaké marking the transfer of power from the comzones to government prefects.
© Kambou Sia/AFP Photo

Reforming the Ranks

PUBLIC SECURITY IN A DIVIDED CÔTE D'IVOIRE

7

INTRODUCTION

On the eve of 2011, Côte d'Ivoire plunged into yet another deep political crisis. On the streets of Abidjan, the internationally recognized winner of the November 2010 presidential elections, former prime minister Alassane Ouattara, took refuge in the Golf Hotel. The incumbent, Laurent Gbagbo, occupied the Presidential Palace, refusing to step down, while repeated clashes between security forces and political supporters in the capital claimed close to 200 lives in December 2010 (Munzu, 2011). The country ushered in the new year with two governments conducting business in parallel and operating in an atmosphere of mutual hostility and violence.

Among the factors influencing the wave of post-election violence is the failure to implement the security provisions of the 2007 Ouagadougou Political Agreement (OPA). The disarmament and demobilization of the rebels and the pro-ruling party militias, scheduled for completion two months prior to the elections, were far behind schedule. Meanwhile, the Integrated Command Center (ICC), a combined government–rebel force designed to serve as a pilot for the future 'New Army' and to provide security during the electoral process, has remained symbolic in nature, to the point that its existence is seriously jeopardized by the current political crisis.

The role of the military in Ivorian politics has become increasingly central since independence in 1960. Not only are today's security forces involved in the current wave of violence, but control over them also represents an important factor in the ongoing political stand-off between Ouattara and Gbagbo. The security sector continues to be key to bringing the crisis to an end and reunifying the country, goals that depend to a large extent on a successful restructuring of the armed forces.

Since 2002, a ceasefire line has divided the Republic of Côte d'Ivoire into a rebel-held area in the north and the government-run south (see Map 7.1). The country is therefore subject to a very peculiar system of governance, with two security apparatuses, two treasuries, and two administrations. As such, Côte d'Ivoire presents a rare opportunity to study not only a complex process of post-conflict security sector reform, but also a dual system of security provision—by rebels on the one hand and an official state administration on the other. The main findings of the chapter include:

- Across Côte d'Ivoire, the population lacks confidence in its security forces; however, people within the Centre Nord Ouest (CNO) zone in the north exhibit a greater level of distrust in their Forces Nouvelles (FN) than do those living in the south with respect to the state security forces.
- The types of insecurity that prevail in the government-controlled area and the rebel-held zone are relatively similar, including banditry and resource-based conflict.
- Although the perception of insecurity in the rebel-run area is higher, civilians in the government zone are as likely to become victims of armed violence.
- While the majority of incidents of armed violence in the CNO zone are perpetrated with assault rifles, most incidents of armed violence in the government zone involve bladed weapons, followed by handguns and assault rifles.

- The deficiencies of the security forces and the level of insecurity encourage civilians to provide their own security through community self-defence and vigilante groups, which in turn create new forms of insecurity. The private security sector has grown rapidly and without regulation.

- So far, security sector reform efforts in Côte d'Ivoire have focused on the reunification of the security apparatus rather than on addressing the lack of democratic oversight, strategic objectives and professionalism, or logistical weaknesses of the security forces.

- The creation of the new, unified armed forces has generated optimism comparable to that projected on the post-colonial military; however, 50 years later, new challenges have reduced the capacity of the military to be an 'agent of modernization'.

Since the war, the private security sector has grown rapidly and without regulation.

The chapter begins with an outline of the political and security evolution of the country since its independence. It continues by analysing the two security sectors and their respective achievements as security providers. The chapter then explores the dynamics of insecurity in both zones and highlights the alternative civilian mechanisms that have been developed to address it. The final section of the chapter provides an assessment of the post-conflict reunification process of the national security sector and draws a parallel between the security transition at the time of independence and the current reform phase, which the country has timidly entered.

While the themes of security sector reform (SSR), disarmament, demobilization, and reintegration (DDR), privatization of security, and regional peacekeeping dynamics have been widely explored in the context of English-speaking Africa, research on francophone Africa is less developed. Literature on the Ivorian security forces, for example, is scarce; not a single reference book is available on the Ivorian military or on the rebellion. The chapter therefore relies largely on field research carried out by the author in Côte d'Ivoire in February and March 2010 and draws on various methodological tools, including a national survey of 2,600 households,[1] focus groups, and interviews conducted by the author with key informants.[2] This chapter is intended to contribute to the developing literature on the topic.

THE PATH TO THE PRESENT

Since gaining independence in 1960, Côte d'Ivoire has maintained strong ties with France, particularly in commerce and security. The tendency to rely on the former colonial power as the 'guarantor of (its) sovereignty' (Bagayoko, 2010, p. 16) through defence agreements, coupled with the determination of the political leadership to suppress any potential challenge to its authority, severely undermined the development of the military system of Côte d'Ivoire. Its limited operational capacity was clearly exposed in 2002, when rebels managed to seize control of a substantial part of the national territory. This section begins with a historical account of the crisis and then examines the evolution of the relationship between the military and the political leadership since independence.

A brief history of the crisis

In contrast to the majority of West African states and sub-Saharan French-speaking countries, Côte d'Ivoire experienced many years of political stability following its independence in 1960. The one-party system and the sizeable global market share of production and export of cocoa and coffee contributed to the country's economic prosperity. Yet with the gradual erosion of the 'Ivorian miracle' brought on by economic recession in the 1980s and the succes-

Map 7.1 **Côte d'Ivoire**

sion of problems that followed the death of long-time president Félix Houphouët-Boigny in 1993, severe political and community tensions emerged, rooted in fierce debates about ethnicity and identity.

Under the presidency of Henri Konan Bédié (1993–99), Houphouët-Boigny's successor, the discourse of ethnic exclusion gave birth to the notion of *Ivoirité,* which sought to distinguish 'true' Ivorians from inhabitants of non-Ivorian origin (particularly from other countries of the sub-region). Tensions culminated in 1999 with the overthrow of Bédié in a military putsch that gave power to Gen. Robert Guéï for ten months (December 1999–October 2000). This

coup constituted the beginning of a period
of turmoil marked by further attempted coups
and 'the radicalization of political repression'[3]
(Banégas, 2010, p. 367).[4] The 2000 elections,
in which Bédié (of the Parti Démocratique
de Côte d'Ivoire) and Alassane Ouattara (of
the Rassemblement des Républicains) were
barred from participating, were eventually
won by Laurent Gbagbo[5] (of the Front
Populaire Ivoirien) after the other major
candidate, Gen. Guéï, fled the country. Post-
election violence claimed the lives of more
than 200 and injured hundreds of people,
the vast majority of whom were reportedly
supporters of Ouattara's party (HRW, 2001).

In September 2002, a group of 700 men,
composed of soldiers exiled in Burkina Faso
and others still serving in the Ivorian Army—
the Forces Armées Nationales de Côte
d'Ivoire (FANCI)—some of whom had participated in Guéï's coup, launched a putsch
to oust Gbagbo (ICG, 2003, p. 1). Having
failed to capture Abidjan, the commercial
capital, the mutineers retreated to Bouaké,
formed the Mouvement Patriotique de Côte
d'Ivoire, and seized control of several cities
in the north of the country. The intervention
of French troops under Opération Licorne
(Operation Unicorn) thwarted a second
attempt to capture Abidjan and the capital
Yamoussoukro (ICG, 2003, p. 1). In November
2002, two other rebel groups emerged in the
western part of the country, where they
managed to seize control of several towns.
Shortly afterwards, the three rebel groups
joined to form the Forces Nouvelles, a movement that continues to control the CNO
zone to this day. While the FN has a hold over
roughly 60 per cent of the national territory

Box 7.1 Chronology of key events since independence

7 August 1960 Independence from France; Félix Houphouët-Boigny
becomes president

28 October 1990 First multi-party elections are held; Houphouët-
Boigny beats Laurent Gbagbo

7 December 1993 Houphouët-Boigny dies and is succeeded by Henri
Konan Bédié

25 December 1999 Bédié is overthrown in a coup d'état led by
Gen. Robert Guéï

26 October 2000 Gbagbo becomes president, defeating incumbent Guéï

19 September 2002 Rebels seize control of the north following an
attempted military coup to oust Gbagbo

October 2002 As part of Opération Licorne, French troops deploy to
intervene in the north-south conflict

26 January 2003 Linas-Marcoussis Agreement: President Gbagbo
and FN sign a compromise in Paris

13 May 2003 UN Mission in Côte d'Ivoire (MINUCI) is established to
implement the Linas-Marcoussis Agreement (UNSC, 2003)

4 July 2003 Government and FN sign an 'End of War' declaration

April 2004 United Nations Operation in Côte d'Ivoire (UNOCI)
replaces MINUCI

November 2004 Ivorian loyalist forces attack French position in
Bouaké; French troops destroy Ivorian military aircraft; UN imposes
an arms embargo (UNSC, 2004)

3–6 April 2005 Peacekeeping talks between government and rebels
are held in Pretoria

4 March 2007 Ouagadougou Political Agreement; Guillaume Soro,
FN leader, becomes prime minister

31 October 2010 First round of presidential elections (already post-
poned six times since 2005)

28 November 2010 Second round of presidential elections: Gbagbo
vs. Ouattara

2 December 2010 Electoral Commission declares victory for
Ouattara (54 per cent of votes)

3 December 2010 Constitutional Council declares victory for
Gbagbo (51.45 per cent)

6 December 2010 Both candidates form respective governments

16 December 2010 First violent, post-electoral clashes between
security forces and Ouattara supporters

(Balint-Kurti, 2007, p. 16), the population density is much higher in the government-controlled areas in the south (RCI, 2009a, para. 291).

In June 2003, six months after the signing of the Linas–Marcoussis Peace Agreement in France (see Box 7.1), a buffer zone was imposed across the country and supervised by the Force Licorne in collaboration with soldiers from the Economic Community of West African States (ECOWAS) peacekeeping mission and the United Nations Mission in Côte d'Ivoire (MINUCI). ECOWAS and MINUCI military personnel were subsequently integrated into the United Nations Operation in Côte d'Ivoire (UNOCI), which was deployed in 2004. Licorne remained under French command but was assigned a support role to UNOCI by UN Security Council Resolution 1721 (UNSC, 2006c).

In July 2003, a ceasefire was signed by both parties; since then, it has been breached repeatedly. In November 2004, for example, the loyalist forces—those who continue to support Gbagbo—launched an air strike on a French military base in rebel-held Bouaké, killing nine French soldiers. In retaliation, French troops destroyed the Ivorian military air fleet, which consisted of two Sukhoi aircraft as well as one MI-8 and five MI-24 helicopters (AFP, 2004), further deteriorating the Franco-Ivorian relationship. A few days later, French troops opened fire on a hostile crowd in Abidjan, killing several people and leading to violent anti-French riots, which hastened the departure of thousands of Western expatriates from the country. That same month, the UN Security Council imposed an arms embargo on the country (UNSC, 2004, para. 7).

While the period of actual armed conflict was relatively short—from September 2002 to July 2003—and did not affect all the regions, an estimated 3,000 people died and a further 750,000 were displaced (Duval Smith, 2010). A large part of the battles took place in the western part of the country, where people are still seriously suffering from the legacy of the conflict. Together with the region of Moyen Cavally, where ethnic tensions turned into deadly clashes after the 2010 presidential elections, Abidjan experienced the worst political violence in Côte d'Ivoire.

> A ceasefire fire was signed in 2003 but it has been repeatedly breached by both parties.

Politics and the armed forces

The Ivorian security apparatus emerged out of a peaceful transition from colonial power to the new independent government, rather than from a struggle for independence. Modelled on the French system, it was largely influenced by the French Fifth Republic's interpretations of the balance of power. That model was characterized by very strong executive power and limited parliamentary control over the security sector, as clearly reflected in the Ivorian constitutional provisions of 1960 and 2000. The Ivorian president 'is the guardian of national independence and of the sovereign integrity of the country', is the Supreme Commander in Chief, and is authorized to take 'exceptional measures' in times of crisis (RCI, 2000, arts. 34, 47, 48). Some analysts claim that this institutional arrangement may encourage authoritarian rule (Bagayoko, 2010, p. 18; N'Diaye, 2010, p. 88).

Houphouët-Boigny maintained a modest army that he paid well to prevent any kind of challenge to his power. He also integrated certain higher-ranking officers into positions of administrative responsibility in the government and in state-owned companies (Conte, 2004, p. 11; Kieffer, 2000, p. 30). On succeeding Houphouët-Boigny in 1993, Bédié relied heavily on the police and the gendarmerie for support and marginalized the army. While Houphouët-Boigny had attempted to achieve a regional balance of military staff, Bédié reserved key positions for officers of his own ethnic group, the Baoulé from central Côte d'Ivoire. This further deepened the gap between the president and an army that was made up largely of people from the west and north of the country (Kieffer, 2000, pp. 33, 36). Security forces thus became increasingly divided into specific ethnic groups, reflecting the mounting identity-related tensions that were rippling through wider Ivorian society at the time. The frustration of the officers culminated in the putsch of 1999 (Kieffer, 2000, p. 36), marking the official entry of the military onto the political stage. During the period of military transition led by Gen. Guéï, there was a notable surge in the establishment of militias within the army, each of which reflected specific political sympathies (R. Ouattara, 2008, pp. 76–7).

After Gbagbo's arrival in power, internal struggles and parallel chains of command continued to develop, exposed by several mutinies and the attempted coup of 2002. To suppress any further potential challenge to the regime, the president, of Bété origin, undertook a *bétéisation* of key positions in the armed forces and sponsored the rapid promotion of young officers to ensure their loyalty (Boisbouvier, 2005). In order to cope with the military weaknesses and to confront the rebellion, the government resorted to private security actors such as foreign mercenaries (Banégas, 2010, p. 361). Furthermore, coercive power was increasingly being handed to both pro-government militias and state paramilitary forces; between 2002 and 2009, the size of the gendarmerie nearly doubled, from 8,000 to 15,000 (Mieu, 2009a) while their operating zone was divided in half along with the division of the country (see Map 7.1).

Other types of paramilitary forces also emerged. One is the Centre de Commandement des Opérations de Sécurité (CECOS), an elite unit of 600 personnel from different security forces; created by presidential decree in 2005, CECOS is commanded by a senior officer from the gendarmerie and armed with FANCI small arms and light weapons (Mieu, 2009a; RTI, 2010a). While CECOS was originally set up to fight organized crime, it has occasionally been assigned

The leader of the Young Patriots, Charles Blé Goudé (right), and Gen. Philippe Mangou (second from the right) arrive for a rally in support of the Ivorian armed forces at Champroux Stadium in Abidjan, January 2011. © Thierry Gouegnon/Reuters

other missions, such as providing security at elections or crowd control at demonstrations, illustrating the lack of a clearly defined mission for the security forces. Finally, as in most former French colonies, the Republican Guard—known as the Presidential Guard in some countries—is also an important asset of the regime. Though well equipped,[6] it is not typically accountable to military authorities and therefore constitutes one of 'the major symptoms of the system of competing security agencies and parallel chains of command which characterise the military in Francophone Africa' (Bagayoko, 2010, p. 29).

The historically rooted rivalries among the different security forces are evident today; while the gendarmerie is regarded as loyalist, the military is still viewed by those in power as a potential threat to the regime (Bagayoko, 2010, p. 40). During the jubilee celebration of independence in 2010, President Gbagbo addressed the last part of his speech to the army officers attending the event, warning them: 'If I fall, [the officers] fall, too. Some people think that a coup is easy! It is a building supported by several pillars. If you try to make it fall, pillars will fall, too' (Gbagbo, 2010).[7] The 2010 post-electoral crisis revealed how control over the army played an important role in the political stand-off between Gbagbo and Ouattara. It also showed that the better-armed and more highly trained 'elite' bodies of the security forces commanded by pro-Gbagbo officers—such as the Republican Guard, CECOS, and the presidential security teams—were crucial to his strategy of staying in power (Airault, Kouamouo, and Meunier, 2011).

Control over elite bodies of state security forces is crucial to Gbagbo's staying in power.

ONE COUNTRY, TWO SECURITY SYSTEMS

Côte d'Ivoire has been divided in two since 2002 and has been subjected to a very peculiar system of security governance made of two distinct security mechanisms. The following sections examine the fragmentation of the security system and its impact on the capacity of the armed forces as security providers.

The decline of the state security sector

The economic recession and the politico-military crisis that the country has endured since 1999 have had an enormous impact on the security sector. Today's security forces lack the capacity and ability to protect people and their assets; corruption and impunity are widespread; and the justice system is non-functional.

The security infrastructure in Côte d'Ivoire is a near carbon copy of the French system. The main security providers are the army and law enforcement agencies—the police and gendarmerie (see Table 7.1). The police operate as a civilian force under the Ministry of the Interior, providing security in urban zones, while the gendarmerie operates as a military body under the auspices of the Ministry of Defence and is responsible for security in rural areas. Yet the mandates of the different bodies overlap and all conduct general policing tasks. In addition, the geographical boundaries that used to separate the police from the gendarmerie have vanished such that it is no longer uncommon to see gendarmes operating alongside police at urban checkpoints.

Previously, each force was supplied with different types of weapons; today these distinctions no longer apply. The crisis ensured a redistribution of weapons among the police, customs, and water and forestry guards, all of whom were armed with assault rifles in support of the army.[8] These weapons are still carried by police on the streets, due in part to the lack of handguns, whose acquisition the 2004 embargo rendered illegal.

The political crisis has largely destabilized the state security sector and has weakened its capacity; security forces have fallen victim to institutional corruption and lack technical and material resources, training, and discipline.[9]

Table 7.1 **Main state security and defence bodies and their weapons***

Body	Supervising ministry	Strength	Status	Armament
Army	Ministry of Defence	18,000	Military	Small arms and light weapons and conventional weapons
Air Force	Ministry of Defence	700	Military	Small arms and light weapons; air fleet not operational since 2004
Navy	Ministry of Defence	900	Military	Small arms and conventional weapons
Gendarmerie	Ministry of Defence	15,000	Military	Small arms and light weapons and conventional weapons (Bagayoko, 2010, p. 40)
Republican Guard	Ministry of Defence	1,350	Military	Small arms and light weapons
Police	Ministry of Interior	17,000 (11% of whom are women)	Non-military	Small arms
Water and forestry	Ministry of Environment	Not available	Military	Small arms (R. Ouattara, 2008, pp. 81-82)
Customs	Ministry of Economy	2,300	Military	Small arms (R. Ouattara, 2008, p. 81)
CECOS	Ministry of Defence	600	Mixed unit	Small arms and light weapons

Note: *This table presents the most recent available estimates for each security body.

Sources: Army: Jeune Afrique (2010); Air Force, Navy, and Republican Guard: IISS (2010, p. 302); Collier (2009, p. 159); gendarmerie: Mieu (2009a); police: UNDP (2010); customs: Bagayoko (2010, p. 44); CECOS: Mieu (2010)

Following the collapse of the economy, military positions are now highly coveted and recruitment is largely based on nepotism (Kieffer, 2000, p. 34). The armed forces are seen as a refuge from unemployment and have thus undergone 'socio-economic ethnicization' (A. Ouattara, 2008, p. 165).

Racketeering and bribery, frequent at security checkpoints, have been identified by the Chamber of Commerce in Abidjan as among the primary causes of a rise in the general costs of living.[10] This criminal activity reduces the competitiveness of Ivorian products and encourages the diversion of international goods to other ports. Racketeering also brings with it serious security implications as agents neglect their primary missions to engage in it; the activity is also associated with violence against women and racism and discrimination against foreigners (Touré, 2008, p. 10). Finally, when security forces demand bribes for completing their tasks, the concept of security as a public good is undermined. The inhabitants of Abidjan, for example, generally have to pay CECOS to respond to their calls for assistance.[11]

While the security sector fails to protect the population, the security of the country as a whole is not assured either. Although the Liberian conflict ended in 2003, customs authorities have not yet redeployed their forces to the central-western border, as this would require the provision of important logistical and financial resources.[12] Customs posts on

Ivorian police drive past burning tyres in Abidjan during a protest by Ouattara supporters, December 2010. © Issouf Sanogo/AFP Photo

the northern border are also non-existent and cross-border cooperation in the sub-region remains limited. Cross-border trafficking, including of weapons, fuels the instability of the country.

While the mandate and performance of the security forces require urgent review, so does the entire justice system. In Côte d'Ivoire in 2010, there was one tribunal for every two million people and one magistrate for every 40,000 inhabitants (RTI, 2010b). The lack of tribunals, coupled with high fees, widespread corruption, and a low level of public awareness of relevant laws, repeatedly calls into question the population's access to justice (RCI, 2009a, p. 25). The result is a greater reliance on traditional mechanisms of justice and mediation. One provincial chief says that 'modern justice is too expensive and slow'; people regularly seek his rulings on cases of adultery, land conflict, and acts of violence.[13]

Further, prisons within the government zone are overcrowded and lack security. For example, although Abidjan's Centre for Detention and Correction was designed to hold approximately 1,500 detainees, it held more than 5,100 inmates by 2009 (LIDHO, 2010, p. 2). In addition, custody periods can be long, with people often waiting several months before going to court.[14] According to the penal authorities, 176 inmates escaped from 22 detention centres in the government zone in 2009 (RCI, 2009b).

The Forces Nouvelles

Since 2002, the Forces Nouvelles have dominated the northern part of the country, which is divided into ten zones, each controlled by a military commander called a *comzone* (*commandant de zone*). In 2004, the FN counted around 25,000 armed members (ICG, 2004, p. 25). According to the census conducted by the National Programme for Reintegration and Community Rehabilitation (Programme national de réinsertion et de réhabilitation communautaire) in 2009, FN strength is approaching 33,000 people, including both combatants and non-combatants.[15] These figures are most probably inflated, as is common in similar post-conflict settings where non-state armed groups seek to gain greater political leverage and positions in the army in the event of integration with the national security forces. Indeed, only one-third of the 33,000 Forces Nouvelles members would be able to fight; the majority are too young, too old, or otherwise unsuitable to fight as combatants (ICG, 2009, p. 10).

The FN authorities comprise a civilian secretariat and a general staff that control the FN Armed Forces (Forces Armées des Forces Nouvelles, FAFN). The ten *comzones* constitute the FN's 'military powerbase' (UNSC, 2009, para. 35). Under his command, each *comzone* has a 'small private army' whose role it is to maintain territorial control over each zone (UNSC, 2009, para. 36; see Table 7.2). A total of 3,000 soldiers make up these different teams (ICG, 2009, p. 10).

While the *comzones* and their respective security outfits are specific to the FAFN, the configuration of the rebel forces in Côte d'Ivoire is to some extent based on the FANCI model. Indeed, FANCI soldiers who defected and formed the FAFN[16] were able to share their knowledge of the regular army; they have comparable military ranks, their arsenals are very similar (see Box 7.2), and the FAFN often use former FANCI barracks and armouries as their own. Moreover, the FN system for the provision of security is almost exclusively based on the military. Yet it seems that the FN police

Comzone Chérif Ousmane (centre) of the Forces Nouvelles speaks with his soldiers in September 2005.
© Issouf Sanogo/AFP Photo

Zone	Location	Comzone	Alias	Military unit name
Table 7.2 The ten comzones and their 'small private armies'				
1	Bouna	Morou Ouattara	Atchengué	Atchengué
2	Katiola	Touré Hervé Pélikan	Vetcho/Che Guevara	Bataillon mystique
3	Bouaké	Chérif Ousmane	Guépard	Les Guépards
4	Mankono	Ouattara Zoumana	Zoua	Various
5	Séguéla	Ouattara Issiaka	Wattao	Anaconda
6	Man	Losseni Fofana	Loss	Cobra
7	Touba	Traoré Dramane	Dramane Touba	Various
8	Odienné	Ousmane Coulibaly	Ben Laden	Various
9	Boundiali	Koné Gaoussou	Jah Gao	Various
10	Korhogo	Martin Kouakou Fofié	Fofié	Fansara 110

Source: UNSC (2009, p. 16)

and gendarmerie may not serve much more than a symbolic purpose, namely to give the impression of a regular, functioning security presence.

Since the FN never intended to secede from the country, the movement has not created a new legislative apparatus. While the FN secretariat is made up of different 'cells', including justice, education, and environment,[17] it appears to be inefficient. For example, it is impossible to obtain administrative documents such as birth certificates or to apply for a gun licence. In contrast, the FN economic system is reasonably organized; the central treasury of the FN administers an efficient taxation system on trade and transport and cities such as Korhogo represent genuine economic centres that attract foreign investment, in particular from Burkina Faso and Mali.

The judicial system has also suffered from the division of the country, which has left many of the security and judicial structures non-functional. Despite the partial redeployment of administrative and judicial authorities to the north of the country in 2010, the FN continues to settle disputes and dispense justice (UNOCI, 2009, p. 16). Many focus group participants in Odienné reported that whenever they faced a problem, they first went to the district chief and, if he was unable to resolve the issue, they turned to the FN brigade. One participant explained: 'The FN brigade sends people [. . .] but you pay a price. You know there'll be no prison sentence so things will be "settled".'[18] Odienné's only prison is not operational. An FN member from the ICC in Odienné reported:

> We are doing everything we can to keep people safe, and we are keeping criminals locked in cells until the prison reopens. The comzone used to help us to transfer people to Bouaké, where there is a prison, but it's not easy because you need a truck, fuel, and personnel.[19]

In the region of 18 Montagnes, one FN *commandant de secteur* (*comsecteur*) commented that:

> When it gets to our level, we come to an amicable agreement. We have no criminal procedure. The civilian prison is not open because there is no criminal procedure, so we 'relocate the problem', for example, by escorting people to the border.[20]

Box 7.2   Weapons used by government and rebel forces

A significant number of the weapons in the armouries of the Ivorian defence and security forces are French, evidence of the two countries' long history of military cooperation.[21] Indeed, up until the collapse of the Soviet Union in 1991, France was Côte d'Ivoire's main commercial partner for military materiel (see Table 7.3). In the aftermath of the cold war, former Soviet states found themselves in possession of important stockpiles of surplus materiel and conflict zones in West Africa offered a lucrative market (Berman, 2007, p. 4). Expert reports confirm that military equipment has been imported into the country in violation of the arms embargo of 2004.[22]

The FAFN arsenal is comparable to that of the state security forces. With the partition of the country in 2002, the FAFN rebels not only laid their hands on armouries in the CNO zone, but also seized weapons left on the battlefield and guns left behind by police officers, gendarmes, and FANCI, who had abandoned them as they fled from the north to the government-held area.[23]

Some differences in arsenals indicate that the FAFN may have acquired weapons from alternative sources. Specifically, Russian, Sudanese, and two other unidentified types of ammunition have been found in FAFN stocks, accounting for a large part of their 7.62 x 39 mm ammunition (UNSC, 2009, paras. 138, 144). The fact that the serial numbers have been removed from thousands of assault rifles in their possession suggests that a foreign state may have been involved in arming the rebels (UNSC, 2009, para. 130). In terms of quantities, however, 'despite a lack of airpower, state security forces have overwhelming superiority in arms, ammunitions and military equipment over the Forces nouvelles' (UNSC, 2009, para. 32).

Finally, both sides suffer from weak stockpile management; several explosions of armouries have been reported and weapons have been diverted from stocks in the north and in the south in the past few years (RASALAO-CI, 2008; Abidjan.net, 2010). Visits to different armouries in 2010 showed that basic security rules were not respected and facilities were in poor condition. Reunification of the armed forces would demand significant efforts in terms of securing and managing stockpiles.

Table 7.3   Key state security weaponry

Arms		Type	Origin
Small arms	Handguns	MAB pistols	France
	Assault rifles	ARM	Bulgaria
		Type 56	China
		Type 83	China
		MAS 49/56	France
		MAT 49	France
		SIG 540 and 543	France
		AKM	Soviet Union/Russian Federation
		AK-47	Soviet Union/Russian Federation
	General-purpose machine guns	PKM	Soviet Union/Russian Federation, China
Light weapons	Heavy machine guns	NSV 12.7 mm	Soviet Union/Russian Federation
		DSHK 12.7 mm	Soviet Union/Russian Federation
		KPV 14.5 mm	Soviet Union/Russian Federation
	Automatic grenade launcher	AGS-17	Soviet Union/Russian Federation
	Rocket launcher	LRAC	France
	Mortars	60 mm and 81 mm	France
	MANPADS	SA-7b	Bulgaria

Source: author interview with a member of the UN Group of Experts on Côte d'Ivoire pursuant to Security Council Resolution 1893 (2009), Abidjan, February 2010

Perceptions of security forces

Human rights organizations in Côte d'Ivoire have repeatedly denounced the atmosphere of impunity in the country.[24] They cite, on the one hand, weak logistical capacity, ethics, and professionalism within the security sector and, on the other, a general reluctance on the part of the population to report crimes, exposing a widespread lack of confidence in the security apparatus.

The national survey and focus groups conducted in Côte d'Ivoire in 2010 have revealed that the population harbours contradictory feelings towards the security forces. A significant part of the population has no trust in the security forces; nevertheless, people generally advocate an increase in the number of security posts throughout the territory as the first step in combating insecurity and illicit weapons circulation.

The low standing of the security forces may be influenced by the abuses committed by both the government and FN forces during the conflict that erupted in 2002. According to Amnesty International, both sides committed serious human rights violations, particularly against women (AI, 2008, p. 24). The decision of the parties not to set up a transitional justice mechanism to deal with the violations committed during the war has contributed to the widespread sense of impunity.

In fact, there is a fundamental lack of confidence in the security forces across the entire territory.[25] Only 45 per cent of those in the government zone who responded to the household survey would report having been the victim of violent crime to the security forces in the hope that something would be done (see Figure 7.1). In the CNO zone, the percentage is significantly lower at 27 per cent, with almost the same percentage of respondents choosing to leave their fate 'in God's hands'. While a lack of trust in the security authorities clearly discourages people from reporting acts of criminality, the survey reveals that the long distances that separate people from their nearest security posts, especially within the CNO zone, is also a major discouraging factor.

For those who participated in the focus groups in the CNO zone, the security situation has largely deteriorated since the eruption of the war and the subsequent partition of the country. When household survey participants living in the CNO zone were asked about ways of combating the illicit circulation of weapons, the most popular responses were to 'complete DDR' (28.2 per cent, n=604) and 'redeploy state authority' (16.8 per cent, n=604).[26] Many participants

Figure 7.1 **Responses to victimization**

Responses (percentage) to the question 'What course of action would you take if you had been the victim of a violent crime?' in the government zone (n=1,782) and in the CNO zone (n=658)

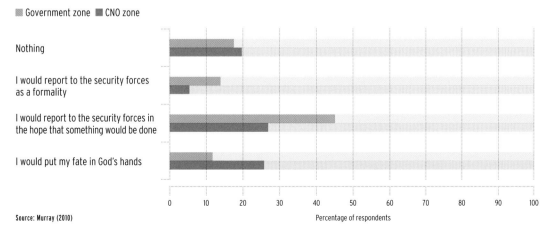

Government zone   CNO zone

in the north also expressed hope for an improved security situation once state security forces had restored control over the rebel-held areas.

The results of the national household survey show that the most trusted authorities nationwide are traditional ones—the chiefs in particular—as they are generally identified as being more efficient and accessible than security forces (Murray, 2010).

INSECURITY IN CÔTE D'IVOIRE

The crisis and the resulting weak security governance have provided fertile ground for armed violence and criminality to flourish. This section compares the scope and types of insecurity in the government- and rebel-held zones. It assesses the validity of the commonly held assumption that insecurity—both real and perceived—are greater in areas held by the rebels.

Figure 7.2 **Principal concerns**

Three principal concerns cited by the respondents to the national survey in the government zone (n=1,808) and in the CNO zone (n=673)

■ Government zone ■ CNO zone

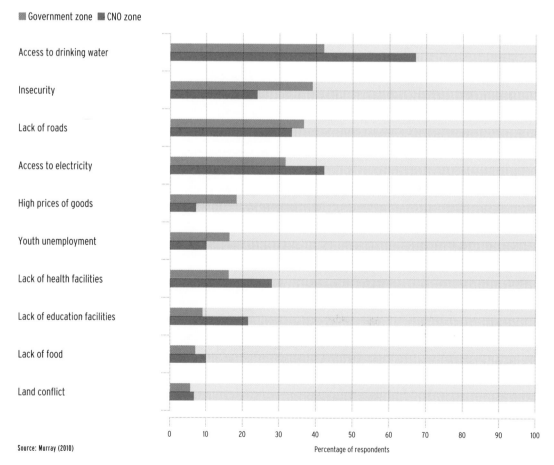

Source: Murray (2010)

Percentage of respondents

Insecurity: a primary concern for the population

An analysis of the data acquired from the 2010 national household survey identifies insecurity as one of respondents' primary concerns, along with a lack of access to basic needs such as drinking water, electricity, and food (see Figure 7.2). Yet, while insecurity is the second most frequently cited problem in the government zone, it ranks fifth for those in the CNO zone. Respondents in the CNO zone primarily reported concerns related to access to water, electricity, a lack of road and transport facilities, and the need for hospitals and medical supplies. Indeed, since independence, the northern part of the country has benefited far less from investment; public infrastructure there has suffered greatly from the partition of the country and the absence of any public administration (RCI, 2009a, pp. 21–23).

People living within the CNO zone do not generally identify insecurity as one of their primary concerns; nevertheless, the overall sense of insecurity is higher in the CNO zone than it is in the government zone, especially outside the home (see Figure 7.3).

The 'victimization' section of the survey indicates that people are as likely to fall victim to armed violence[27] in the government-run areas as in those held by rebels.[28] As in many other countries, criminality is higher in towns and cities than in the countryside. With an estimated six million inhabitants, Abidjan experiences types of criminality common in big cities (RCI, 2009a, p. 64; see Table 7.4).

Both of the major hotspots in terms of insecurity are in the southern part of the territory: Abidjan and the Moyen Cavally. Western Côte d'Ivoire, which suffered more from the war and its aftermath than other parts of the country, is divided into the two regions of 18 Montagnes (rebel-held) and Moyen Cavally (state-held). Insecurity appears to be particularly high in Moyen Cavally (ICG, 2010a, pp. 10–11); the region has a high incidence of violent armed highway robberies around Guiglo and aggravated burglaries in towns (HRW, 2010, pp. 31, 34). Moyen Cavally's security situation is threatened by active armed militias and by tensions between communities of different origins. Weapons are in high circulation and law and order is almost non-existent (p. 6). The region is also one of the main producers of cocoa, wood, and coffee and is therefore a major hub for the transport of cash and goods, which attracts highway criminals.

Figure 7.3 Perceptions of insecurity

Percentage of respondents who said that they 'did not feel at all secure' by location (inside the home or outside), period (day or night), and zone (government zone: n=1,823; CNO zone: n=673)

■ Government zone ■ CNO zone

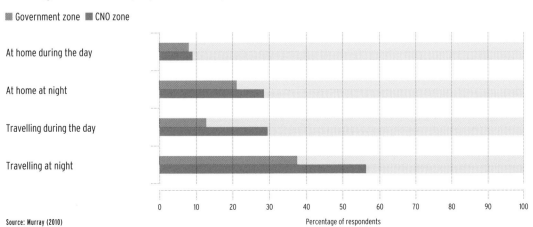

Source: Murray (2010)   Percentage of respondents

Table 7.4 Crimes reported to the Criminal Police[29] in the district of Abidjan, 2004–09

Crimes	2004	2005	2006	2007	2008	2009	Total
Total number of crimes	**5,045**	**5,479**	**2,889**	**3,472**	**3,267**	**2,379**	**22,531**
Voluntary homicides	46	55	48	71	57	46	323
Theft–all categories	4,409	5,294	2,454	2,731	2,740	2,067	19,695
Armed robberies or violent robberies	3,539	4,653	1,813	1,900	2,067	1,490	15,462
Armed burglaries	322	343	255	215	190	286	1,611
Attacks on private vehicles	1,576	2,312	865	803	1,011	732	7,299
Attacks on taxis	1,519	1,869	642	744	638	328	5,740

Sources: Criminal Police (2005-10)

Criminal statistics are scarce in Côte d'Ivoire, making it difficult to build any comprehensive security picture. Indeed, while there is a comparatively large amount of institutional data on Abidjan, the only available data in the CNO zone comes from UN agencies. The UN Police, for example, collates criminal statistics from its 21 stations across the country. While their monitoring system does not provide an exhaustive record of criminal activity, it is nevertheless able to identify trends. As such, it indicates that the incidence of crimes is particularly high in Abidjan and in the western part of the country (see Table 7.5).

Types of insecurity

Côte d'Ivoire is subject to many types of insecurity that are common in post-conflict countries that have not yet completed their transition to peace: economic and criminal violence, sexual violence, post-war displacements and disputes, political violence, and violence related to law enforcement (Geneva Declaration Secretariat, 2008, p. 53). Types of insecurity found in the government zone do not differ significantly from those found in the rebel-held area. The typology of insecurity presented here draws on results from the household survey as well focus groups.

Banditry. Available institutional data and the national survey identify banditry as the primary source of insecurity. Armed robbery—including theft at home and outside, and organized armed banditry—represents a plague in Côte d'Ivoire; even though the phenomenon existed well before the war, the conflict certainly exacerbated it. The difficult economic situation,[30] the security vacuum, and the unrestricted circulation of weapons all influence the incidence of crime. The nature of criminal acts committed within the government zone mirrors, to a considerable extent, those

Table 7.5 Acts of armed violence reported to UN Police stations in 2009

	Abidjan	Western zone*	Eastern zone*	Total
Homicides	216	87	61	364
Armed robberies	406	102	49	557

Note: * UN Police-designated western and eastern zones are mixed areas controlled by both government and rebel forces.

Source: UNPOL (2009)

Figure 7.4 **Most prevalent crimes reported**

Most prevalent crimes reported in the government zone (n=1,786) and the CNO zone (n=640), (percentage)

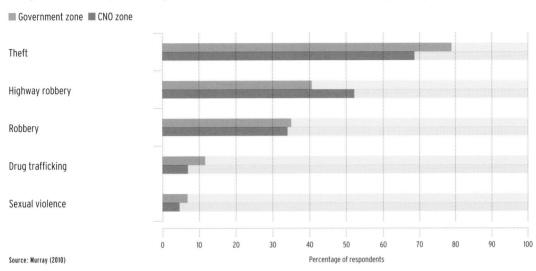

Source: Murray (2010)

committed in the CNO zone; basic theft is the most frequently reported crime in both zones, followed by highway robberies (see Figure 7.4).

When individuals who had reported a member of their household as a victim of armed violence in 2009 were asked about the types of weapons used in those acts, participants in the CNO zone indicated that assault rifles were primarily used (65 per cent of cases, n=113). In the government zone, where 29 per cent (n=350) cited assault rifles, bladed weapons (50 per cent) and handguns (38 per cent) were reported more used.

Gender-based violence. According to several organizations working in this area, sexual and gender-based violence takes various forms in Côte d'Ivoire, from domestic violence to female genital mutilation (FGM), human trafficking, sexual exploitation, and sexual violence.[31] While examples of all of these can be found in both the north and the south of Côte d'Ivoire, there are two main distinctions to be noted: cases of FGM are much more prevalent in the northern territories than in the southern areas[32] and armed sexual violence committed in gangs seems to be particularly prevalent in the south-western part of the country, as an extension of highway robberies. In the western part of the country in particular, the population complains about the absence of security forces.[33]

Resource-based conflict. Land is a perennial focus of dispute in Côte d'Ivoire. Tensions over resources are common within families and communities in every part of the country (Chauveau, 2000, p. 99); the most extreme manifestations can occasionally turn into deadly clashes. While the spotlight may be on the fertile western part of the country, site of serious violence and ethnic tensions both before and during the crisis, the remainder of the country has not escaped resource-based conflict. In the CNO zone, for example, tensions between farmers and cattle breeders revolve around access to water and land (Diallo, 2007).

Violence linked to the electoral process. The series of violent events sparked by the electoral disputes[34] of February 2010 destroyed the relative calm that had settled over the country following the signing of the OPA. Riots flared up in many towns after the dissolution of both the electoral commission and the government; in some places the security forces brutally crushed these uprisings, leaving 13 people dead and more than 100 injured (including at least 76 civilians and 18 members of the state security forces) (UNOCI, 2010a, p. ii). Both the north and south of the country also experience other forms of political violence, including arbitrary detentions of political activists and clashes between members of different political parties (UNOCI, 2009, pp. 4, 6). Political violence climaxed in Côte d'Ivoire in December 2010 in the south of the country during the immediate post-electoral period, when more than 170 people were killed and many more injured in five days (AFP, 2010b).

Security forces as vectors of insecurity. Both sets of security forces committed significant human rights violations during the crisis (AI, 2008, p. 22). Even though the level of violence is significantly lower than during the conflict, national and international human rights organizations regularly denounce the violations committed by state and rebel security forces and the impunity they enjoy (HRW, 2010, p. 21).[35]

The state forces—army, gendarmerie, and police—all report to the military tribunal; data released by this institution provides an account of the kind of crimes committed by members of the state security forces. Between 2002 and 2008, 1,448 cases of significant human rights violations were established by the military tribunal, including rapes, homicides, grievous bodily harm, and illegal confinement. In addition, the tribunal has dealt with 9,014 offences against personal property, including racketeering,[36] theft, swindling, and abuse of trust (Djipro, 2009; Kohon, 2009). This data reflects only reported cases in the government-held area; the reality is likely to be much worse. As noted above, a lack of trust in the security forces deters civilians from reporting most crimes to the police; with respect to crimes committed by the security forces, reporting is probably even more infrequent.

Evidence indicates that security forces use force in a disproportionate manner. Statistics from the national Criminal Police reveal that between 2006 and 2009 security forces in Abidjan and the surrounding areas killed 416 bandits,

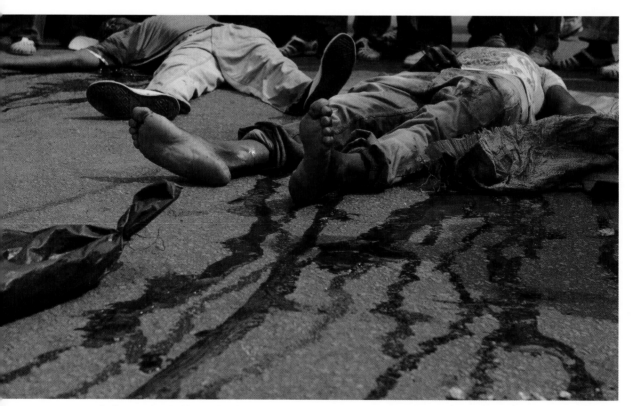

Ouattara supporters stand by the bodies of two men reportedly killed by Ivorian police and armed forces loyal to Ivorian leader Gbagbo, Abidjan, December 2010. © Issouf Sanogo/AFP Photo

more than 30 per cent of whom were killed by CECOS (Criminal Police, 2007–10). This figure is almost twice as high as the number of reported homicides committed by civilians in Abidjan during the same period (Criminal Police, 2007–10). The types of weapons used by law enforcement agencies are of concern as well; the director of the National Police asserted that assault rifles were not appropriate for police missions as they cause 'immeasurable damage'.[37] In its 2009 report, the National Human Rights Commission of Côte d'Ivoire denounced the 'ease' with which the state security forces used firearms as having 'nothing to do with their concern to protect human life in the heart of the action' (CNDHCI, 2010, p. 14). The suppression of the demonstrations in February 2010 and December 2010 are good examples of this.

Many of the same types of abuses are perpetrated in the FN-held part of the country; however, some drivers of violence are specific to the area. UNOCI's Human Rights Section reports on cases of illegal occupation of homes, homicide, extortion, arbitrary arrest, and mistreatment (UNOCI, 2009, pp. 6–10). The near absence of any judicial system—evidenced by the lack of tribunals and the closing of most prisons—has encouraged the FN to expedite justice. CNO zone survey respondents identified demobilized FAFN members among the main perpetrators of armed violence. Specifically, 32.4 per cent cited ex-combatants as primary perpetrators of armed violence, second after bandits (67.9 per cent, n=564).

Finally, the exploitation of resources in the northern part of the country is also a cause of violent tensions. Over the past two years, competition within the FN has repeatedly led to deadly local clashes (Zobo, 2009; AFP, 2010a). Some observers refer to the *comzones* as 'war lords' who reign over natural resources and are reluctant to end the crisis as it would result in a loss of revenue and status (UNSC, 2009, paras. 34–35).

> Human rights organizations have denounced the reckless use of firearms by state security forces.

COPING MECHANISMS: RESORTING TO NON-STATE SECURITY PROVIDERS

The security vacuum, in terms of both national defence and domestic order, encourages individuals and indeed whole communities to use their own means of countering insecurity. A wide range of coping mechanisms and armed actors have emerged, calling into question the state monopoly on force and the role of arms control measures. To some extent, these initiatives constitute new sources of insecurity. This section looks at the different types of non-state security providers.

Community self-defence: vigilantes and mob justice

The large distances between security posts and the inefficiency of the security forces have led communities to organize vigilante groups and to dispense mob justice. Vigilantes emerged long before the crisis; today they are responsible for security in villages or urban areas in every region in Côte d'Ivoire (R. Ouattara, 2008, p. 83). They are often equipped with firearms and usually hail from the community that they are charged to protect. Village vigilantes occasionally work in collaboration with the nearby security forces; the village chief provides a list of group members to representatives of the security forces.

Despite their ostensible role in providing community security, authorities have condemned certain elements of these armed groups for abuses, such as the administration of arbitrary justice.[38] Mob justice is indeed widespread in Côte d'Ivoire (Galy, 2004, p. 118); however, it seems that the practice is more prevalent in the more urbanized government zone, where traditional authorities play a lesser role (see Figure 7.5). This finding highlights that where they are prevalent, traditional authorities continue to represent an important regulatory force.

Figure 7.5 **Dealing with crime suspects**

Response (percentage) to the question 'In your community, what happens to people who are suspected of having committed a crime?' in the government zone (n=1,780) and in the CNO zone (n=633)

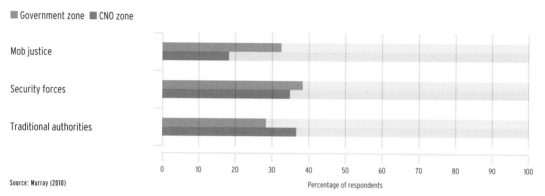

Source: Murray (2010)

The *dozos*–traditional hunters

Dozos are a traditional brotherhood of hunters; like other traditional hunters in the West African sub-region under Mandé influence (Hellweg, 2006, pp. 466–67), they claim to possess mystical powers and to be tasked with protecting the population. They are sometimes organized into vigilante groups and employed by communities. *Dozos* generally consider it part of their identity to carry a weapon and therefore reject calls for gun licences. A census conducted in 1999 provided an estimate of at least 42,000 hunters with 32,000 rifles among them (Basset, 2004, p. 43; see Box 7.3).

Dozos mainly belong to social groups from northern Côte d'Ivoire and were credited with playing a significant role in maintaining public order and fighting criminality during the recession of the 1990s, making them relatively popular (Basset, 2004, p. 32). When criminality rose, villages formed their own vigilante groups and urban communities called on the *dozos* to protect their districts (Basset, 2004, p. 31; Badou, 1997). The association representing them at that time campaigned, in vain, for the *dozos* to be awarded the status of 'auxiliary security forces'[39] (Badou, 1997).

Throughout history, political leaders have either marginalized the *dozos* or attempted to assimilate them and benefit from their popularity (Basset, 2004, p. 39). In 2002, numerous *dozos* joined the rebellion (Hellweg, 2006, p. 467), controlling small towns and villages and preventing infiltration by loyalists.

Today, many *dozos* continue to work as guardians throughout the country. In Moyen Cavally, the *dozos* also guard and protect communities where the majority of inhabitants are originally from the north (ICG, 2010a, p. 11). Others are also employed by private security companies (Kouamé, 2009). Yet certain *dozos* have been known to abuse their positions, for example by arresting people arbitrarily and mistreating them (UNOCI, 2009, pp. 1, 6).

The militias

The outburst of hostilities in 2002 spawned the creation of a multitude of militias in the southern part of the country, particularly in the south-west and in Abidjan. These bands have evolved within the *galaxie patriotique,* a system of groups that support the ruling party of President Gbagbo, and 'intersect depending on the spirit of the moment'[40] (*Les milices hors Abidjan,* 2009). There are two main types of militias: the armed militias of the western part of the country and the groups of 'young patriots' based in Abidjan, usually armed with bladed and blunt weapons.

The 20,000-strong militias from the west of the country (ICG, 2009, p. 1) were armed and funded by the ruling party and trained by state security forces whom they supported during the civil war and who helped them defend the south (Banégas, 2008, p. 7; ICG, 2010a, p. 11; Konadjé, 2008, p. 8). These militias also received the support of combatants from the anti-Charles Taylor Liberians United for Reconciliation and Democracy (Ero and Marshall, 2003, p. 96). In 2005, the most important Wé militias from the west[41] formed the Force de Résistance du Grand Ouest (Resistance Forces of the Great West).

Despite peace agreements stipulating that they should be disarmed and dismantled—through the 'disarmament and dismantlement of militias' (DDM) programme—before the elections, the militias are still armed, posing a serious threat to security in the post-electoral period (ICG, 2009, p. 1). 'The chains of command are still intact', affirmed the head of one of the main militias, adding that 'even if a member has been disarmed, he still considers himself part of a self-defence group'.[42] So far, the DDM programme is far from being a success (see Box 7.4).

The absence of economic opportunities causes frustration among groups of young people at risk; indeed, almost 90 per cent of militia members are reportedly 24–35 years old (*Les milices hors Abidjan,* 2009). Several civil society organizations have suggested that there is a link between the prevalence of armed violence in the western part of the country and the presence of numerous armed young people with no source of income.[43] Within this context, observers have also pointed to the abject poverty of militia members (Banégas, 2008, p. 15; UNSC, 2009, para. 112).

Denis Maho Gofliehi, head of the Forces de Libération du Grand Ouest (Liberation Forces of the Great West, FLGO), the pro-Gbagbo militia in western Côte d'Ivoire, talks to the press in March 2008 in Guiglo. © Kambou Sia/AFP Photo

Today, the militias claim to play their part by punishing criminals,[44] protecting the population from potential attack by the FN, and defending them against *allochtones* (Ivorians from other regions) and *allogènes* (foreigners).[45] The immediate post-electoral period in 2010 witnessed serious ethnic tensions resulting in deadly clashes between militia members and *allogènes*. Some analysts have attributed the failure to dismantle the militias to the government's strategy of retaining these forces in case of future conflict with the FN (UNSC, 2009, para. 121).

The *galaxie patriotique* also includes urban groups of young people such as the Congrés Panafricain des Jeunes et des Patriotes (Pan-African Congress of Young Patriots, COJEP). In 2006, the UN Security Council imposed sanctions on the group's leader, Charles Blé Goudé, for calling for violence against foreigners and UN staff and for directing and participating in acts of violence committed by street militias (UNSC, 2006a). Human rights organizations have also denounced violent acts committed by the Ivorian Student Federation (FESCI) and the impunity that its members enjoy.[46] While these groups do not generally carry firearms, they represent a real threat as a vast force that can be mobilized quickly and that are potentially difficult for the international peacekeeping forces to contain or suppress in case of confrontation.[47]

Security personnel stand by their motorcycles at the headquarters of the private security company Risk, in the Marcory commune of Abidjan, August 2009.
© Kambou Sia

Private security companies

While some private security companies (PSCs) had already appeared by the 1970s, the insecurity caused by the political crisis and civil war has been a driving force for the expansion of the sector. Indeed, since the wave of violence that targeted Westerners in 2004, the sector has grown rapidly and without regulation. Côte d'Ivoire has witnessed an increasing privatization of its security sector, which now includes more than 400 companies employing more than 55,000 people[48] and accounting for more than XOF 50 billion (USD 100 million) in revenue in 2008 (Kan, 2009). PSCs are located in every region of Côte d'Ivoire but the overwhelming majority are found in the southern urbanized regions of the country. The primary responsibilities of these PSCs include the protection of private houses, shops, companies, and banks, as well as the guarding of mines and the provision of bodyguards.

PSCs fill part of the security vacuum; the principal companies are better equipped in communications and transportation than the police and are therefore able to react faster and more efficiently (Mieu, 2009b). While PSCs have the right to possess non-lethal weapons, legislation stipulates that only cash-in-transit personnel and bodyguards (two types of staff that only a few Ivorian PSCs provide) are entitled to carry firearms. Nevertheless, most PSC personnel carry various kinds of firearms illegally,[49] given that PSCs make more of a profit by providing the services of armed security guards than unarmed ones. Illegal weapon carrying persists largely due to a lack of logistical capacity and the reluctance of the authorities to enforce the law. Indeed, while it is difficult to identify the owners of the PSCs, it is clear that some politicians and former military officers are among those involved in this very profitable business (Kougniazondé, 2010, p. 12).

Box 7.3   Civilian weapons ownership

Arms possession in Côte d'Ivoire is widespread and is to a large extent considered 'traditional', particularly in rural zones where hunting is a primary means of subsistence. Hunting is particularly popular in the savanna region, in the CNO zone. Unlike other countries of the region, such as Liberia or Burundi, where the majority of the population regards weapons as a source of danger (Gilgen and Nowak, 2011; Pézard and de Tessières, 2009, pp. 108–09), close to half of the sample interviewed in Côte d'Ivoire (n=2,424) consider firearms primarily a means of protection (49.3 per cent).

A lack of accurate data makes rates of ownership difficult to estimate; between 1989 and 2010, the authorities only issued 2,598 gun licences. Fieldwork suggests that firearms are much more widespread and that the vast majority are held illegally. While people from the CNO zone identify hunting as the primary motivation for gun carrying, insecurity and criminal activities together are strong motivating factors in the government-run area (see Figure 7.6).

Figure 7.6   Motivations for gun carrying

First answer given (percentage) to the question 'What are the principal reasons why people in your locality (except security forces) carry firearms?' in the government zone (n=1,602) and in CNO zone (n=605)

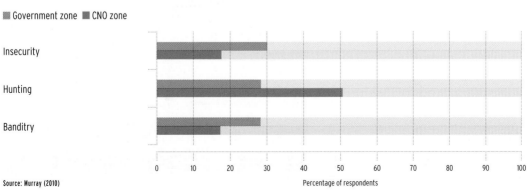

Source: Murray (2010)   Percentage of respondents

International forces

As a result of the defence agreements signed by France and Côte d'Ivoire after independence, the ineptitude of the Ivorian security forces, and the deployment of various peacekeeping forces during and after the crisis, foreign troops continue to play a key role in ensuring the security of Côte d'Ivoire. As the recently appointed commander of the Force Licorne explained: 'The mission of the impartial forces is to secure the country and facilitate the process of normalization' (ONUCI-FM, 2010). Specifically, these 'impartial forces' have been supporting the Ivorian forces in ensuring the security of the electoral process. Indeed, due to a lack of resources, the joint government–rebel ICC has not been able to deploy 8,000 personnel as planned (ICG, 2009, p. 10). These shortcomings and the volatile atmosphere in the run-up to the elections encouraged UNOCI to strengthen its presence from 8,650 to 9,150 military and police personnel in 2010 (UNSC, 2010, para. 1).

A TWO-STEP PROCESS: RESTRUCTURE THEN REFORM

The security sector in Côte d'Ivoire has attracted intense international attention. In line with the security provisions of the OPA, the government, the rebels, and international stakeholders are currently engaged in the process of reunifying the national security sector; however, while the OPA sets the framework for the rehabilitation of the sector, it does not provide the basis for thorough SSR. The following section examines the process of reunification of the security sector and looks at the hopes projected on the future New Army.

Jordanian ONUCI soldiers ride past an electoral poster of President Laurent Gbagbo, which reads, 'I am 8 years old. I was born during the war. I want to grow up in peacetime.' Abidjan, December 2010. © Kambou Sia/AFP Photo

Reunification of the national security apparatus

The OPA defines a series of measures that aim to bring an end to the conflict and reunify the country. These conventional post-conflict initiatives include the organization of free and transparent democratic elections, the integration of rebel fighters into the new national security forces, and the disarmament, demobilization, and reintegration of the remainder of the rebel forces and militias, as well as the restoration of state authority across the nation. Yet three years since the signing of the OPA, little progress has been made in implementing the clauses that were intended to lead Côte d'Ivoire towards peace.

Towards new defence and security forces

In 2007, the working group for restructuring and reforming the army was created. Composed of different actors from the rebel and state security sectors, it is tasked with restructuring the armed forces and paving the way for SSR (UNDP, n.d.). While the group conducted an assessment of the security sector in 2010, a national reform strategy has yet to be designed; current initiatives focus mainly on the technical reunification of the forces.[50] The OPA outlines the precise steps of the reunification process. Some measures have been adopted, such as the deployment of a combined government–rebel unit (the ICC), which is charged with implementing the restructuring of the armed forces (RCI, 2007, art. 3.1.1); the harmonization of ranks; and the cantonment of some FAFN to be integrated in the national army. Nevertheless, several obstacles threaten to disrupt the process.

One issue concerns the mixed composition of the ICC, which serves as a pilot for the future reunification of the army. While providing security during the 2010 elections allowed the two armies to foster cooperation, the process was not entirely smooth, not least because both lacked trust in the peace process. While the FAFN were supposed to contribute half of the 8,000 ICC personnel, they provided fewer than 2,000 men. The International Crisis Group posits that the FAFN fear that injecting such a large number of men into the combined unit would weaken their capacity (ICG, 2010a, p. 14). Secondly, the development of the ICC has been hindered by a lack of finance and logistical resources (CNDHCI, 2010, p. 11). The units spread across the country lack vehicles and equipment, most notably anti-riot gear, which is essential for handling potential civil unrest linked to the electoral process.[51]

Yet the most alarming problem is the disparity in treatment between the FAFN and members of the state security forces working in the same units. Soldiers are still dependent on their own hierarchies, which means that the state soldiers receive a salary while the FAFN do not. 'For the time being, we're not receiving the allowances that we were promised and so we're relying on our *comzone,* who gives us food',[52] explained one FAFN soldier who served as part of a combined ICC unit operating in the CNO zone. This imbalance in treatment is the source of much frustration and does little to foster the spirit of camaraderie that is so essential during this crucial period of transition.[53] As one FAFN soldier put it: 'It's a symbolic mission—a strong gesture to show the country and the international community that we are united.'[54]

Rifles and sub-machine guns that belonged to the Forces Nouvelles are displayed in a stadium, pending their destruction. Bouaké, July 2007.
© Kambou Sia/AFP Photo

Box 7.4 The failure of DDR and DDM

DDR has been on the peace process agenda since the signing of Côte d'Ivoire's first peace accord in 2003. Although several attempts were made to launch the programme after 2004, all were aborted because of a lack of funding and political will. DDR and DDM were originally conceived as preconditions for the electoral process, but the provisions have not been honoured and the continued existence of non-state armed groups still jeopardizes the stability of the post-electoral period.

In total, around 33,000 FAFN members have been screened for DDR (PNRRC, 2010). As per the OPA, 5,000 volunteers will be integrated into the New Army, in addition to 600 who were already selected at the beginning of the ICC deployment and a further 4,000 FAFN members will join forces with the police and the gendarmerie.[55] Nearly 24,000 people are thus scheduled to disarm, demobilize, and reintegrate. Of these, more than 12,000 have already been disarmed and demobilized.

Yet the FN disarmament exercises have been relatively fruitless to date. Only 237 weapons and 22 grenades have been handed in (UNOCI, 2010b; 2010c), leaving observers to draw the conclusion that there are still a large number of weapons in the hands of the demobilized. Indeed, one former FAFN combatant confirmed that:

> A large number of ex-combatants have held on to their weapons. I've kept some. Some ex-combatants have pistols and Kalashnikovs, which they're keeping for their own security [. . .]. As long as the current situation remains unstable, things could always erupt again.[56]

In the context of DDM, almost 38,000 members have been screened by the national programme of reintegration;[57] by May 2010, more than 17,000 people had been disarmed. Fewer than 1,200 weapons were collected, of which the vast majority were unusable (Bidi, 2007; UNSC, 2006b, paras. 39-45).

In short, the chains of command in the militias and the FAFN have survived and the reintegration phase lacks financial support and an effective coordination mechanism. Whether in the CNO zone or in the west of the country, the financial insecurity and inactivity of youths previously involved in the rebellion and in the militias worries the population. The national survey and focus groups highlight the importance of reintegration and the creation of jobs for at-risk young people.[58] Currently, their post-military economic opportunities are scarce, and their frustration is slowly building.[59]

Nevertheless, the population is generally supportive of the ICC; focus group participants repeatedly expressed their desire to see the ICC take on policing missions. Many regard ICC agents as more disciplined and less corrupt than other security forces, partly because they do not appear to have set up checkpoints for extortion.[60]

Finally, not everyone stands to benefit from the reunification of the country and the establishment of the New Army. Indeed, it is in some people's interest to obstruct the process. Both rebel and government authorities collect a large amount of tax from cocoa production and trade. *Comzones* in particular are likely to have much to lose from the reunification as they currently produce and export resources, levy taxes on the transport of goods, and control the political administration of their zones (UNSC, 2009, para. 36). Global Witness estimates that between 2004 and 2007 the FN authorities collected more than USD 30 million in cocoa proceeds (Global Witness, 2007, p. 4). As previously noted, the *comzones* run well-armed private armies[61] and potentially constitute one of the main obstacles to the reunification of the country. The 'reintegration package' for the *comzones* and their security team is still under discussion; incentives seem difficult to identify. Both Gbagbo and Ouattara avoided the topic in their 2010 political campaigns (ICG, 2010b, p. 7).

Strengthening the judicial system

Not only do Ivorian authorities have to oversee the restoration of the judicial administration throughout the territory, but they also have to address the current internal deficiencies of the system. Supported by the European Commission, the Ivorian government is in the process of preparing a new policy on the sector that includes reinforcing judicial actors (supporting the National Institute of Legal Training and the National Police Training College); strengthening the efficiency of certain legal departments (through computerization of procedures and modernization of infrastructure);

fighting corruption; facilitating access to the judicial system (by restructuring courts and legal aid); and improving prison conditions, including in youth detention centres (European Commission, 2009).

Since the OPA, the Ivorian authorities have timidly advanced towards the goals of reuniting the security forces and strengthening the capacity of the judiciary. National and international stakeholders are focused on quick-fix measures that are meant to address immediate security threats, including the integration of rebel elements into the New Army and the disarmament and demobilization of combatants. Yet the OPA does not only call for a technical overhaul of the armed forces, but also insists on reaching longer-term security objectives. The construction of a security system that is 'committed to integrity and republican values' (RCI, 2007, part III) and capable of assuring both the protection of the people and the security of the territory will require significant financial resources, technical know-how, and, above all, strong political will to break free from the past.

Fifty years on: a second security renewal?

To date, SSR efforts have not address core weaknesses of the security sector.

In post-conflict Côte d'Ivoire, the period of reunification seems an appropriate time for a review and overhaul of the security system. An efficient, legitimate system that respects human rights is one of the keys to the normalization of the country. To some extent, the enthusiasm behind the New Army echoes the optimism that moulded the post-colonial armies as agents of modernization in the independence period (Janowitz, 1964). As this chapter demonstrates, however, the role of the Ivorian military as a modernizer has proved very limited. Expected reforms do not appear to address the core deficiencies that face the security sector in any way.

The security sector today is crippled by some of the same deficiencies that emerged after independence, including limited democratic control over the armed forces and a lack of strategic objectives. Fifty years later, the reform dynamic faces additional challenges: the existence of two armed forces in one country, a reduction in the quality of personnel, weak capacities, divisive ethnic and regional tensions exacerbated by post-electoral conflict, a lack of consensus on political leadership, and an inability to deliver public security. Today, security forces can almost be described as predatory.

In the aftermath of independence, the army was considered a force for social modernization due largely to the 'modernity' of the institution itself (Lavroff, 1972, p. 988). In light of the decline in the quality of recruitment and training of soldiers during the crisis[62] and the rise of nepotism and corruption, this description no longer applies. While it is possible to draw parallels between the role of the nation-builders of the 1960s and the central role that the military is expected to play in national reconciliation, today's divided, post-conflict Côte d'Ivoire faces an increased level of ethnic and regional tensions. In addition, the balance of power has changed. Whereas Houphouët-Boigny had neglected the army—which didn't represent a base for his power—his successors increasingly relied on the military.

The ongoing political crisis has shown that control over the military continues to be a very important guarantee of power. The state forces are pivotal in the political struggle between Ouattara and Gbagbo and may even determine the eventual outcome of the crisis; while the incumbent is trying to keep control over the military, Ouattara claims that an increasing number of FANCI officers are supporting him (*Le Panafricain,* 2010). Finally, as in many other countries, the experience of civil war represents an important development in terms of the military budget, personnel, and equipment, making the army much larger and perhaps more difficult to sidestep, which may influence the power-sharing relationship between the political and military leaderships.[63]

CONCLUSION

Although the armed conflict ended more than five years ago, the ongoing 'no war, no peace' situation continues to have an enormous impact on the security of the country. People generally feel unsafe, insecurity is widespread and varied, and the inadequacy of the security sector only reinforces this sense of insecurity. The crisis has not only affected the security provision in the rebel-held area but also in the southern, government-held part of the country. While the capacity of the security forces is generally more trusted in the south, the state forces do not perform much better than the rebels; resorting to private security actors as a coping mechanism is common in both zones. To a large extent, both sets of armed forces face the same deficiencies: violations of human rights are widespread, corruption is high, democratic oversight is lacking, accountability is poor, and resources are insufficient.

Security sector reform is a key element of post-conflict institutional engineering and the need for such reform in Côte d'Ivoire is indisputable. The current situation calls for more than a simple reshuffling of ranks, however; the authorities need to devise a genuine reform strategy that goes beyond the pre-crisis status quo. This would involve the adoption of an integrated approach and a wider definition of the security sector—one that also takes into account the judiciary and non-state security actors, such as PSCs.

While reform must eventually take place, the post-2010 election turmoil has put the sensitive initiative on hold. The ongoing post-electoral crisis brings with it additional challenges: it will have a detrimental impact on the SSR process by further reducing the level of trust between the two sides and intensifying the population's lack of confidence in the security forces. The violent repression of civilian demonstrations is further polarizing the political system and reducing the chances of a compromise. Perhaps most worrisome, the political stalemate has increased the politicization of the armed forces and removed any democratic oversight of the military. The process of redefining political-military relations in Côte d'Ivoire is some way off. ∎

LIST OF ABBREVIATIONS

APDH	Action pour la protection des droits de l'homme
CECOS	Centre de Commandement des Opérations de Sécurité
CNO	Centre Nord Ouest
Comsecteur	Forces Nouvelles *commandant de secteur* (sector commander)
Comzone	Forces Nouvelles *commandant de zone* (zone commander)
DDM	Disarmament and dismantling of militias
DDR	Disarmament, demobilization, and reintegration
ECOWAS	Economic Community of West African States
FAFN	Forces Armées des Forces Nouvelles
FANCI	Forces Armées Nationales de Côte d'Ivoire
FGM	Female genital mutilation
FN	Forces Nouvelles
FLGO	Les Forces de Libération du Grand Ouest
ICC	Integrated Command Center

LIDHO	Ligue ivoirienne des droits de l'homme
MIDH	Mouvement ivoirien pour les droits humains
MINUCI	United Nations Mission in Côte d'Ivoire
OPA	Ouagadougou Political Agreement
PSC	Private security company
SSR	Security sector reform
UNOCI	United Nations Operation in Côte d'Ivoire
WANEP	Réseau ouest-africain pour l'édification de la paix
XOF	Franc de la Communauté Financière Africaine (African Financial Community franc)

ENDNOTES

1   The national household survey was commissioned by the Ivorian National Commission against the Proliferation and Illicit Circulation of Small Arms and Light Weapons with the financial and technical support of the UN Development Programme. It was designed and administered by the Small Arms Survey in February 2010. Note that the views expressed in this chapter are those of the Small Arms Survey and are not necessarily those of the National Commission or the UN Development Programme.

2   The author organized 12 focus groups (made up of 10–15 people each) in the regions of 18 Montagnes, Denguélé, Lacs, Lagunes, Moyen-Cavally, Vallée du Bandama, and Zanzan. Focus groups addressed issues related to perceptions of security and security providers, armed violence, the prevalence and use of weapons in the community concerned, non-state armed groups, and possible solutions to the problems of armed violence and the illicit circulation of firearms. More than 70 key informant interviews were conducted by the author in the government-held zone and in the CNO zone with high-ranking officers and other members of the security forces, the FAFN, and heads of armed militias, as well as representatives of civil society organizations, public health workers, traditional hunters, ex-combatants, and official and traditional authorities.

3   'La radicalisation de la répression politique' (author's translation).

4   Yet violence also existed in the political sphere prior to this period; indeed, the presidencies of Houphouët-Boigny were themselves marked by political violence (Banégas, 2010, p. 367).

5   Bédié, Ouattara, and Gbagbo were also the main presidential candidates in the 2010 elections.

6   Authorities have been very reluctant to allow inspections of Republican Guard installations by the UNOCI Embargo Cell or the UN Group of Experts. Since the implementation of the embargo, these bodies have not been granted access to the Republican Guard's armouries, in clear violation of UN Security Council resolutions (UNSC, 2009, paras. 49–53). As far as weapons are concerned, the Republican Guard is likely to be particularly well equipped due to its status as an elite and privileged corps.

7   'Si moi je tombe, ils tombent aussi. Il y en a qui croient qu'un coup d'Etat est facile! C'est un édifice qui a plusieurs poteaux. Si tu le fais tomber, tes poteaux tomberont aussi' (author's translation).

8   Author interview with the director of the National Police, Abidjan, February 2010.

9   The arms embargo and the subsequent lack of ammunition contributed to the deficit of appropriately trained new officers (author interview with representatives of the security forces, Côte d'Ivoire, February 2010).

10   Checkpoints cost the Ivorian private sector USD 300–600 million per year and affect the price of consumer goods; for one, the price of coal, which is delivered to Abidjan from relatively nearby, quadruples between the factory and the city (Reuters, 2010).

11   Focus group, Abidjan, February 2010.

12   Author interview with a customs inspector, National Customs Office, Abidjan, March 2010.

13   Author interview with a province chief, February 2010.

14   Author interview with the head of a human rights organization, Abidjan, February 2010.

15   See PNRRC (n.d.).

16   In contrast to 2002, when estimates put the figure at around 800 people (Florquin and Berman, 2005, p. 239), in 2010 it was estimated that 467 FAFN personnel were ex-FANCI and therefore entitled to retroactive pay like other government troops in 2007, following the signature of the peace accords (Labertit, 2010).

17   See the organizational chart of the FN secretariat (FN, n.d.).

18   'La brigade [FN] envoie des gens mais tu dois payer. Tu sais qu'il n'y a pas de prison donc que ça va être "arrangé"' (author's translation). Focus group with civilians, Odienné, February 2010.

19   'On fait tout pour que les personnes soient hors d'état de nuire, on garde les criminels en cellule jusqu'à ce que la prison ré-ouvre. Avant le comzone donnait les moyens pour transférer les gens à Bouaké où il y a une prison, mais c'est pas facile car il faut un engin, du carburant et du personnel' (author's translation). Author interview with an FN officer, Odienné, February 2010.

20   'Quand ça vient à notre niveau, on règle à l'amiable. Nous n'avons pas de procédure juridique. La prison civile n'est pas ouverte parce qu'il n'y a pas de procédure pénale. Donc "on déplace les problèmes"; par exemple, on escorte la personne jusqu'à la frontière' (author's translation). Author interview with a *comsecteur*, 18 Montagnes region, February 2010.

21   In April 1961, a few months after the declaration of Ivorian independence, France signed defence agreements with Côte d'Ivoire, Niger, and Dahomey (modern Benin), stipulating, among other things, that France would provide 'the assistance required for the establishment of their armed forces' (RCI et al., 1961, art. 6). Four years later, France and Côte d'Ivoire drew up a convention detailing the logistical support the former would provide to the latter, including the sale of military equipment. See RCI Presidency (n.d.).

22   For more information on this, see the reports of the Group of Experts on Côte d'Ivoire (SCC, n.d.).

23   Interview with Gen. Bakayoko, FAFN Chief of Staff, Bouaké, February 2010; interview with the director of the National Police, Abidjan, February 2010.

24   Author interviews with representatives of the main human rights organizations in Côte d'Ivoire—Ligue ivoirienne des droits de l'homme (LIDHO), Réseau ouest-africain pour l'édification de la paix (WANEP), Mouvement ivoirien des droits humains (MIDH), and Action pour la protection des droits de l'homme (APDH)—Abidjan, February–March 2010; focus group with representatives of civil society, Bouaké, February 2010.

25   Author interviews with representatives of the main human rights organizations in Côte d'Ivoire—LIDHO, WANEP, MIDH, and APDH—Abidjan, February–March 2010; focus group with representatives of civil society, Bouaké, February 2010.

26   Other answers included: 'conduct sensitization campaigns' (12.7 per cent), 'strengthen the number of security posts' (11.4 per cent), and 'reinforce border control' (7.5 per cent).

27   Armed violence is defined as 'the intentional use of illegitimate force (actual or threatened) with arms or explosives, against a person, group, community, or state, that undermines people-centred security and/or sustainable development' (Geneva Declaration Secretariat, 2008, p. 2).

28   Eighteen per cent (n=673) of respondents in the CNO zone and 20 per cent (n=1,819) in the government zone reported that at least one member of their household had been the victim of an act of armed violence in 2009. 'Household' was understood by respondents as 'people who live under the same roof'.

29   The Criminal Police (*Police criminelle*) is a special unit that deals with major crimes. These figures probably under-represent the actual levels of violence. For example, the homicide rate in Abidjan in 2009 would be 0.77 per 100.000—if the city's population of six million people (RCI, 2009a, p. 64) were taken into account; that figure seems unusually low. According to estimates provided by the UN Office on Drugs and Crime, West African countries generally experience a homicide rate closer to 20 per 100,000 (Geneva Declaration Secretariat, 2008, p. 70).

30   The poverty rate has shot up in the last ten years, from 33.6 per cent in 1998 to 48.9 per cent in 2008 (RCI, 2009a, p. xi). The north of the country is more deeply affected, with four out of five people classified as poor. Unemployment is high and affects young people in particular. Nearly one-quarter of 15–24-year-olds are unemployed (RCI, 2009a, para. 282); according to criminal statistics, they are responsible for much of the country's banditry.

31   Author interviews with several different human rights organizations based in Abidjan, February 2010; author interview with representatives of the Duékoué Social Centre, February 2010. See also the annual report of the Ivorian National Commission on Human Rights (CNDHCI, 2010) and the reports of the Human Rights Section of UNOCI (UNOCI, 2009; 2010a).

32   In spite of the 1998 law against FGM, a UNICEF study shows that the practice is still widespread in Côte d'Ivoire, with the national figure for circumcized women at 36 per cent. In the north and west of the country, where the populations practice FGM as a matter of tradition, the rate approaches 90 per cent (UNICEF, 2008).

33   Focus group, Duékoué, February 2010; focus group, Bangolo, February 2010.

34   The compilation of the electoral list triggered a strong social reaction. It was not only the right to vote that was at risk, but also access to an official Ivorian identity, which brought with it access to land ownership, jobs within the civil service, and the free circulation of people, particularly at checkpoints. These tensions were linked to the concept of *Ivoirité*, which had fuelled the dynamics leading to the conflict.

35   Interviews with representatives of the main human rights organizations of Côte d'Ivoire—LIDHO, WANEP, MIDH, and APDH—Abidjan, February–March 2010.

36   See endnote 10.

37   Author interview with the director of the National Police, Abidjan, February 2010.

38   Author interviews with security authorities, Côte d'Ivoire, February 2010.

39   'Auxiliaries des forces de l'ordre' (author's translation).

40   'S'interpénètrent en fonction de l'intérêt du moment' (author's translation).

41   The following militias were created at the beginning of the conflict in Moyen Cavally: les Forces de Libération du Grand Ouest (FLGO), le Mouvement Ivoirien pour la Libération de l'Ouest de la Côte d'Ivoire (MILOCI), l'Alliance Patriotique du Peuple Wé (Ap-Wé), l'Union des Patriotes pour la Résistance du Grand Ouest (UPRGO), and les Forces Spéciales pour la Libération du Monde Africain (FS Lima) (Banégas, 2008, pp. 3–4).

42   'Les chaînes de commandement sont encore là. Même si l'élément a été désarmé il se pense encore parti d'un groupe d'autodéfense' (author's translation). Author interview with the head of one the major militias of the west, Guiglo, February 2010.

43   Focus group with representatives of civil society, Guiglo, February 2010.

44   Author interview with a militia member, Abidjan, March 2010.

45   Group discussion with heads of militias, Guiglo, February 2010.

46   Author interview with representatives of different Ivorian human rights organizations, Abidjan, February 2010.

47   In November 2004, French soldiers opened fire outside Abidjan's Hôtel Ivoire on a hostile crowd made up largely of 'Young Patriots', killing several people. The event made headlines in the international media. Human rights organizations accused French soldiers of losing control of the situation and shooting at unarmed people (ICG, 2005, p. 8).

48   Author interview with the head of the private security and cash-in-transit unit, Direction de Surveillance du Territoire (Directorate of Territorial Surveillance), Abidjan, March 2010.

49   Author interview with the head of the private security and cash-in-transit unit, Direction de Surveillance du Territoire, Abidjan, March 2010.

50   Author interview with a representative of the cabinet of the prime minister, Abidjan, February 2010.

51   Author interview with a senior military officer of the ICC, Yamoussoukro, March 2010.

52   'Pour le moment on a toujours pas reçu les primes qu'on nous avait promises et donc on dépend du comzone qui nous donne de la nourriture' (author's translation).

53   Author interview with the head of an ICC unit, 18 Montagnes, February 2010.

54   'C'est une mission symbolique, un geste fort pour montrer au pays et à la communauté internationale que nous sommes unis' (author's translation).

55   Author interview with a representative of the cabinet of the prime minister, Abidjan, March 2010.

56   'Beaucoup d'ex-combattants ont gardé des armes. Moi j'en ai gardées. Certains ex-combattants ont des pistolets et des kalaches en particulier, ils les gardent pour leur sécurité [. . .]. Tant que la situation n'est pas stable les choses peuvent reprendre' (author's translation). Author interview with a demobilized ex-combatant, Bouaké, March 2010.

57   According to the National Programme for Reintegration and Community Rehabilitation, 5,000 additional militia members are yet to be screened (PNRRC, 2010).

58   Focus group with civil society representatives, Bouaké, March 2010; focus group, Duékoué, February 2010.

59   Author interview with a senior military officer of the ICC, Yamoussoukro, March 2010. In theory, DDR beneficiaries can choose between a reinsertion project or XOF 500,000 (USD 1,000) as a 'final settlement'.

60   Focus group, Bangolo, February 2010; focus group, Danané, February 2010.

61   In October 2009, the report of the UN Group of Experts noted that some of the *comzones* were rearming themselves (UNSC, 2009, para. 122).

62   Author interviews with representatives of the security forces, Côte d'Ivoire, February and March 2010.

63   Military expenditures more than doubled between 1997 and 2009, rising from XOF 171 million (USD 300,000) to more than XOF 369 million (USD 700,000), or 1.5 per cent of the gross domestic product in 2009 (SIPRI, n.d.). While it is difficult to identify the exact strength of the army, different sources show that both the police and the gendarmerie have doubled in size in the past ten years (UNDP, 2010; Mieu, 2009a).

BIBLIOGRAPHY

Abidjan.net. 2010. 'Séguéla: Une explosion dans la poudrière de la caserne des Forces nouvelles fait 2 blessés graves.' 5 January.

AFP (Agence France-Presse). 2004. 'Côte d'Ivoire: tous les aéronefs ivoiriens neutralises.' 7 November.

—. 2010a. 'Côte d'Ivoire: trois morts lors d'affrontements entre ex-rebelles'. 29 April.
      <http://www.france24.com/fr/20100429-cote-divoire-trois-morts-lors-daffrontements-entre-ex-rebelles>

—. 2010b. 'Côte d'Ivoire: 173 morts selon l'ONU, le camp Ouattara en appelle a la CPI.' 23 December.

AI (Amnesty International). 2008. *Blood at the Crossroads: Making the Case for a Global Arms Trade Treaty*. ACT 30/011/2008. London: AI.
      <http://www.amnesty.org/en/library/info/ACT30/011/2008/en>

—. 2010. *Amnesty International Report 2010: The State of the World's Human Rights*. London: AI.

Airault, Pascal, Théophile Kouamouo, and Marianne Meunier. 2011. 'Les mystères d'Abidjan.' *Jeune Afrique*, No. 2609. 10 January.
      <http://www.jeuneafrique.com/Article/ARTJAJA2609p020-026.xml1/>

Badou, Jerôme. 1997. 'Côte d'Ivoire: les chasseurs dozos traquent les bandits.' Syfia Info. 1 July.
      <http://www.syfia.info/index.php5?view=articles&action=voir&idArticle=407>

Bagayoko, Niagalé. 2010. *Security Systems in Francophone Africa*. Brighton: Institute of Development Studies.
      <http://www.ntd.co.uk/idsbookshop/details.asp?id=1175>

Balint-Kurti, Daniel. 2007. *Côte d'Ivoire's Forces Nouvelles*. Africa Programme Armed Non-State Actors Series. London: Chatham House.

Banégas, Richard. 2008. *La République oublie-t-elle ses enfants? Milicianisation et démilicianisation du champ politique en Côte d'Ivoire*. Colloquium on 'Regards croisés sur les milices d'Afrique et d'Amérique latine en situation de violence'. Paris: Centre d'Etudes de Relations Internationales (CERI). September. <http://www.lasdel.net/cours%20ue/milicianisation.pdf>

—. 2010. 'Génération "guerriers"? Violence et subjectivation politique des jeunes miliciens en Côte d'Ivoire.' In Nathalie Duclos, ed. *L'adieu aux armes: Parcous d'anciens combattants.* Paris: Khartala, pp. 359–97.

Basset, Thomas. 2004. 'Containing the Donzow: The Politics of Scale in Côte d'Ivoire.' *Africa Today*, Vol. 50, No. 4, pp. 31–49.

Berman, Eric G. 2007. 'Illicit Trafficking of Small Arms in Africa: Increasingly a Home-Grown Problem.' Presentation made at the GTZ–OECD–UNECA Expert Consultation of the Africa Partnership Forum Support Unit, Addis Ababa. 14 March. <http://www.oecd.org/dataoecd/33/25/38647866.pdf>

Bidi, Ignace. 2007. 'Désarmement à l'ouest: 1027 armes rendues par le FRGO.' Koffi.net. 21 May.
    <http://www.koffi.net/koffi/actualite/7506-Desarmement-a-l-ouest-1027-armes-rendues-par-les-FRGO.htm>

Boisbouvier, Christophe. 2005. 'Gbagbo et l'armée: qui menace qui?' *Jeune Afrique.* 22 August.
    <http://www.jeuneafrique.com/Article/LIN14085gbagbecanem0/actualite-afriquegbagbo-et-l-armee-qui-menace-qui.html>

Chauveau, Jean-Pierre. 2000. 'Question foncière et construction nationale en Côte d'Ivoire.' *Politique Africaine*, No. 78, pp. 94–125.

CNDHCI (Commission Nationale des Droits de l'Homme de Côte d'Ivoire—Ivorian National Commission on Human Rights). 2010. *L'Etat des droits de l'homme de Côte d'Ivoire: Rapport Annuel 2009.* Abidjan: CNDHCI.

Cocks, Tim. 2010. 'Interview: Ivory Coast Checkpoints Cost Millions a Year—Business.' Reuters. 3 May.
    <http://www.reuters.com/article/2010/05/03/idUSLDE6421QF>

Collier, Paul. 2009. *Wars, Guns and Votes: Democracy in Dangerous Places.* London: Bodley Head.

Conte, Bernard. 2004. *Côte d'Ivoire: clientélisme, ajustement et conflit.* Bordeaux: Centre d'économie du développement.
    <http://ideas.repec.org/p/mon/ceddtr/101.html>

Criminal Police. 2005. *Statistiques élaborées pour la période du 1er janvier au 31 décembre 2004.* Abidjan: Direction de la police judiciaire.

—. 2006. *Statistiques élaborées pour la période du 1er janvier au 31 décembre 2005.* Abidjan: Directorate of the Judicial Police.

—. 2007. *Statistiques élaborées pour la période du 1er janvier au 31 décembre 2006.* Abidjan: Directorate of the Criminal Police.

—. 2008. *Statistiques criminelles élaborées pour la période du 1er janvier au 31 décembre 2007.* Abidjan: Directorate of the Criminal Police.

—. 2009. *Statistiques criminelles élaborées pour la période du 1er janvier au 31 décembre 2008.* Abidjan: Directorate of the Criminal Police.

—. 2010. *Statistiques criminelles élaborées pour la période du 1er janvier au 31 décembre 2009.* Abidjan: Directorate of the Criminal Police.

Diallo, Youssouf. 2007. 'Les Peuls, les Sénoufo et l'Etat au nord de la Côte d'Ivoire: Problèmes fonciers et gestion du pastoralisme.' *Bulletin de l'Association Euro-Africaine pour l'Anthropologie du Changement Social et du Développement*, No. 10.

Djipro, Koukou Frimo. 2009. 'Tribunal militaire d'Abidjan, Ange Kessi (Commissaire du gouvernement)—"Voici la raison de notre existence."' *Le Temps.* 24 April.

Duval Smith, Alex. 2010. 'Côte d'Ivoire: Des partenaires en Côte d'Ivoire s'efforcent d'approvisionner en eau potable des milliers de personnes.' 20 September. <http://www.unicef.org/french/infobycountry/cotedivoire_56081.html>

Ero, Comfort and Anne Marshall. 2003. 'L'Ouest de la Côte d'Ivoire: un conflit libérien?' *Politique Africaine*, Vol. 89. March, pp. 88–101.

European Commission. 2009. *Fiche d'Identification pour une Approche de Projet: Appui à la réforme et modernisation du système judiciaire et pénitentiaire en Côte d'Ivoire.* <ec.europa.eu/europeaid/documents/aap/2009/af_aap_2009_civ.pdf>

Florquin, Nicolas and Eric G. Berman. 2005. *Armed and Aimless: Armed Groups, Guns, and Human Security in the ECOWAS Region.* Geneva: Small Arms Survey.

FN (Forces Nouvelles). n.d. 'Organigramme Secrétariat Général.' <http://www.forcesnouvelles.info/site/organigramme-fn.gif>

Galy, Michel. 2004. 'Côte d'Ivoire: la violence, juste avant la guerre.' *Afrique contemporaine*, Vol. 209, pp. 117–39.
    <http://www.cairn.info/article.php?ID_ARTICLE=AFCO_209_0117>

Gbagbo, Laurent. 2010. *Discours du Président Laurent Gbagbo à l'occasion du cinquantenaire anniversaire de l'indépendance de la CI.* 7 August.
    <http://www.cotedivoirepr.ci/?action=show_page&id_page=12890>

Geneva Declaration Secretariat. 2008. *Global Burden of Armed Violence.* Geneva: Geneva Declaration Secretariat.

Gilgen, Elisabeth and Matthias Nowak. 2011. *From Conflict to Crime: Patterns of Security Perceptions in Post-war Liberia.* Liberia Armed Violence Assessment Issue Brief No 1. Geneva: Small Arms Survey.

Global Witness. 2007. *Chocolat chaud: comment le cacao a alimenté le conflit en Côte d'Ivoire.* Washington, DC: Global Witness.
    <http://www.liberationafrique.org/IMG/pdf/cotedivfrench.pdf>

Hellweg, Joseph. 2006. 'Manimory and the Aesthetics of Mimesis: Forest, Islam and State in Ivorian Dozoya.' *Africa*, Vol. 76, No. 4, pp. 461–84.

HRW (Human Rights Watch). 2001. *Côte d'Ivoire: le nouveau racisme—La manipulation politique de l'ethnicité en Côte d'Ivoire.* New York: HRW.
    <http://www.hrw.org/en/reports/2001/08/28/le-nouveau-racisme>

—. 2010. *Terrorisés et abandonnés: l'anarchie, le viol et l'impunité dans l'ouest de la Côte d'Ivoire.* New York: HRW.
    <http://www.hrw.org/fr/reports/2010/10/22/terroris-s-et-abandonn-s>

ICG (International Crisis Group). 2003. *Côte d'Ivoire: The War Is Not Yet Over.* Africa Report No. 72. Brussels: ICG.

—. 2004. *Côte d'Ivoire: No Peace in Sight.* Africa Report No. 82. 12 July. Brussels: ICG. <http://allafrica.com/peaceafrica/resources/view/00010234.pdf>

—. 2005. *Côte d'Ivoire: Le pire est peut-être encore à venir.* Rapport Afrique No. 90. 24 March. Brussels: ICG.

—. 2009. *Côte d'Ivoire: Les impératifs de sortie de crise.* Briefing Afrique No. 62. 2 July.

—. 2010a. *Côte d'Ivoire: Sécuriser le Processus Electoral.* Rapport Afrique No. 158. 5 May. Brussels: ICG.

—. 2010b. *Côte d'Ivoire: sortir enfin de l'ornière*. Briefing Afrique No. 77. 25 November. Nairobi/Brussels: ICG.

IISS (International Institute for Strategic Studies). 2010. *The Military Balance 2010*. London: Routledge.

Janowitz, Morris. 1964. *Military Institutions and Coercion in the Developing Nations*. Chicago: University of Chicago Press.

Jeune Afrique. 2010. 'La fracture militaire au centre de l'affrontement Gbagbo-Ouattara.' 13 December.
<http://www.jeuneafrique.com/Article/ARTJAWEB20101213090905/laurent-gbagbo-forces-nouvelles-armee-guillaume-sorola-fracture-militaire-au-centre-de-l-affrontement-gbagbo-ouattara.html>

Kan, Armand. 2009. 'Intelligence Economique et Sécurité Privée en Côte d'Ivoire: vers une nouvelle concurrence.' Intelligence Economique & Prospective (blog). <http://arnesta.over-blog.com/article-intelligence-economique-et-securite-privee-vers-une-nouvelle-concurrence-38460957.html>

Kieffer, Guy-André. 2000. 'Armée ivoirienne: le refus du déclassement.' *Politique Africaine*, No. 78, pp. 26–44.

Kohon, Landry. 2009. 'Tribunal militaire: Les audiences ouvertes hier en présence du Chef de l'Etat.' Connectionivoirienne.net. 22 April.
<http://www.connectionivoirienne.net/on-dit-quoi-au-pays-actualites/tribunal-militaire-les-audiences-ouvertes-hie/>

Konadjé, Jean-Jacques. 2008. *Bruits de bottes et feu sur la Côte d'Ivoire: armes légères et groupes armés dans le conflit ivoirien*. Unpublished background paper. Geneva: Small Arms Survey.

Kouamé, Allah. 2009. 'Désarmement—Bamba Mamoutou (Président des dozos de Côte d'Ivoire): "Celui qui nous désarme ne durera pas au pouvoir."' *Nord-Sud*. 6 July.

Kougniazondé, Christophe. 2010. 'L'Etat des lieux de la privatisation de la sécurité en Afrique Francophone: une revue de littérature.' Working Paper Series No. 1. Santiago de Chile: Global Consortium on Security Transformation. June.
<http://www.securitytransformation.org/images/publicaciones/158_Working_Paper_1_-_Securite_privee_en_Afrique.pdf>

Labertit, Guy. 2010. 'Côte d'Ivoire: l'heure de vérité.' Website contribution. 17 May. <http://www.cotedivoirepr.ci/?action=show_page&id_page=11967>

Lavroff, Dimitri-George. 1972. 'Régimes militaires et développement politique en Afrique noire.' *Revue française de science politique*, Vol. 22, No. 5, pp. 973–91. <http://www.persee.fr/web/revues/home/prescript/article/rfsp_0035-2950_1972_num_22_5_418945>

Le Panafricain. 2010. 'Côte d'Ivoire: Ouattara affirme avoir beaucoup de soutiens dans l'armée.' 18 January.
<http://www.lepanafricain.com/2011/01/14/cote-d-ivoire-ouattara-affirme-avoir-beaucoup-de-soutiens-dans-l-armee/>

Les milices hors Abidjan: une menace à prendre en compte dans le processus de paix. 2009. Unpublished document.

LIDHO (Ligue Ivoirienne des Droits de l'Homme). 2010. *Communication du rapport d'enquête sur la MACA*. Unpublished report.

Mieu, Baudelaire. 2009a. 'Côte d'Ivoire: un pays, deux armées.' *Jeune Afrique*. 8 September.
<http://www.jeuneafrique.com/Article/ARTJAJA2538p028-030.xml0/rebellion-laurent-gbagbo-armee-alassane-ouattaraun-pays-deux-armees.html>

—. 2009b. 'Côte d'Ivoire: Security Business.' *Jeune Afrique*. 8 September.
<http://www.jeuneafrique.com/Article/ARTJAJA2538p060-061.xml0/securitecote-d-ivoire-security-business.html>

—. 2010. 'Le shérif d'Abidjan.' *Jeune Afrique*. 11 February. <http://www.jeuneafrique.com/Article/ARTJAJA2560p038.xml0/>

Munzu, Simon. 2011. 'Côte d'Ivoire: 210 morts suite aux violences postélectorales.' ONUCI-FM. 6 January.
<http://www.unmultimedia.org/radio/french/detail/118133.html>

Murray, Ryan. 2010. *Results of the National Survey on Small Arms and Light Weapons and Security Perceptions Conducted in Côte d'Ivoire in 2010*. Unpublished background paper. Geneva: Small Arms Survey.

N'Diaye, Boubacar. 2010. 'Beyond Keenness: The Structural Obstacles to Security Sector Transformation in Francophone Africa.' In Axel Augé and Patrick Klaousen, eds. *Réformer les armées africaines: En quête d'une nouvelle stratégie*. Paris: Karthala, pp. 85–110.

ONUCI-FM. 2010. *Général Jean-Pierre Palasset, Commandant de la Force Licorne*. 15 November.
<http://www.onuci.org/onucifm/spip.php?article3133>

Ouattara, Azoumana. 2008. 'L'armée dans la construction de la nation ivoirienne.' In Jean-Bernard Ouédrago and Ebrima Sall, eds. *Frontières de la citoyenneté et violence politique en Côte d'Ivoire*. Dakar: CODESRIA, pp. 149–68.

Ouattara, Raphael. 2008. 'Côte d'Ivoire.' In Alan Bryden, Boubacar N'Diaye, and 'Funmi Olonisakin, eds. *Challenges of Security Sector Governance in West Africa*. Geneva: Geneva Centre for the Democratic Control of Armed Forces (DCAF), pp. 75–92.
<http://www.ssrnetwork.net/uploaded_files/5141.pdf>

Pézard, Stéphanie and Savannah de Tessières. 2009. *Insecurity Is Also a War: An Assessment of Armed Violence in Burundi*. Geneva: Geneva Declaration Secretariat.

PNRRC (Programme national de réinsertion et de réhabilitation communautaire). 2010. 'Profilage des ex-combattants des forces armées des Forces Nouvelles et des ex-membres des groupes d'autodefense.' 19 May.
<http://www.pnrrc-ci.org/index.php?option=com_content&view=article&id=24&Itemid=82>

—. n.d. Website. <http://www.pnrrc-ci.org/index.php?option=com_content&view=article&id=24&Itemid=82>

RASALAO-CI (Réseau d'action sur les armes légères en Afrique de l'ouest Section Côte d'Ivoire). 2008. 'Déclaration du RASALAO-CI relative au cambriolage de la poudrière de la garde républicaine de Yamoussoukro.' 26 August. <http://rasalao-ci.org/archives_communique.php3?date=11/2009>

RCI (Republic of Côte d'Ivoire). 2000. *Loi n° 2000-513 du 1er août 2000 portant constitution de la Côte d'Ivoire*.

—. 2007. Accord politique de Ouagadougou. Ougadougou, Burkina Faso. March.

—. 2009a. *Stratégie de Relance du Développement et de Réduction de la Pauvreté (DRSP)*.
<http://www.ci.undp.org/uploadoc/docs/DSRP%20FINAL%20RCI_janvier%2009.pdf>

—. 2009b. *Mouvement des détenus au titre de l'année 2009*. Abidjan: Penitentiary Administration, Ministry of Justice and Human Rights.

— et al. 1961. Accord de défense entre les Gouvernements de la République de Côte d'Ivoire, de la République du Dahomey, de la République Française et de la République du Niger. Paris, 24 April. <http://www.cotedivoirepr.ci/files/pdf/Accords_de_defense_entre_la_Cote_d_Ivoire_et_la_France.pdf>

RCI Presidency. n.d. 'Accords de défense entre la Côte d'Ivoire et la France.' <http://www.cotedivoirepr.ci/?action=show_page&id_page=581>

Reuters. 2010. *Interview: Ivory Coast Checkpoints Cost Millions a Year-Business*. 3 May. <http://www.reuters.com/article/2010/05/03/idUSLDE6421QF>

RTI (Radiodiffusion Télévision Ivoirienne). 2010a. 'Sécurité: 3 jours d'exercice de maintien d'ordre du CECOS à Adiaké.' 8 May.

—. 2010b. 'Face à Face: Gbagbo et Ouattara—Election Côte d'Ivoire, Part 2.' <http://news.abidjan.net/v/4366.html>

SCC (Security Council Committee Established Pursuant to Resolution 1572 (2004) Concerning Côte d'Ivoire). n.d. 'Reports of the Group of Experts Submitted through the Security Council Committee Established Pursuant to Resolution 1572 (2004) Concerning Côte d'Ivoire.' <http://www.un.org/sc/committees/1572/CI_poe_ENG.shtml>

SIPRI (Stockholm International Peace Research Institute). n.d. 'SIPRI Military Expenditure Database.' <http://milexdata.sipri.org/result.php4>

Touré, Moustapha. 2008. *Etude du racket sur les routes en Côte d'Ivoire*. Abidjan: World Bank and Republic of Côte d'Ivoire. <http://www.cgeci.org/cgeci/docs/documents/racket.pdf>

UNDP (United Nations Development Programme). 2010. *Le genre dans les institutions de sécurité*. <http://www.ci.undp.org/actu_detail.php?id_news=118>

—. n.d. *La question de la réforme du secteur de la sécurité en Côte d'Ivoire*. <www.ci.undp.org/docs/ARTICLEMJO%20ely.pdf>

UNICEF. 2008. *Communiqué de presse: Journée Internationale de lutte contre les mutilations génitales féminines—Tolérance zéro contre l'excision*. 6 February. <http://www.unicef.org/french/media/media_42765.html>

UNOCI (United Nations Operation in Côte d'Ivoire). 2009. *Human Rights Report to the Sanctions Committee: July–September*. Abidjan: Human Rights Section, UNOCI.

—. 2010a. *Rapport de l'enquête sur les violations des droits de l'homme liées aux événements de février 2010*. Abidjan: Human Rights Section, UNOCI.

—. 2010b. *Table/Progressive totals on FAFN regrouping operations (up. 31/03/10)*. Abidjan: Disarmament, Demobilization, and Reintegration Section, UNOCI.

—. 2010c. *Effectifs réalisés et armes/munitions collectées lors des opérations conduites du 15 au 30 juin dans la région des Savanes*. Abidjan: Disarmament, Demobilization, and Reintegration Section, UNOCI.

UNPOL (United Nations Police). 2009. 'Criminal Statistics Compiled by UNPOL Stations in 2009 in RCI.'

UNSC (United Nations Security Council). 2003. Resolution 1479. S/RES/1479 of 13 May.

—. 2004. Resolution 1572. S/RES/1572 of 15 November. <http://daccess-dds-ny.un.org/doc/UNDOC/GEN/N04/607/37/PDF/N0460737.pdf?OpenElement>

—. 2006a. *Security Council Committee Concerning Côte d'Ivoire Issues List of Individuals Subject to Measures Imposed by Resolution 1572 (2004)*. SC/8631 of 7 February. <http://www.un.org/News/Press/docs/2006/sc8631.doc.htm>

—. 2006b. *Report of the Group of Experts Submitted Pursuant to Paragraph 9 of Security Council Resolution 1643 (2005) Concerning Côte d'Ivoire*. S/2006/735 of 5 October.

—. 2006c. Resolution 1721. S/RES/1721 of 1 November.

—. 2009. *Final Report of the Group of Experts on Côte d'Ivoire Pursuant to Paragraph 11 of Security Council Resolution 1842 (2008)*. S/2009/521 of 9 October. <http://www.un.org/ga/search/view_doc.asp?symbol=S/2009/521>

—. 2010. Resolution 1942. S/RES/1942 of 29 September. <http://daccess-dds-ny.un.org/doc/UNDOC/GEN/N10/559/52/PDF/N1055952.pdf?OpenElement>

Zobo. 2009. 'Affrontement entre Forces nouvelles: 3 morts à Man.' *Fraternité matin*. 3 February.

ACKNOWLEDGEMENTS

Principal author

Savannah de Tessières

A child cries as he is questioned by a police officer after he witnessed a gunfight in the La Saline slum in Port-au-Prince, March 2010. © Ramon Espinosa/AP Photo

Securing the State
HAITI BEFORE AND AFTER THE EARTHQUAKE

8

INTRODUCTION

On 12 January 2010, a devastating earthquake killed an estimated 158,000 people in Haiti's capital, Port-au-Prince, and displaced 1.3 million more.[1] In its wake, the nation's first cholera epidemic killed more than 3,700 and infected another 185,000.[2] The international community pledged more than USD 10 billion towards rebuilding the country. As of January 2011, however, less than one-tenth of this sum had been disbursed in Haiti.[3]

The costs of the natural disaster extended well beyond death and injury. Port-au-Prince and surrounding towns were left in ruins. Virtually every government building was damaged, and many civil servants—including police officers—were killed. Wary of the potential for escalating crime and violence in the capital, multilateral agencies, regional organizations, and bilateral donors rapidly focused on promoting increased policing capacities and wider security sector reforms.

The international focus on improving security sector capacity in Haiti is not new. Since declaring independence 200 years ago, the country has contended with periodic outbursts of political violence and international efforts to influence Haitian governance through the establishment of structural adjustment programmes and reform of the justice, military, policing, and corrections systems. In spite of billions of dollars poured into enhancing conventional security promotion, these approaches are routinely criticized for generating marginal returns in terms of improved safety on the ground in Haiti.[4]

Criticism aside, outsiders have a limited understanding of the dynamics of security and insecurity experienced by Haitian communities and households. While foreign and nationally based human rights agencies and researchers have alternately blamed international and domestic actors for repression or inaction, there is virtually no evidence-based research into how Haitians actually experience day-to-day security—or what kinds of violence prevention and reduction efforts are effective.[5]

In order to bridge this information gap, this chapter considers the context of security promotion efforts in the years preceding Haiti's 2010 earthquake and emerging trends in its aftermath. It draws on the findings of three household surveys administered before and after the earthquake in order to highlight crime victimization, access to basic needs, and attitudes about gun use and policing. A central objective of the chapter is to give voice to the real threats facing Haitians, in their own words.

Key findings indicate that:

- Haiti lacks both human resources and infrastructural capacity to police its country. Its ratio of 1.05 police officers for every 1,000 inhabitants is among the lowest in the world.
- Household survey data generated since 2004 suggests that security has improved in Haiti over the past decade and has continued to improve since the earthquake. Police involvement in criminal activity, as reported by crime victims, decreased sharply after the transition to an elected government in 2007.

- Findings from surveys show that, in 2010, more than two-thirds of the general population would turn first to the police if faced with a threat to their person or property.

- The distribution of firearms may be much lower than commonly believed. In 2010, just 2.3 per cent of Port-au-Prince area households reported owning firearms. Among the wealthy, 'personal protection' was most often cited as the reason for gun ownership, while the poor most often declared they held weapons 'for work'.

- In 2010, more than three-quarters of all respondents—both in the general population and residents of internally displaced person (IDP) camps—said that more control over the issuing of firearms licences would make their communities safer.

- Despite considerable challenges in advancing police reform over the past decade, popular confidence in the Haitian National Police (HNP) has increased since the earthquake.

Divided into three sections, this chapter begins by reviewing the state of the security sector before the 2010 earthquake. The focus is primarily on the Haitian National Police since former President Jean-Bertrand Aristide demobilized the armed forces—long associated with repressive practices—in 1994. The chapter discusses the current state of

Map 8.1 **Haiti**

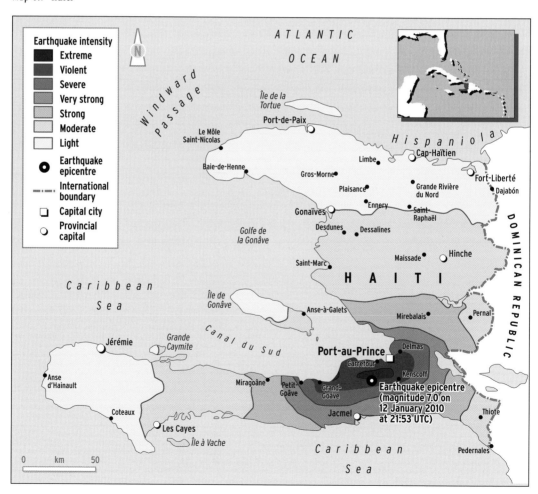

security as well as challenges facing the criminal justice system and HNP in the post-earthquake period. Lastly, it draws on findings from household surveys completed before and after the earthquake, to examine both the prevalence of crime victimization among Haitians since 2004 as well as recent changes in public opinion regarding gun ownership and security provision.

HAITI AND ITS DISCONTENTS

Haiti has been characterized by outsiders as a fragile, failing, or failed state since at least the 1980s.[6] The country has experienced considerable political volatility over the past two centuries, with more than 30 coups since independence in 1804 and no fewer than nine UN missions since 1990. While geopolitical interference in Haiti has played a significant role, particularly since the 1990s, certain analysts point to the country's extreme concentration of authority and wealth in the hands of the elite—elected and otherwise—as a source of persistent instability (Muggah, 2008; Maguire, 2009a; 2009b).

Haiti has been seen as a fragile, failing, or failed state since at least the 1980s.

For some of Haiti's diaspora and certain foreign governments, the heavy-handed dictatorships and associated paramilitary rule from the 1950s to the 1980s afforded a degree of stability. Yet, from the perspective of the vast majority of Haitians, especially those eking out an existence in the country's popular zones or shantytowns in and around Port-au-Prince and other major cities, the Duvaliers—father and son—terrorized the population into submission. They did this both through the arming of the so-called Tonton Macoute militia and by empowering Haiti's police force, then part of the Haitian armed forces, to use indiscriminate killings, torture, and arbitrary detention to enforce their power.[7]

In the latter half of the 1980s, the country experienced a rocky transition to democracy during which President Jean-Bertrand Aristide became the country's first democratically elected leader in 1991. In addition to promoting political participation by the impoverished majority of the population—a first in the country's history, which won him supporters and critics both in Haiti and abroad—Aristide demobilized the Haitian armed forces by presidential decree in 1994 and created the country's first civilian national police force, the HNP (Dupuy, 2005).

Haitians were initially hopeful that this new body, the HNP, would effectively control crime and increase safety, especially in the larger cities. During the 1990s, property crime and violence were widespread, in sharp contrast to Haiti's historically low crime rates. Business owners and the wealthy relied on privately hired armed guards—who were frequently implicated in vigilante-style violence—to provide basic security. Despite considerable investments in capacity development and training of the nascent force, the HNP was unable to address community-level criminal violence adequately during its early years (Hayes and Wheatley, 1996).

The political and economic situation in Haiti began to deteriorate dramatically during the late 1990s and early 2000s. Growing instability tested the HNP's ability to fight criminal violence and respond to organized armed violence committed by political groups. Trafficking in persons, weapons, and drugs, reportedly connected to Haiti's business elite, continued unabated, bringing financial support to the few while generating political unrest. As tensions mounted between the Haitian government and certain members of the international donor community, such as the United States and France, former members of the disbanded Haitian armed forces created the so-called 'rebel army', also known as the National Revolutionary Front for the Liberation of Haiti (Front pour la Libération et la Reconstruction Nationales) (Muggah, 2005a). The army was composed of paramilitary thugs active during the 1991–94 military coup years and had recruited politically motivated armed gangs into their ranks.

With foreign backing and support from the national elite, as well as supporters in key positions within the HNP itself, the rebel army proved to be a surprisingly resilient opponent. Heavily armed with assault rifles and a firm supply network, the force began launching quiet but efficient attacks against border towns and urban centres between 2000 and 2004, with the goal of overthrowing the elected Haitian government. HNP officers struggled to respond. International and US-led restrictions against arms sales to the government since the early 1990s had never been fully lifted,[8] effectively prohibiting the HNP from legally purchasing weapons (Muggah, 2005a; 2005b). The HNP officers who remained committed to upholding the rule of law had few arms and little chance of surviving direct armed conflict with the rebel army.

By 2004, following successful rebel army attacks in the towns of St. Marc and Gonaïves, the HNP was overcome and scattered. The insurgent army rapidly advanced on the capital. With Aristide removed from power by foreign diplomats and with US marines occupying the National Palace, the insurgents were free to take the capital. Indeed, one of the insurgents' first actions after entering Port-au-Prince was to march two blocks past the National Palace to the National Penitentiary, where they freed hundreds of convicts.[9]

Despite the circumstances surrounding the interim government's establishment, the international community stepped in to support it with a stated goal of reshaping the fragile security sector.[10] Much like the US-led de-Bathization process in Iraq, the HNP was purged of 60 per cent of its officers, many of whom fled to other areas of the country

An anti-Aristide rebel beats on the corpse of a police officer with a machete in Gonaïves, some 100 km north of Port-au-Prince, February 2004.
© Rodrigo Abd/AP Photo

or to the Dominican Republic, fearing that remnants of the rebel army might exact revenge. Some 540 members of the rebel army, many of whom had been soldiers in Haiti's long demobilized armed forces, were integrated into the 'new' HNP. Few of them, if any, were required to undergo the formal training and graduation from the police academy required of new recruits (Hallward, 2008, p. 128; ICG, 2005; Mendelson-Forman, 2006).[11]

At the request of the new interim government, the United Nations Security Council established the UN Stabilization Mission in Haiti (MINUSTAH) in June 2004.[12] Led by Brazil, Canada, the European Union, and the United States and involving more than 40 countries, the large-scale deployment of international peacekeepers and police support marked an important turning point. With nearly 9,000 blue helmets and 3,000 international police deployed, the mission focused on ensuring stability by enhancing HNP capacities, extending the rule of law through improved delivery of justice services, and rebuilding the country's dilapidated judicial system (Muggah, 2010b, pp. 451–52). Though initially formed to uphold a coup government widely viewed by Haitians as illegitimate and repressive, MINUSTAH was successful in establishing strong support among both Haitian policy-makers and segments of the general population. Thus, MINUSTAH was able to maintain its presence even after the transition to a democratically elected president was made in late 2006.

A STATE OF INJUSTICE

Many international organizations and institutions—from the UN to the Organization of American States and the International Organisation of La Francophonie—devoted considerable energy to the reform and strengthening of Haiti's security and judicial system (Baranyi and Fortin, forthcoming). Support has ranged from financial assistance to the provision of technical expertise in policing, investigation, customs, and corrections reform. Donor-supported efforts to promote judicial reform since the mid-1990s have included the restructuring and revision of judicial procedures, legal codes, and protocols.

Many international organizations have been involved in reforming Haiti's security and judicial system.

Since 1998, efforts to codify and implement improved criminal and corrections laws have yielded few returns.[13] The most significant development in reforming the judicial sector during the past decade was arguably the passage of laws by Parliament in 2007 to create the Superior Judicial Council, mandated to devise rules for the training, recruiting, and disciplining of magistrates and the regulation of Haiti's magistrates school. In 2007, the Ministry of Justice and Public Security published a 'roadmap' identifying a range of key priorities to enhance the quality and quantity of justice, and, in particular, service delivery (UNDP, 2009).

To enhance implementation and improve access to justice for the population, an approach to justice reform emerged, focusing on simultaneous reforms across the judicial, policing, and corrections sectors and linking these to enhanced accountability.[14] For example, a Citizen's Forum (Comité Coordonnateur du Forum Citoyen) was created both to enhance citizen engagement and to monitor government transparency. Nevertheless, the country continued to feature outdated and disregarded laws, weak human resources, and practically non-existent infrastructure to manage cases (Baranyi and Salahub, 2010).

Over the past two decades, a major obstacle to high-quality judicial, police, and corrections service delivery was their illegitimacy in the eyes of Haitian civilians. This was particularly true during the years when unelected governments were in power. For instance, during the military dictatorship (1991–94), police officers were frequently implicated in the illegal arrest and torture of ordinary citizens (O'Neill, 1995). This changed from October 1994 to February

Haitian police officers protect opponents of former President Jean-Bertrand Aristide from being shot during a march in Port-au-Prince, March 2004.
© Pablo Aneli/AP Photo

2004, as the country was in a period of struggling democratic governance, and state leaders rejected the use of the police force to exert political control. However, the post-coup interim government of President Boniface Alexandre and Prime Minister Gérard Latortue (2004–06) flirted with using the police as a tool to suppress popular dissent and punish political opponents (Dupuy, 2005).

As a result of the HNP's inefficiency and susceptibility to corruption, but also, in many cases, officer involvement in a wide range of human rights violations during the 2004 coup and its two-year aftermath, both international donors and local populations lost faith in the police force's capacity and willingness to deliver services. To bridge this legitimacy gap, donors invested heavily in police reform, recruitment, and human rights training, as well as community policing from 2004 onwards (Baranyi and Fortin, forthcoming; CIGI, 2009).

The Haitian National Police: before 2010

Although the 1987 Haitian Constitution sets out the terms for a national police force—including provisions for its composition and purpose[15]—the official HNP force was not established until the mid-1990s. Formally launched by President Aristide in 1995, the HNP was intended to serve as the exclusive armed entity responsible for maintaining law and order and protecting the life and property of citizens. Following the dissolution of the armed forces (the Forces Armées d'Haïti) the previous year, the HNP enjoyed wide jurisdiction. Haitians initially greeted the formation of the HNP with considerable enthusiasm. Despite investment from the United States and Canada and two UN missions (MIPONUH I and II), however, popular support for the police began to erode between 1996 and 2003 (Muggah, 2005a).

Administratively, the HNP is overseen by the Ministry of Justice and Public Security. Similarly, the Prisons Administration Directorate and the emergency fire brigade (*sapeurs pompiers*) fall under the jurisdiction of the HNP. According to an internal review led by the Haitian authorities, the HNP has faced a host of inadequacies and problems since 1995, including limited staff, weak training, unpredictable funding, limited senior personnel, systemic corruption, poor inspection capacities, and a history of violating human rights.[16] To the average Haitian, the police were seen as having limited effectiveness at best.

> Haiti has 1.05 officers for every 1,000 inhabitants— among the lowest ratios in the world.

The size and distribution of the HNP has oscillated since its creation, and it has never met international standards with respect to strength or capacity. As of late December 2009, public authorities claimed there were some 9,520 enlisted police officers (including some 746 female officers), most of them deployed in urban areas.[17] Added to this were 705 police and administrative staff of the Prisons Administration Directorate, for a grand total of 11,458 personnel.[18] This figure was lower than the projected strength of 14,000 officers set out in an HNP reform plan to be reached by late 2011 (GoH, 2006; US, 2009). With a population of at least 9.7 million, this accounts for a ratio of 1.05 officers for every 1,000 inhabitants—among the lowest ratios in the world.[19]

With assistance from the UN and bilateral donors, the institutional infrastructure of the HNP experienced considerable reforms beginning in the mid-1990s. As of 2009, there were more than 236 HNP facilities throughout the country.[20] Yet an estimated 39 of these—including seven precincts and 32 sub-precincts—were still considered non-operational prior to the January 2010 earthquake. Moreover, the overall size and configuration of the country's motor fleet was limited—with an estimated 600 vehicles of varying make and quality—resulting in major maintenance challenges.[21]

As noted above, the HNP is also mandated to oversee corrections facilities. Indeed, the Penitentiary Administration, the first civilian prison system in the country's history, was established in 1995 by former President Aristide. It was reconstituted as the Prisons Administration Directorate under the auspices of the HNP in 1997, and rudimentary investments were made until the country's descent into extreme violence in 2004. Prior to 2004, there were some 21 prisons with an estimated 6,440 m² of cell space for 3,640 detainees, or 1.76 m² per detainee. By 2009, however, the remaining 17 functioning prisons (overseen by 705 officers) reported a capacity of 4,894 m² for some 8,686 detainees, or 0.4 m² per detainee—well below international standards.[22] Haitian corrections authorities contended in 2009 that the number of detainees could surpass 16,000 by 2012.[23] Just as disconcerting, according to Haitian prison authorities, more than three-quarters (76 per cent) of detainees were 'pre-trial', with an average pre-trial detention period of approximately 20 months (see Box 8.1).[24]

Some international organizations credit MINUSTAH with having improved security across Haiti, particularly between 2007 and 2009 (Muggah, 2010b). However, human rights groups and researchers also heavily criticized MINUSTAH's early tactics, particularly its repressive handling of gangs (Hallward, 2008, pp. 272–76, 398–404). Specifically, between 2004 and 2006, heavy-handed interventions pursued by the HNP with tacit support from

MINUSTAH were designed specifically to arrest and neutralize armed elements. In some cases, these activities—described bluntly as 'disarm or die' campaigns—resulted in the accidental shooting deaths of dozens of citizens, including children.[25]

Against a backdrop of MINUSTAH-led stabilization efforts, UN civilian agencies were busy crafting a reform plan for the HNP with local counterparts in 2006.[26] In view of the frequency of natural disasters and the legacy of political unrest in Haiti, donors placed an emphasis on improving HNP capacity to counter floods, fires, and hurricanes throughout the country.[27] By 2009, there was growing confidence among international actors in the potential of the HNP to provide security, with the UN Security Council acknowledging key gaps but also citing real improvements.[28]

The Haitian National Police: after the earthquake

The impact of the earthquake on the human and physical infrastructure of the justice and security sector—and particularly the HNP—was extensive. Almost 80 HNP personnel were killed and another 253 injured directly by the earthquake. By UN estimates, almost one-quarter of Haiti's police capacity was rendered non-operational.[29] MINUSTAH records show that 55 buildings used by the HNP were affected, including some 28 facilities suffering 'major damages', such as collapse, and another 27 experiencing 'minor damages'.[30] If these structures are added to the 39 facilities that were already non-operational at the time of the earthquake, almost 40 per cent of HNP capacities could not be used at this stage.[31]

In the immediate aftermath of the earthquake, the focus of the UN and international donor community was on rapidly ensuring the delivery of life-saving supplies, personnel, and equipment and restoring police communication, coordination, and response capabilities, particularly in anticipation of increased gang violence. International observers were concerned that damage and displacement generated by the earthquake—coupled with the impact of the global fiscal crisis on food prices—could generate a humanitarian disaster and an upswing of crimes against property and violence.[32]

In the first six months after the natural disaster, fears that escapees from prisons would perpetrate targeted attacks, extortion, and kidnappings were commonplace among NGOs and international organizations working in Haiti (Muggah, 2010b). International aid providers were worried that, if such violence were to occur, it would

Members of a Haitian police special unit arrest a suspect during a sweep through Cité Soleil, the largest slum in Port-au-Prince, February 1997. © Thony Belizaire/AFP Photo

Table 8.1 Characteristics of Port-au-Prince residents arrested by the HNP, 2004–10

Category	Characteristics	Percentage
Sex	Male	86.9
	Female	13.1
Age	Under 18	19.0
	18–27 years old	42.9
	28–37 years old	16.7
	38 or older	21.4
Religion	Catholic or Protestant	48.9
	Voodoo	16.7
	Christian but also practises Voodoo	29.8
	Rastafarian	3.6
	Other or no religion	1.2
Employment at time of arrest	Self-employed–sales or service	35.7
	Self-employed–trade or professional	11.9
	Employed part-time or sporadically	7.9
	Employed full-time	2.4
	Unemployed adult	33.8
	Unemployed child under 18	8.3

Table 8.2 Characteristics of Port-au-Prince residents arrested by the HNP, 2004–10

		2004	2005	2006	2007	2008	2009	2010
Place of detention (%)	Police station	38.4	41.5	40.6	32.7	24.1	22.8	60.0
	Prison	38.2	45.8	53.9	67.3	75.9	77.2	10.0
	Other	23.4	12.7	5.5	0.0	0.0	0.0	30.0
Reason for arrest (%)	Violent crime	12.1	9.0	4.8	13.7	18.3	20.2	*
	Non-violent crime	21.4	21.5	32.2	50.5	61.2	66.7	*
	Delinquency	22.1	17.4	16.2	13.8	9.7	10.5	*
	Unknown	44.4	52.1	46.8	22.0	10.8	2.6	*
Percentage who saw a judge		1.2	3.4	6.4	8.1	7.0	6.8	*
Percentage convicted of a crime		9.2	8.9	5.8	6.0	4.1	2.1	*
Average length of detention (in days)	Held in police station	23.2	14.8	11.7	4.3	3.1	2.6	1.1
	Held in prison	402.1	423.8	355.0	301.2	280.6	284.2	*
	Held in another place	11.0	9.2	**	**	**	**	2.5

Notes:
* Insufficient information was available for 2010, and these values could not be calculated.
** No individuals indicated being held in a place other than a police station or prison between 2006 and 2009.

Members of the local police force relax at their makeshift headquarters in Port-au-Prince in January 2010.
© Fred Dufour/AFP Photo

hamper relief efforts in Haiti and exacerbate instability if humanitarian assistance did not successfully reach affected populations. In certain cases, US officials turned away flights delivering supplies and medical personnel so that planes with US combat troops could land instead.[33]

Throughout this period, MINUSTAH military and police personnel supported domestic efforts alongside US and Canadian troops.[34] Fears of food riots, fleeing prisoners, and growing disorder were matched with massive investments in restoring public security. The so-called 'security umbrella' generated by this international presence is credited with enhancing humanitarian aid distribution, search and rescue operations, and the gradual return of national police to challenging areas. Meanwhile, a growing number of large, foreign private security companies began to explore opportunities in the country.[35]

VIOLENCE AND CRIME IN HAITI: BEFORE AND AFTER THE EARTHQUAKE

Despite considerable investment in justice and security before and after the 2010 earthquake, little is known about whether real and perceived safety have improved for most Haitians. Rather, media accounts continue to emphasize lawlessness, chaos, and brutality in the capital and surrounding regions. The problem of sexual violence is repeatedly cited as evidence that the security sector is in shambles, as is the alleged surge of criminal banditry in the shanty-towns of the country's major cities (MADRE et al., 2011).

Box 8.2 Household surveys in Haiti

Researchers affiliated with the Small Arms Survey conducted studies in Haiti to assess the experiences and opinions of Haitian citizens over the past five years. Similar sampling procedures and data collection instruments were used in each of the surveys. Households were randomly sampled from the population using Random GPS Coordinate Sampling, and adult household members were then randomly selected to participate in the study. This process allowed for a representative sample that can be generalized to the entire population, providing invaluable insight into the experiences, ideas, and opinions of ordinary Haitians.

The 2010 surveys, funded by the UN Development Programme and the Ottawa-based International Development Research Centre, were undertaken primarily to inform the Post-Disaster Needs Assessment process and support government authorities in determining priorities for security promotion (see Table 8.3). These surveys allow for a careful reading of mortality, injury, and victimization as well as attitudes towards public institutions and security providers, including the HNP and foreign actors.

Table 8.3 Port-au-Prince area household surveys

Year	Study	Population and focus
2005	Wayne State University Study of Health and Human Rights in Haiti	1,260 Port-au-Prince area households regarding experiences in the 22 months following the February 2004 coup against Aristide (Kolbe and Hutson, 2006; Wayne State University, 2005).
2009	University of Michigan Neighbourhood Survey	Nearly 1,000 households from three highly populated and impoverished neighbourhoods in Port-au-Prince; focused on crime, gun use, and opinions about security provision (Small Arms Survey, 2009b).
	National Assessment of Health & Harm in Haiti	2,800 households from urban and peri-urban communities throughout Haiti, of which 1,800 were from the Port-au-Prince area; focus on physical and mental health, experiences with human rights violations from 2004 to 2009, substance use, gun use, quality of life, and opinions regarding security and national events (Kolbe et al., 2010; University of Michigan, 2009). Qualitative interviews were conducted with an additional 150 crime victims, their household members, and community leaders regarding the need for access to community services (University of Michigan, 2009).
2010	Post-Earthquake Assessment–General Population	Recontacted the 1,800 capital-area households interviewed in 2009 for an additional interview; focus on current location and status of all previously included household members, provision of basic human needs, food insecurity, sanitation, physical and mental health, mortality, opinions about service provision, gun use, access to information, and experiences with crime. An additional 150 qualitative interviews were conducted with crime victims, household members of victims, and leaders in the community (Kolbe and Muggah, 2010; Kolbe et al., 2010; Small Arms Survey, 2010).
	Post-Earthquake Assessment–IDPs	1,147 households residing in 30 IDP camps (25 of which were randomly chosen, five of which were identified as the largest camps in the capital area); focus on assessing basic human needs, health, access to information, gun use, opinions about security, and experiences with human rights violations (Kolbe and Muggah, 2010; Kolbe et al., 2010; Small Arms Survey, 2010).

In fact, comparatively little data or analysis is available on actual rates of criminal violence over time and across communities. National government capacity to conduct vital and criminal statistics remains startlingly weak. Despite some descriptive studies undertaken in recent years by Canadian and US research institutes and universities, Amnesty International, and Médecins Sans Frontières, the availability of valid and reliable evidence is limited.[36] To fill these data gaps, several surveys of Haitian households were conducted in 2005, 2009, and 2010 (see Box 8.2). The 2010 post-earthquake surveys found that, contrary to popular belief, crime rates were much lower than suggested by the global media.

Other surprising findings emerged from the 2010 surveys. First, violent crime was considerably less pervasive in the six weeks after the earthquake than was indicated by the media. Despite major concerns among international donors about the risk of property-related crime, it was also surprisingly low. Another notable finding in 2010 was that the preferred security provider for addressing crime and victimization was, overwhelmingly, the HNP.

Violent crime

Crime and insecurity were widely considered problems long before the earthquake of January 2010. Indeed, almost two-thirds (62.9 per cent) of those surveyed in the general population in 2009 asserted that, before the earthquake, crime was a serious problem. After the earthquake, however, just one in five (20.0 per cent) of those surveyed in the general population said that crime or insecurity constituted a major problem (see Figure 8.1). While this drop could be due to a reprioritization of needs in the immediate aftermath of the disaster, it could also reflect reduced experiences of crime as rates of property crime, kidnapping, physical assault, and murder decreased in comparison to previous years.

In terms of violent crime, a clear pattern emerges that mirrors the country's democratic transitions: the incidence of violent and non-violent crime was low in the first two months of 2004 (when the country was ruled by an elected leader), rose significantly from March 2004 to late 2006 (when an unelected leader ruled following a coup), and decreased steadily between early 2007 (after democratic elections) and 2010 (Kolbe et al., 2010).

Figure 8.1   **How serious a problem is crime/insecurity?**

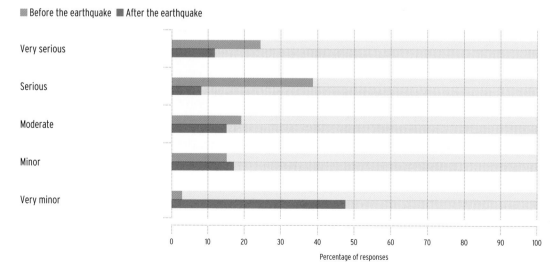

■ Before the earthquake ■ After the earthquake

Sources: Small Arms Survey (2010); University of Michigan (2009)

Sexual violence

Notwithstanding reductions in reported violent or politically motivated crimes, reports of sexual assault increased dramatically in the post-earthquake period. Considerable media attention was devoted to the rising incidence of sexual violence in Port-au-Prince and around displaced person camps in the months following the earthquake.[37] While Amnesty International pointed to the likelihood of gender-based victimization, its reports relied on qualitative assessments that could not be used to calculate rates of victimization.[38] The findings generated by this study do indeed support claims of a sharp increase in sexual assaults made in these reports; survey data shows that an estimated 10,813 individuals[39] in Port-au-Prince were sexually assaulted in the six weeks after the earthquake, with almost 70 per cent of the attackers identified by the respondents as (anonymous) 'criminals' (Kolbe et al., 2010).[40] The number of victims is significantly higher than in the previous three years, when an estimated 30,000–50,000 individuals were sexually assaulted per year.

Property crime

While NGO reports regarding sexual violence were supported by the surveys, media claims of widespread looting and organized theft were not. The vast majority of Port-au-Prince residents reported that neither they nor any members

The corpse of a 55-year-old man lies on the ground of his shack as residents look on in the slum of Cité Soleil in Port-au-Prince, February 2005. Witnesses said he was accidentally killed by police. © Ariana Cubillos/AP Photo

Table 8.4 **Percentage of Port-au-Prince households reporting property crime**								
	General property theft	**Food only**	**Food and property**	**Money and property**	**Water only**	**Money only**	**Vandalism**	**Broke in but did not steal anything**
2004	1.5	0.5	0.2	0.2	0.1	1.3	0.2	0.1
2005	2.0	0.0	0.0	0.6	0	0.6	1.2	0.2
2006	0.4	0.2	0.1	0.2	0.2	0.4	2.2	0.1
2007	0.4	0.2	0.1	0.2	0.1	0.2	0.0	0.0
2008	0.5	0.1	0.1	0.0	0.0	0.3	0.2	0.0
2009	0.6	0.1	0.1	0.1	0.3	0.1	1.3	0.0
2010	0.3	0.8	0.7	0.0	1.4	0.1	0.2	0.3

Sources: Small Arms Survey (2010); University of Michigan (2009)

of their household had property stolen from them or intentionally destroyed by others after the earthquake. Only an estimated 4.1 per cent of all Port-au-Prince households experienced some form of theft, vandalism, or intentional destruction of property in the first six weeks after the earthquake. Indeed, the most common thefts reported related to water or food, unsurprising given the high levels of food insecurity.

These incidents tended to be geographically concentrated in certain neighbourhoods and usually involved relatively modest values. Notably, in comparison to neighbourhoods ranked as dangerous by survey respondents, Cité Soleil and Bel Air were identified as 'average'. A comparative analysis reveals that property crime decreased as a whole; while 4.2 per cent of surveyed households reported property crime during January and February 2005, only 1.3 per cent of households did so in the same months of 2009 (see Table 8.4).

Attitudes towards security providers

The household surveys highlight that the general population and IDP camp residents viewed the HNP as the preferred security provider in 2010. What is perhaps most interesting is to what extent appreciation of police had improved since the previous year. When asked, 'Who would you turn to first if you were robbed or someone threatened to hurt or kill you?', more than two-thirds (66.7 per cent) of all respondents in 2010 (both general and displaced) identified the police. This figure stands in sharp contrast to that of 2009, when just 38 per cent of the population listed the HNP as a first recourse in the case of threats to person or property. Other responses included relatives or neighbours, heads of household, and community elders (see Table 8.5). Possible explanations for the increased confidence in the HNP include heightened confidence in public institutions, a decrease in the UN's credibility, and disruptions to alternative routes for personal security (such as relying on family or hiring private security guards) in the wake of the earthquake.

There also appears to be widespread agreement among Haitians that the HNP should be the primary security entity in the country. When asked in 2010, 'Ideally, who should be responsible for security?', almost two-thirds (63.6 per cent) of the general public named the police. Meanwhile, the 'community' was cited by more than one-quarter (27.2 per cent), and the remainder chose MINUSTAH, the family, local government, the Ministry of the Interior, or

Table 8.5 Who would you turn to first if you were robbed or someone threatened to hurt or kill you?

Response	2009		2010	
	Robbed (%)	Threatened (%)	Robbed (%)	Threatened (%)
Relative, friend, or neighbour	12.0	18.1	38.5	13.5
Police	40.7	38.0	56.6	66.7
Former members of the Haitian army	0.7	0.4	0.1	0.0
Foreign military	9.7	28.9	0.3	0.0
Private security company or similar	0.3	0.7	0.0	0.0
Community elders	3.7	2.5	2.3	8.4
Head of the family	0.6	2.5	0.8	9.2
An armed group	0.4	1.2	0.2	0.0
Nothing/no point in doing anything	29.9	6.6	0.6	0.1
Other/don't know	2.0	1.1	0.6	0.0

Sources: Small Arms Survey (2010); University of Michigan (2009)

private security firms. Not one respondent opted for the former members of the armed forces (see Table 8.6). This is an important finding, since there has been persistent debate in some quarters since 2004 around resurrecting the disbanded Haitian armed forces (Hallward, 2008).

Table 8.6 General population: Ideally, who should be responsible for security?

Response	2009		2010	
	Frequency	Percentage	Frequency	Percentage
Local government	136	7.8	17	1.0
Ministry of the Interior	85	4.9	7	0.4
MINUSTAH	374	21.6	61	3.5
Police	859	49.5	1,102	63.6
Former members of the armed forces	19	1.1	0.0	0.0
Private security firms	50	2.9	14	0.8
The community	109	6.3	471	27.2
Family	46	2.7	60	3.5
An armed group	31	1.8	0.0	0.0
Other	25	1.4	0.0	0.0
Total	**1,734**	**100.0**	**1,732**	**100.0**

Sources: Small Arms Survey (2010); University of Michigan (2009)

A private security guard stands outside a burning store in downtown Port-au-Prince, January 2010.
© Carlos Barria/Reuters

Similar themes emerged from an analysis of in-depth qualitative interviews conducted with randomly selected survey participants in 2009 and 2010. Respondents across all socio-economic backgrounds expressed increasing confidence that the police could and would respond to their requests for assistance (see Box 8.3). As one interviewee put it:

> *The police have changed and now they are getting better at doing their job. In the past the police sat around doing nothing. If you approached them to complain about a crime you would be lucky to get any response. Now they are more active because they know the eyes of the foreign police are on them.*[41]

Other respondents attributed improved policing to increased funding, better training, and technical assistance provided by MINUSTAH and foreign consultants; they also reported that the police were no longer engaged in politically motivated mistreatment of particular segments of society. Several respondents claimed that police officers treat residents with more respect because they are recruited from within the neighbourhoods that they are policing. As one elderly resident of Carrefour explained:

> *It's not an 'us versus them' situation anymore. We've known some of [the officers] since they were children playing here in our streets. So when they come here to do police work we treat them with respect and they treat us with respect as well.*[42]

Box 8.3 Perceptions of policing in 2010: 'getting better all the time'

Respondents routinely used terms such as 'changed', 'better', and 'more professional' to describe the HNP in 2010 as compared to five years ago. Political changes, as well as professionalization and training of the HNP, could account for this shift; five years ago, the police were regularly used by the national government to curb free speech and maintain unelected power in the face of widespread popular discontent. Today's elected government largely avoids using the police force in such a way, and some police officials who were involved in those practices in the past have left the force.

While respondents described some shortcomings, they expressed overall confidence that the police would respond when needed, that they were more active in policing specific neighbourhoods, that they were not systematically used by the current government to target political dissentients, and that they were less corrupt than in the past. According to those surveyed, police presence was robust following the earthquake, with about half of all respondents having seen police within the last 48 hours (see Figure 8.2). Police appeared to be readily available in most neighbourhoods; they continued with regular policing duties, sometimes despite the destruction of their neighbourhood police station.

In the course of the 2010 survey and qualitative interviews, respondents shared personal stories that highlighted their attitudes towards and experiences with the police:

During that time [2006] if you told the police you had been raped, they might take that as an invitation to have you [sexually] as well. They wouldn't protect you or arrest the rapist. But when my daughter was violated [in 2009] I was confident in the police. They had a policewoman interview my child and they arrested [the rapist]. The police said: 'Don't worry, we will help you. You don't need to shed any more tears.' And they were right.

—37-year-old woman, market vendor

Two police officers are always stationed at the end of our street. My wife went to them when our home was robbed. They called the boss and made a report. The boss agreed that robbery was occurring too frequently in our area so he sent some other police to track down the robbers. They spoke to everyone and found witnesses so they could arrest the criminals. In the past only the wealthy received this kind of service from the police; now even an ordinary man can expect to have his report treated with importance.

—45-year-old man, taxi driver

Figure 8.2 **When was the last time you saw the police?**

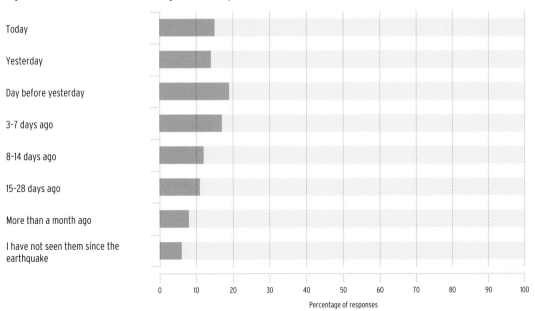

Percentage of responses

Sources: Small Arms Survey (2010); University of Michigan (2009)

Residents also highlighted the increased responsiveness on the part of the police when responding to situations such as domestic violence, sexual abuse, crimes against children, and conflicts involving the mentally ill. Most respondents said that, in the past, HNP officers had refused to intervene in some situations, such as domestic violence, claiming it was not illegal for a man to hit his wife. They described how, in recent years, particularly since 2008, the police had become more responsive in addressing violence against women.

The 2010 survey revealed a host of anecdotes regarding enhanced police sensitivity. For example, one man recounted seeking police assistance for a spousal abuse incident involving his neighbour:

> *We were reluctant to get involved, but finally my wife told me to go to the police. The police came and arrested [my neighbour]. A female officer talked to his wife and warned her that a judge might release her husband soon. So she decided to leave and return to her parents' home in the provinces. The police helped her pack her things and drove her to the taptap [private transport] station.*[43]

HNP involvement in human rights violations

In 2010, Haitians expressed a new trust that the HNP would provide security.

Each of the household surveys examined the role of the HNP in committing human rights violations. A frequent complaint voiced by respondents concerned the HNP's use of excessive force and arbitrary arrest during various periods since 2004. Respondents described how officers had indiscriminately arrested passers-by during demonstrations, beat and shot market women and children during anti-gang manoeuvres, and engaged in unethical conduct, such as theft, vandalism, or demanding bribes while on duty. Reports from the International Crisis Group recount similar incidents involving the HNP (ICG, 2005; 2008).

Empirical evidence reveals a pattern of HNP involvement in criminal activity, particularly when examining property crimes reported by survey respondents between January and February of each year from 2004 to 2010. During those months in 2004, HNP officers were not held responsible for any of the property crimes; however, during the same months of 2005 and 2006, HNP officers were named as responsible for 17.7 and 10.8 per cent of property crimes, respectively. Similarly, HNP and other government security forces were identified as responsible for 5.7 per cent of an estimated 32,000 property crimes committed in the Port-au-Prince area in the 22 months after the departure of President Aristide on 28 February 2004 (Kolbe and Hutson, 2006). This pattern of HNP involvement in criminal activity tracks the overthrow and return of democracy exactly.

Yet the past three years have witnessed a reversal of this trend. Indeed, for both property crimes and crimes against persons, police were seldom found responsible for perpetrating criminal acts from early 2007 (see Table 8.7). In the six weeks after the earthquake, HNP and foreign soldiers were blamed for some property crimes. Yet, on closer inspection, respondents affirmed that while their home had been broken into, nothing had actually been stolen. Acts that residents called vandalism were probably conducted in the course of post-earthquake search and rescue operations rather than for criminal purposes.

Instead of regarding the police as perpetrators of violence who are to be feared, respondents interviewed for the 2010 surveys were more likely to describe the HNP as 'protectors'. Data from both surveys reflects this perception, with police significantly less likely to be named as the perpetrator of a crime in 2008–10 as compared to 2004–07. Indeed, qualitative interview respondents who had previously relied on private security firms or had felt compelled to arm themselves to ward off crime expressed a new trust that the HNP could and would provide security and that the aforementioned coping strategies were less necessary than in the past. Nevertheless, a significant debate continues

Table 8.7 Perpetrators of property crimes in January and February, 2004-10

Perpetrators	Percentage of responses						
	2004	2005	2006	2007	2008	2009	2010
Criminals	67.8	39.0	21.4	36.5	44.9	68.2	17.1
HNP	0.0	14.4	13.8	0.0	0.0	0.0	2.0
Foreign soldier	0.0	11.2	13.9	5.8	1.0	0.0	4.4
Gang member	1.1	6.2	9.5	6.0	8.7	0.0	1.5
Armed political group	0.0	12.7	6.2	0.0	0.0	0.0	0.0
Neighbour	7.1	1.4	1.8	7.8	15.7	8.7	29.4
Crowd of desperate people	0.0	1.4	0.0	0.0	0.0	0.0	5.9
Current or ex-friend/partner	4.8	1.0	6.0	8.4	18.3	21.7	11.8
Unknown	19.2	12.7	27.4	35.5	11.4	1.4	27.9

Sources: Small Arms Survey (2010); University of Michigan (2009)

in Haiti, as elsewhere, over the merits of publicly versus privately administered security (Jones, 2010). When asked in 2010 who should be responsible for security, however, the overwhelming majority of IDP camp respondents (more than two-thirds) indicated local government. The army (long disbanded) was a distant second at 8.7 per cent.

Arming for self-defence

As citizens develop more confidence in their police force, they are arguably less likely to obtain and use weapons to protect themselves. When asked in 2010 whether they held a weapon, only 2.3 per cent of Port-au-Prince area households reported owning firearms. Respondents may be reluctant to discuss sensitive topics such as gun ownership, or they may appear cooperative although they are dishonest when responding. To increase accuracy, interviewers repeatedly reminded respondents that the survey was confidential, and that they could decline to answer any question they wished.

Despite this, few respondents declined to answer, and most were not only willing to answer, but also to show their weapons and gun permits as evidence. Haitian society does not have cultural taboos regarding gun ownership that would prevent respondents from disclosing their ownership of firearms. Since respondents were forthright in other segments of the interviews when providing sensitive information (for instance, when discussing substance abuse, experiences of sexual abuse, and, in a few cases, their own illegal 'employment'), the low ownership rate of 2.3 per cent may be treated as reasonably reliable. It is slightly higher than the percentage of Port-au-Prince residents with permits to own a firearm (which was 1.9 per cent in 2009), but is lower than the figures provided by MINUSTAH and other international actors, which range from 8 to 22 per cent of all Port-au-Prince area households (Small Arms Survey, 2010; University of Michigan, 2009).

Among those who reported owning a weapon, there were an average 2.7 firearms per home. Handguns, such as revolvers, were the most commonly reported, followed by rifles and pistols. Shotguns and 'other arms' (including grenade launchers and machine guns) were the least commonly reported. Those who reported possessing a firearm

Table 8.8 Why do you own a firearm?

Response	Popular zones (%)	Other areas (%)
Personal protection	14.2	46.6
Property protection	16.6	12.6
Political security	19.4	25.7
Work	30.7	15.2
Left over from the army	9.0	0.0
Tradition	4.9	0.0
Valued family possession	5.2	0.0

Source: Small Arms Survey (2010)

were asked why they first obtained the weapon and when, as well as the reason for its last use; it was thus possible to disaggregate the reasons why gun owners chose to arm themselves.

Responses provided by residents of popular zones (densely populated low-income areas with higher crime rates) differ markedly from those of residents of neighbourhoods with greater economic diversity and lower population density (see Table 8.8). The survey reveals that, in the popular zones, the most common reason given for weapons possession was 'work' (30.7 per cent); in each of these households, at least one adult was employed as a security guard or police officer, and the reported weapon was either a pistol or a shotgun (the two firearms most commonly used in these professions). In other areas, however, the most common reason offered was 'personal protection' (46.6 per cent). 'Political security' was the second most common reason for gun possession for both geographic groups.

Further nuances emerge between households in different income brackets. Specifically, it appears that wealthier households were more likely to own weapons than middle- or lower-income households, whether in 2009 or 2010. Wealthier households also owned a greater number of weapons than low- and middle-income households and were less likely to state that their reason for weapons ownership was work-related. There is no discernable socio-cultural explanation for why wealthy households would be more willing to disclose gun ownership than middle- or lower-income households in this context, so it may be assumed that the findings are accurate.

In Haiti, as elsewhere, political groups and factions are generally assumed to be more likely to hold weapons than those claiming to be unaffiliated. Each of the surveys in 2009 and 2010 examined political party affiliation and gun ownership, finding that political party membership had no statistically significant impact on whether a household owned a firearm. Rather, qualitative interviews indicated wide-ranging reasons for gun acquisition and ownership. Respondents from wealthier households, for example, often cited 'protection of persons and property' as the primary reason they had chosen to obtain a gun.

Among the wealthy, firearm ownership was frequently accompanied by the use of private security companies to protect one's home and business (though weapons owned or used only by private security personnel were not included in the household's roster of weapons or in calculating gun ownership by the respondent). One resident described his choice to amass a small arsenal as 'taking fate into one's own hands'. He added: 'If I never have to use it, so be it. But for me, it is peace of mind. If criminals break in, I'll shoot them. Then I'll call the police.'

Figure 8.3 **Post-earthquake: Does owning a firearm make your family more or less safe?**

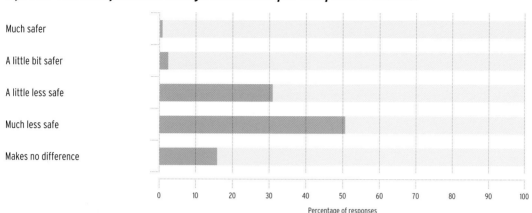

Source: Small Arms Survey (2010)

Nevertheless, owning firearms as a means to 'increase personal security' was seldom identified as a coping strategy among survey respondents (see Figure 8.3). Indeed, more than half of the respondents from the general population surveyed after the earthquake said that owning a weapon made one less safe. Only 15.7 per cent said that it makes no difference, and 3.0 per cent said that owning a weapon makes one either much safer (0.7 per cent) or a little bit safer (2.3 per cent).

OPTIONS FOR SECURITY PROMOTION IN HAITI

Despite considerable improvement in local perceptions of police capacity and effectiveness, Haitians have identified areas of security promotion that need further attention. One goal relates to enhancing the capacity and responsiveness of HNP to all Haitians, given the pervasive concern that security provision remains intolerably unequal. In particular, individuals still need to 'know someone' in order to ensure a rapid reaction to complaints. Respondents also acknowledged that certain officers are 'lazy' when responding to requests for assistance.[44] In addition, they pointed to the need for better regulation of firearms and the reining in of armed groups.

Reinforcing the HNP

In 2010, the Haitian government and the international community focused on facilitating practical improvements to security on the ground. A key question related to balancing efforts to reconstitute the 'formal' legal and procedural systems in the capital on the one hand, with investments in more 'informal' and locally targeted community policing, the deployment of private security, informal mediation, and grassroots violence prevention in specific neighbourhoods on the other. Should interventions be shaped by fundamental changes in law and justice, investment in magistrates and judges, support for police and corrections systems, or community-driven mediation and peace-making efforts? During qualitative interviews conducted as part of the post-earthquake assessment, respondents proposed local options for security promotion, a few of which are reproduced in Box 8.4.

Box 8.4 Perceptions matter: practical suggestions to improve policing in Haiti

Haitians appear to be resolute that support for the HNP is an investment worth making. Indeed, a large majority of respondents from the general population surveyed in 2010 believe that strengthening the capacity of police would make their community safer (94.4 per cent). This suggests a growing faith in the ability of the police to protect local interests. Numerous interviewees suggested ways of improving policing:

The police should go after the criminals by pretending to be weak so the criminals attack them and then the police could arrest them. The police shouldn't just wait for the crime to happen, they should be looking for the criminals under every rock until they find them and put them in jail.

–19-year-old man, student

The police need better equipment, like trucks, and they should patrol both by walking around and also from the trucks. Instead of having ten police in the back of the truck, they should get out and walk around. They should go in the corridors and get to know the women of the neighbourhood so that people will come to them and say, 'Hey, there was a crime here and I saw who did it.'

–48-year-old woman, nurse

You should be able to call or text message the police and tell them if you see a crime or need help. And then they should come right away, not later when it is more convenient.

–26-year-old man, plumber

Two weeks after the 2010 earthquake, a Haitian police officer holds his rifle as he stands guard in the business district of Port-au-Prince. © Joe Raedle/Getty Images

A major concern registered by respondents to surveys in 2009 and 2010 related to the issues of arrest and incarceration. Specifically, respondents observed that criminals were often released without being charged or even seeing a judge. While this issue reflects deficiencies in the wider justice sector as a whole, many respondents blame the HNP in particular. In fact, this dynamic puts pressure on the police to take justice into their own hands, for example by reverting to past practices of punishing or even executing suspects because they know the courts cannot hold criminals accountable. This illustrates one of the many ways in which security sector improvements are partly reliant on justice sector reforms and vice versa.

Interviewees complained that police were sometimes reluctant to deal with juvenile delinquents, drunks, and violent mentally ill people, claiming that there was 'no place to put them'. As noted above, Haiti currently has few facilities in which to house detainees. Moreover, despite a complete lack of legal aid in most neighbourhoods, it is widely recognized that police station cells are, in fact, holding centres for individuals who may present a danger to themselves or others although they are not necessarily criminals.

Steps are under way to improve the capacity of the HNP, which has a strength of about 11,500. For one, the HNP reform plan for 2006–11 projects a police force of 14,000 officers. The reform places major emphasis on modernizing and upgrading existing police structures and ensuring that procedures are in compliance with international standards. Accountability and respect for human rights are also key features. Nevertheless, the recruitment and training process has repeatedly been delayed.[45] In 2009, the certification and vetting process of HNP personnel was ongoing. As of December 2009, some 7,154 applicant investigation files were opened by joint UN–HNP teams, with some 3,496 of these under active investigation. Between 2006 and 2009, a total of 3,503 files were handed over to the UN–HNP teams with recommendations on certification (UNDP, 2009).

Enhancing the regulation of firearms

The debate on enhancing firearm controls in Haiti extends back at least two decades. Indeed, the failure of the United States to disarm the military and paramilitary during the US military intervention in 2004, the UN decision not to implement a disarmament campaign during the 1995–98 mission, and the limited number of weapons collected from 'gang members' by repeated UN- and HNP-led anti-gun campaigns have all highlighted some of the challenges inherent in collecting and destroying weapons already in circulation (Muggah, 2005a). Nevertheless, respondents to the 2009 and 2010 surveys reveal that Haitians would welcome more firearm licensing control, more robust penalties for illegal firearm possession, and corresponding legislation to outlaw militias. Although many Haitians in both the general population and the IDP camps reported feeling safer in 2010 as compared to 2009, there are still widespread concerns about the particular influence of firearms in shaping community security and safety.

Table 8.9 **Greater control of firearms licences would make my community safer**

	2010 IDP camp residents (%)	2009 general population (%)	2010 general population (%)
Strongly agree	44.5	39.3	38.7
Agree	44.1	44.1	40.2
Disagree	4.1	5.4	7.4
Strongly disagree	6.8	11.3	5.6
Don't know	0.6	0.0	8.1
Total	**100.0**	**100.0**	**100.0**

Sources: Small Arms Survey (2010); University of Michigan (2009)

Table 8.10 **Harsher punishment for illegal weapons possession would make my community safer**

	2010 IDP camp residents (%)	2009 general population (%)	2010 general population (%)
Strongly agree	53.0	28.0	30.1
Agree	40.1	43.2	50.0
Disagree	5.0	11.1	7.7
Strongly disagree	1.8	17.7	5.9
Don't know	0.2	0.0	6.3
Total	**100.0**	**100.0**	**100.0**

Sources: Small Arms Survey (2010); University of Michigan (2009)

Table 8.11 Collecting illegal guns from their owners would make my community safer			
	2010 IDP camp residents (%)	2009 general population (%)	2010 general population (%)
Strongly agree	49.3	43.9	49.0
Agree	46.0	45.2	47.6
Disagree	2.8	2.3	2.2
Strongly disagree	1.6	1.4	1.2
Don't know	0.4	0.0	0.0
Total	**100.0**	**100.0**	**100.0**

Sources: Small Arms Survey (2010); University of Michigan (2009)

Specifically, respondents frequently expressed support for government-led measures to regulate access and to use of firearms. More than three-quarters of all respondents in 2010 (whether IDP camp residents or the general population) either agree or strongly agree that more control over the issuing of firearm licences would make their communities safer (see Table 8.9). Likewise, more than 80 per cent of all respondents in 2010 also agreed or strongly agreed that harsher punishments for illegal weapons possession would improve community safety (see Table 8.10). Finally, almost all asserted that arms collection programmes would make their community safer (see Table 8.11). In qualitative interviews conducted in 2009 and 2010, many expressed frustration that wealthier segments of society are not held to the same standards as the rest of the population where firearms are concerned.

Addressing armed gangs and groups

As noted above, the issue of firearms and weapons misuse is widely considered a major security issue in Haiti. Indeed, roughly half of all respondents said there were too many guns in society today. But these weapons are unevenly distributed throughout society, and it matters fundamentally which groups are perceived to be armed. When the general population was asked in 2010 which segments of society had too many guns, they most often named 'criminal groups' (74.1 per cent), 'business people' (65.1 per cent), and 'ex-soldiers' (45.7 per cent). The least commonly named included 'politicians' (2 per cent), 'households' (1.8 per cent), and 'armed political groups' (4 per cent).[46]

Table 8.12 Outlawing armed groups would make my community safer (2009)	
	Percentage
Strongly agree	35.0
Agree	38.3
Disagree	6.5
Strongly disagree	20.2
Total	**100.0**

Source: University of Michigan (2009)

Table 8.13 Peace accords between armed gangs would make my community safer (2009)	
	Percentage
Strongly agree	25.4
Agree	21.6
Disagree	30.3
Strongly disagree	21.6
Refused to respond	1.1
Total	**100.0**

Source: University of Michigan (2009)

Although 'armed gangs' and 'political groups' were seldom identified as responsible for violence in recent years, crimes were often attributed to them in the past. In order to ascertain public assumptions about violence attribution and responsibility, and what steps could make communities safer, respondents were asked whether outlawing 'armed groups' would improve security. In 2009, 73.3 per cent either agreed or strongly agreed that it would (see Table 8.12); after the earthquake this figure has decreased slightly, with 69.6 per cent of the general population either agreeing or strongly agreeing.

One approach to reducing violence among 'armed groups' was pioneered by the Brazilian NGO Viva Rio in Port-au-Prince. Focusing on Bel Air, informal 'peace accords' were agreed between warring factions in order to reduce homicidal violence. Communities reporting a decline in homicide rates were rewarded with primary school scholarships and neighbourhood parties. When asked whether these peace accords between armed gangs were effective at increasing community safety, 47 per cent either agreed or strongly agreed in 2009 (though this percentage was significantly higher among survey respondents living in or around Bel Air; see Table 8.13). Since the earthquake, this figure increased slightly; in 2010, 55.4 per cent either agreed or strongly agreed.

Figure 8.4   **Since 2004, supporters of Aristide have committed a lot of violence in my community**

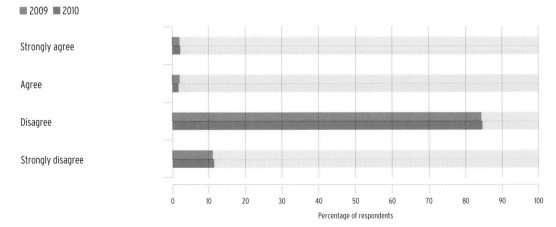

Sources: Small Arms Survey (2010); University of Michigan (2009)

Approaches to addressing so-called 'political groups' are more complex. For the purposes of this chapter, information about the particular role of political groups in perpetrating violence and crime was collected through public opinion questions, from the attributed 'perpetrator' for crimes committed against household members in the previous five years, and through additional qualitative questions to respondents. Overall, both quantitative and qualitative data collected in 2009 and 2010 indicates that political groups were engaged in violence and crime during 2004, 2005, and 2006, though this involvement appears to have steadily tapered off since early 2007 (Moestue and Muggah, 2010; Muggah, 2010b).

Political groups were not named for any of the crimes reported by respondents from 2008 (political groups could have committed crimes during these later years but in such small numbers that the survey design was unable to detect them). Indeed, oft-quoted statements about the Lavalas party being involved in violence were not supported by these surveys. In both 2009 and 2010, fewer than four per cent of the general population agreed with the statement, 'Since 2004, supporters of Aristide have committed a lot of violent acts in my community' (see Figure 8.4).

CONCLUSION

This chapter challenges the conventional wisdom on pre- and post-earthquake security in Haiti. It shows that Haiti remained stable in the months following the earthquake despite impaired policing capacity and deficiencies in key justice institutions. Far from what was expected, crime rates were also dramatically lower than predicted during the period just after the earthquake. In fact, only 4.1 per cent of all Port-au-Prince households experienced property violation such as theft, vandalism, or the intentional destruction of property in the six weeks after the earthquake; the property crime that did take place was concentrated in just a few neighbourhoods and caused only modest losses.

The chapter detects what appears to be an important shift in attitudes towards the Haitian security sector since the earthquake. While the HNP perpetrated acts of violence against the population during the 2004–06 crisis, the transition to a democratically elected government in 2007 was accompanied by a change in government policy and a cessation of state-ordered organized violence against civilians. This change created an opportunity for the HNP to steadily gain the trust of ordinary Haitians. While the police force remains far from perfect, in 2010 both the general population and the residents of IDP camps identified the HNP as their preferred security provider. What is more, the vast majority of respondents believed that strengthening police capacity would make their communities safer.

Almost 64 per cent of the general population referred to the police as the primary actors responsible for security in 2010—up from roughly 50 per cent in 2009. Likewise, IDP camp residents echoed this sentiment, with 63 per cent stating they would turn to the police to safeguard their security. In a country where the police were implicated in widespread human rights violations and where confidence in public institutions was extremely low, these findings offer some grounds for optimism. ◾

LIST OF ABBREVIATIONS

HNP	Haitian National Police
IDP	Internally displaced person
MINUSTAH	United Nations Stabilization Mission in Haiti

ENDNOTES

1   Kolbe et al. (2010) estimates that 158,679 people in the Port-au-Prince area died during the earthquake or in the six-week period afterwards owing to injuries or illness (95% confidence interval: 136,813–180,545). The official estimate established by the Haitian authorities is 222,500 people reported dead and some 1.3 million rendered homeless throughout the country. See GoH (2010a).

2   See OCHA (2011).

3   See Muggah (2010a).

4   See, for example, CIGI (2009; 2010) and ICG (2009).

5   Three recent efforts to begin evaluating the impact of security promotion include Moestue and Muggah (2010), Alda and Willman (2009), and World Bank (2010).

6   See Maguire (2009a); Muggah (2010b; 2008); and Perito (2009).

7   This trend continues into 2011. For example, on 16 January 2011, 'Baby Doc' Duvalier returned to Haiti. His effect on the country was electric. See, for example, Cunningham and Kennedy (2011).

8   In 2006 the United States 'eased', but did not lift, its arms embargo on Haiti. See BBC (2006).

9   See Prengaman (2005).

10   In this context, the United States, France, and Canada led the international community's activities.

11   By 2008, fewer than 100 of these former soldiers remained in the force, with most retiring, voluntarily moving on to other jobs, or being dismissed for various reasons.

12   UN Security Council Resolutions 1529 (2004) and 1542 (2004) set out the mandate of MINUSTAH; see UNSC (2004a; 2004b).

13   Author communication with rule of law programme officers, UN Development Programme, Port-au-Prince, December 2009.

14   The approach to justice system reform is drawn from the following strategic documents: the *Growth and Poverty Reduction Strategy Paper* (GoH, 2007a); the Ministry of Justice 'roadmap' (GoH, 2008); the *HNP Reform Plan* (GoH, 2006); and the *Strategic Plan for the Reform of the Prisons Administration Directorate* (GoH, 2007b)—all of which outline key priorities and areas of investment for the coming years.

15   See GoH (1987, ch. II, arts. 269-1; 272). See also the law of 29 November 1994, which describes the objectives of the police: '*maintenir l'ordre en général et de prêter force à l'exécution de la loi et des règlements*' ('maintain order in general and assist in the execution of the law and regulations') (GoH, 1994, art. 7).

16   See GoH (2010b).

17   Author correspondence with HNP authorities, December 2009.

18   Author correspondence with HNP authorities, December 2009.

19   By way of comparison, some officially reported ratios of officers to civilians include: Japan (1:443), Nigeria (1:400), and South Africa (1:318); European countries report a ratio of roughly 1:250–300. Other developing countries' police-to-citizen ratios are similar to Haiti's, including Ghana (1:1,421), Kenya (1:1,150), Mozambique (1:1,279), and East Timor (1:1,040). See UNDP (2009).

20   The HNP-allocated national operating budgets for 2006–07, 2007–08, and 2008–09 were about USD 89 million, USD 105 million, and USD 118 million, respectively. The budget for 2009–10 suffered a 20 per cent reduction in comparison with the previous budget and was insufficient to meet the operational and capacity development requirements. Multilateral and bilateral donors, such as Brazil, Canada, Spain, and the United States, invested heavily in rebuilding key physical and social infrastructure, training, and promoting more responsible models of policing. See UNDP (2009).

21   Author communication with MINUSTAH civilian personnel, January–March 2010.

22   The international norm is 2.5 m^2 per detainee. See UNSC (2008a, para 38).

23   The Prisons Administration Directorate issued this dire warning: '*La Direction de l'Administration pénitentiaire gère actuellement la situation la plus délicate de son histoire. Elle fait face à une réduction et une fragilité des infrastructures carcérales entraînant une diminution de l'espace cellulaire*' ('The current situation is the most fragile one the Prisons Administration Directorate has ever had to manage. The Directorate is facing the decline and weakness of the prison infrastructure, which leads to a reduction in cell space') (GoH, 2007b).

24   Author interview with Haitian prison authorities, December 2009.

25   See, for example, AI (2006), NYT (2005), Perito (2007), and Stimson Center (2008).

26   See the HNP reform plan (UNSC, 2006). The plan was developed with the technical support of MINUSTAH, the UN Development Programme, the International Organization for Migration, and other agencies and bilateral partners engaged in efforts to strengthen the HNP.

27   See UN (2010).

28   The UN Secretary-General's report on MINUSTAH, dated 1 September 2009, highlights that 'although the capacity of the National Police is gradually improving, it still lacks the force levels, training, equipment and managerial capacity necessary to respond effectively to these threats without external assistance' (UNSC, 2009, para. 21).

29   Author correspondence with a representative of the UN Development Programme, March 2010.

30   For example, the General Directorate was completely destroyed, depriving the high command of the ability to coordinate efficiently an immediate public security response. Author communication with MINUSTAH personnel, March 2010.

31   UN (2010).

32   See, for example, Muggah (2010c).

33   See, for example, UN (n.d.).

34   For example, on 19 January 2010, the UN Security Council endorsed the recommendation made by the Secretary-General to increase MINUSTAH's overall force levels to support immediate recovery, reconstruction, and stability efforts. MINUSTAH was authorized 2,000 additional military troops and 1,500 more international police (UNSC, 2010).

35   See Homeland Security Newswire (2010).

36   See, for example, MSF (2006; 2007) and research generated by the coalition Interuniversity Institute for Research and Development (INURED, n.d.).

37   See, for example, NYT (2010).

38   See, for example, AI (2010).

39   With a 95 per cent confidence interval: 6,726–14,900.

40   Others named as responsible for sexual assaults include neighbours and individuals with whom the victim had a prior relationship or non-sexual friendship.

41   This quote and the following direct quotes attributed to respondents were taken from qualitative interviews conducted as a part of the post-earthquake assessment survey of the general population. This particular respondent interview was conducted in February 2010 in Port-au-Prince.

42   Author interview with a respondent, Port-au-Prince, March 2010.

43   Meanwhile, another woman praised police for their sensitive response to a local man with mental illness: 'When a crazy man broke into our house and scared my children, my son ran to the police station. They came straight away and, seeing that the man was not in his right mind, they were gentle with him. They took him to [a clinic] where his wife could come for him' (author interview with a respondent, Delmas, February 2010).

44   As described by a taxi driver from Port-au-Prince's Delmas neighbourhood: 'My cousin is in the police so I can call him any time and say, "Hey, this guy stole from me" and they'll come quick. But for the man who doesn't have a friend in the police, he's just out of luck. The police are like any business. You get the best service from those you know personally' (author interview with a respondent, Port-au-Prince, March 2010).

45   Even before the earthquake, the recruitment of the 22nd, 23rd, and 24th promotions were under way. Each batch includes 800 cadets, approximately 15 per cent of whom are female. See UN (2010).

46   Respondents were asked to identify the top two segments of society. Percentages include only those who responded positively when asked if society had too many guns.

BIBLIOGRAPHY

AI (Amnesty International). 2006. *Annual Report for Haiti 2006.* <http://www.amnestyusa.org/annualreport.php?id=ar&yr=2006&c=HTI>

—. 2010. 'Sexual Violence in Haitian Camps of the Displaced, beyond the Numbers.'
      <http://livewire.amnesty.org/2010/03/22/sexual-violence-in-haitian-camps-of-the-displaced-beyond-the-numbers/>

Alda, Erik and Alys Willman, ed. 2009. *Bottom-up Statebuilding: Preventing Violence at the Community Level.* World Bank Research Paper.
      <http://siteresources.worldbank.org/WBI/Resources/213798-1253552326261/do-oct09-alda.pdf>

AP (Associated Press). 2004. 'Rebels Take Haiti's Second Largest City.' 23 February.
      <http://www.jamaicaobserver.com/news/56108_Rebels-take-Haiti-s-second-largest-city>

Baranyi, Stephen and Isabelle Fortin. Forthcoming. 'La réforme de la Police Nationale et la construction démocratique en Haïti.' *Canadian Journal of Development Studies,* Vol. 31, Nos. 3–4.

Baranyi, Stephen and Jennifer Salahub. 2010. 'Police Reform and Democratic Development in Lower-Profile Fragile States.' Montreal: Congress of the Canadian Association for the Study of International Development. June.

BBC. 2006. 'US Eases Arms Embargo on Haiti.' BBC News. 11 October. <http://news.bbc.co.uk/2/hi/6040270.stm>

CIGI (Centre for Governance and Innovation). 2009. *Security Sector Reform Monitor: Haiti,* No. 3. November.
      <http://www.cigionline.org/sites/default/files/SSRM%20Haiti%20v3.pdf>

—. 2010. *Security Sector Reform Monitor: Haiti,* No. 4. April. <http://www.cigionline.org/sites/default/files/SSRM%20Haiti%20v4_April_20%202010.pdf>

Cunningham, Jennifer and Helen Kennedy. 2011. 'Anxiety Mounts over Fear of Civil War as Deposed Haiti Dictator "Baby Doc" Duvalier Prepares to Talk.' *New York Daily News.* 18 January. <http://www.nydailynews.com/news/world/2011/01/18/2011-01-18_tensions_rise_as_dictator_returns_anxiety_mounts_over_fear_of_civil_war_as_baby_.html#ixzz1BOmOx7FC>

Dupuy, Alex. 2005. 'From Jean-Bertrand Aristide to Gerard Latortue: The Unending Crisis of Democratization in Haiti.' *Journal of Latin American and Caribbean Anthropology,* Vol. 10, Iss. 1, pp. 186–205.

Fortin, Isabelle and Yves-François Pierre. 2009. *Haïti et la réforme de la PNH.* Ottawa: North-South Institute.

GoH (Government of Haiti). 1987. Constitution of Haiti. <http://www.oas.org/juridico/MLA/en/hti/en_hti-int-const.html>

—. 1994. Loi relative à la Police nationale. 29 November. <http://www.oas.int/juridico/mla/fr/hti/fr_hti_mla_gen_police.html>

—. 2006. *Plan de réforme de la PNH: 2006–2011.* Port-au-Prince: GoH.

—. 2007a. *Document de la Stratégie Nationale de Croissance et de Réduction de la Pauvreté: DSNCRP (2008–2010).* Port-au-Prince: GoH. November.
      <http://www.ht.undp.org/public/publicationdetails.php?idpublication=4>

—. 2007b. *Plan de Réforme Stratégique de la Direction de l'Administration Pénitenciaire, 2007–2012*. Port-au-Prince: GoH.

—. 2008. *Feuille de Route Syndicale pour la Reconstruction et le Développement d'Haïti*. Port-au-Prince: Ministry of Justice.

—. 2009. *Groupe de travail sur la réforme de la Justice*. Port-au-Prince: Presidency.

—. 2010a. *Haiti Earthquake PDNA: Assessment of Damage, Losses, General and Sectoral Needs*. Annex to the Action Plan for National Recovery and
    Development of Haiti. Port-au-Prince: GoH. <http://www.refondation.ht/resources/PDNA_Working_Document.pdf>

—. 2010b. *Plan d'Action pour le Relèvement et le Développement d'Haïti: Les grands chantiers pour l'avenir*. Port-au-Prince: GoH. March.

—. 2010c. *Groupe Sectoriel 'Gouvernance': Sous-groupe Police/Réduction de la Violence*. Draft version. Port-au-Prince: GoH.

Hallward, Peter. 2008. *Damming the Flood: Haiti, Aristide, and the Politics of Containment*. London: Verso Press.

Hayes, Margaret and Gary Wheatly. 1996. *Interagency and Political-Military Implications of Peace Operations: Haiti–Case Study*. Washington, DC:
    National Defense University. <http://www.dtic.mil/cgi-bin/GetTRDoc?AD=ADA310900&Location=U2&doc=GetTRDoc.pdf>

HNP (Haitian National Police). 2007. *Programme de Réformes de la P.N.H: Le respect des droits humains*. Port-au-Prince: HNP.

Homeland Security Newswire. 2010. 'Private Security Firms Eyeing Haiti Contracts.'
    <http://homelandsecuritynewswire.com/private-security-firms-eyeing-haiti-contracts>

ICG (International Crisis Group). 2005. *Spoiling Security in Haiti*. Latin America/Caribbean Report No. 13. 31 May.

—. 2008. *Reforming Haiti's Security Sector*. Latin America/Caribbean Report No. 28. 18 September.

—. 2009. *Haiti 2009: Stability at Risk*. Latin America/Caribbean Briefing No. 19. 3 March.
    <http://www.crisisgroup.org/~/media/Files/latin-america/haiti/b19_haiti_2009___stability_at_risk.ashx>

IHSI (Institut Haïtien de Statistique et d'Informatique). 2009. *Population Totale, Population de 18 ans et plus Ménages et Densités Estimés en 2009*.
    Port-au-Prince: IHSI.

INURED (Interuniversity Institute for Research and Development). n.d. 'Research Laboratory on the Study of Violence.' <http://inured.org/rec_en.html>

Jones, Michael. 2010. 'Private Security Companies Prepare to Consume Haiti.' Change.org. 4 March.
    <http://humanrights.change.org/blog/view/private_security_companies_prepare_to_consume_haiti>

Kolbe, Athena and Royce Hutson. 2006. 'Human Rights Abuse and Other Criminal Violations in Port-au-Prince. Haiti: A Random Survey of Households.'
    Lancet, Vol. 368, Iss. 9538, pp. 864–73.

Kolbe, Athena and Robert Muggah. 2010. 'Surveying Haiti's Post-Quake Needs: A Quantitative Approach.' *Humanitarian Exchange Magazine,* Iss.
    48. London: Overseas Development Institute. October.

Kolbe, Athena, et al. 2010. 'Mortality, Crime and Access to Basic Needs before and after the Haiti Earthquake: A Random Survey of Port-au-Prince
    Households.' *Medicine, Conflict and Survival*, Vol. 26, Iss. 4, pp. 281–97.

MADRE et al. 2011. *Our Bodies Are Still Trembling: Haitian Women Continue to Fight against Rape*. New York: MADRE, International Women's Human
    Rights Clinic at the City University of New York School of Law, Institute for Justice & Democracy in Haiti, Bureau des Avocats Internationaux,
    and Lawyers' Earthquake Response Network. <http://www.madre.org/images/uploads/misc/1294686468_Haiti_Report_FINAL_011011_v2.pdf>

Maguire, Robert. 2009a. *Transcending the Past to Build Haiti's Future*. Peace Brief. Washington, DC: United States Institute for Peace.
    <http://www.usip.org/publications/transcending-the-past-build-haiti-s-future>

—. 2009b. *What Role for the United Nations in Haiti?* Peace Brief. Washington, DC: United States Institute for Peace.
    <http://www.usip.org/publications/what-role-the-united-nations-in-haiti>

Mendelson-Forman, Johanna. 2006. 'Security Sector Reform in Haiti.' *International Peacekeeping*, Vol. 13, No. 1. March, pp. 14–27.

MICIVIH (International Civilian Mission in Haiti). 1996. *MICIVIH Activity Report*. New York: United Nations.

Moestue, Helen and Robert Muggah. 2010. *Social Integration, Ergo, Stabilization: Viva Rio in Port-au-Prince*. Rio de Janeiro/Geneva: Viva Rio and
    Small Arms Survey. <http://www.smallarmssurvey.org/fileadmin/docs/E-Co-Publications/SAS-VIVA%20RIO-2009-Port-au-Prince.pdf>

MSF (Médecins Sans Frontières). 2006. 'New Wave of Violence Hits Port-au-Prince.' <http://www.doctorswithoutborders.org/news/article.cfm?id=1847>

—. 2007. 'Severe Injuries to Civilians Follow Recent Violence in Haiti's Cité Soleil Slum.'
    <http://www.msf.ca/news-media/news/2007/02/severe-injuries-to-civilians-follow-recent-violence-in-haitis-cite-soleil-slum/>

Muggah, Robert. 2005a. *Securing Haiti's Transition: Reviewing Human Insecurity and the Prospects for Disarmament, Demobilization, and
    Reintegration*. Occasional Paper 14. Geneva, Small Arms Survey.

—. 2005b. 'More Arms No Solution for Haiti.' *Globe and Mail*. 7 April.

—. 2008. 'The Perils of Changing Donor Priorities in Fragile States: The Case of Haiti.' In Jennifer Welsh and Ngaire Woods, eds. *Exporting Good
    Governance: Temptations and Challenges in Canada's Aid Program*. Waterloo: Wilfrid Laurier University Press, pp. 169–202.

—. 2009. 'Du vin nouveau dans de vieilles bouteilles? L'analyse de l'impasse de la gouvernance en Haïti.' *Telescope*.

—. 2010a. 'The World's Broken Promises to Haiti.' *Guardian*. 31 December.
    <http://www.guardian.co.uk/commentisfree/cifamerica/2010/dec/31/haiti-development>

—. 2010b. 'The Effects of Stabilisation on Humanitarian Action in Haiti.' *Disasters,* Vol. 34, S3, pp. 444–63.

—. 2010c. 'Dealing with Haiti's Gangs.' *Ottawa Citizen*. 18 January.

— and Keith Krause. 2006. 'A True Measure of Success? The Discourse and Practice of Human Security in Haiti.' *Whitehead Journal of Diplomacy and
    International Relations*, Vol. 57, No. 2, pp. 153–81.

NYT (*The New York Times*). 2005. 'Up to 25 Die in Police Raid on Haiti Slums.' 5 June.
    <http://query.nytimes.com/gst/fullpage.html?res=9E0CE4DB1538F936A35755C0A9639C8B63>

OCHA (Office for the Coordination of Humanitarian Affairs). 2011. *Haiti Cholera Situation Report 32.* 12 January.
  <http://www.reliefweb.int/rw/RWFiles2011.nsf/FilesByRWDocUnidFilename/VVOS-8D4MZS-full_report.pdf/$File/full_report.pdf>
O'Neill, William. 1995. 'Human Rights Monitoring vs. Political Expediency: The Experience of the OAS/U.N. Mission in Haiti.' *Harvard Human Rights Journal*, Vol. 8, pp. 101–28. <http://heinonline.org/HOL/LandingPage?collection=journals&handle=hein.journals/hhrj8&div=7&id=&page=>
PBS (Pubic Broadcasting System). 2011. *Frontline Documentary: The Battle for Haiti.* <http://video.pbs.org/video/1737171448/>
Perito, Robert. 2007. *Haiti: Hope for the Future.* USIP Special Report. Washington, DC: United States Institute of Peace, p. 4.
  <http://www.usip.org/publications/haiti-hope-future>
—. 2009. *Haiti after the Storms: Weather and Conflict.* Peace Brief. Washington, DC: United States Institute for Peace.
  <http://www.usip.org/publications/haiti-after-the-storms-weather-and-conflict>
Prengaman, Peter. 2005. 'Scores of Prisoners at Large in Haiti after Prison Attack.' Associated Press. 21 February.
Saligon, Pierre. 2005. 'Violence Intensifies in Port-au-Prince.' <http://www.doctorswithoutborders.org/news/article.cfm?id=1561>
Small Arms Survey. 2009a. *Port-au-Prince Neighbourhood Survey: Violence, Security, and Health in Three Highly Populated Zones.* Unpublished background paper. Port-au-Prince/Geneva: Small Arms Survey.
—. 2009b. 'University of Michigan/Small Arms Survey Neighbourhood Survey.' Data file. Geneva: Small Arms Survey.
—. 2010. 'University of Michigan/Small Arms Survey Study of Health and Harm in Post-Earthquake Haiti.' Data file. Geneva: Small Arms Survey.
START (Stabilization and Reconstruction Task Force). 2010. *Policy Fast-Talk: Canada's Early Recovery and Reconstruction Efforts in Security Reform in Haiti.* Ottawa: Canadian Department of Foreign Affairs and Trade/START.
Stimson Center. 2008. 'MINUSTAH.' Peace Operations Fact Sheet Series. Washington, DC: Stimson Center. 16 July, p. 2.
UN (United Nations). 2010. *United Nations Integrated Strategic Plan for the Assistance to the Haitian National Police.* Draft document. Port-au-Prince: UN.
—. n.d. 'MINUSTAH Background.' New York: Department of Public Information.
  <http://www.un.org/en/peacekeeping/missions/minustah/background.shtml>
UNDP (United Nations Development Programme). 1999. *Justices en Haïti ('Rapport Barthelemy').*
—. 2009. 'Renforcement de L'Etat de Droit en Haïti.' Port-au-Prince: UNDP and MINSUTAH.
  <http://www.ht.undp.org/public/projetdetailsprint.php?idprojet=35&PHPSESSID=f3c1dc8fbe63ab823f42371bc4fadd63>
University of Michigan. 2009. 'University of Michigan Haiti Violence Assessment: A National Study of Health, Human Rights, and Small Arms Violence.' Data file. Ann Arbor, MI: University of Michigan.
UNSC (United Nations Security Council). 2004a. Resolution 1529. S/RES/1529 of 29 February. <http://minustah.org/pdfs/res/1529_en.pdf>
—. 2004b. Resolution 1542. S/RES/1542 of 30 April. <http://minustah.org/pdfs/res/1542_en.pdf>
—. 2006. *Haitian National Police Reform Plan.* S/2006/726 of 12 September.
  <http://daccess-ods.un.org/access.nsf/Get?Open&DS=S/2006/726&Lang=E&Area=UNDOC>
—. 2008a. *Rapport du Secrétaire général sur la Mission des Nations Unies pour la stabilisation en Haïti.* S/2008/586 of 27 August.
  <http://daccess-ods.un.org/access.nsf/Get?Open&DS=S/2008/586&Lang=F&Area=UNDOC>
—. 2008b. Resolution 1840. S/RES/1840 of 14 October. <http://minustah.org/pdfs/res/res1840_en.pdf>
—. 2009. *Report of the Secretary-General on the United Nations Stabilization Mission in Haiti.* S/2009/439 of 12 September.
  <http://www.un.org/ga/search/view_doc.asp?symbol=S/2009/439>
—. 2010. 'Security Council Boosts Force Levels for Military, Police Components of United Nations Stabilization Mission in Haiti.' SC/9847 of 19 January.
  <http://www.un.org/News/Press/docs/2010/sc9847.doc.htm>
US (United States). 2009. *Human Rights Report: Haiti.* Washington, DC: US State Department.
  <http://www.state.gov/g/drl/rls/hrrpt/2009/wha/136116.htm>
Wayne State University. 2005. 'Wayne State University Study of Health and Human Rights in Haiti.' Data file. Detroit, MI: Wayne State University.
World Bank. 2010. *Violence and the City: Understanding and Supporting Community Responses to Urban Violence.* Washington, DC: World Bank.

ACKNOWLEDGEMENTS

Principal authors

Athena Kolbe and Robert Muggah

Contributors

Royce A. Hutson, Leah James, Naomi Levitz, Bart W. Miles, Jean Roger Noel, Marie Puccio, Harry Shannon, and Eileen Trzcinski

How many more have to DIE?

A young girl holds a banner featuring the pictures of most of the 35 Port Arthur massacre victims during a demonstration for tougher gun laws in Sydney's Hyde Park, May 1996.
© Megan Lewis/Reuters

Balancing Act
REGULATION OF CIVILIAN FIREARM POSSESSION

9

INTRODUCTION

In all but a handful of countries around the world,[1] civilians are permitted to purchase and possess firearms—with restrictions. While only a fraction of the world's civilians own guns, they possess a total of some 650 million—representing nearly three-quarters of the global firearm arsenal or approximately three times the number held by national armed forces and law enforcement (Small Arms Survey, 2007, p. 43; 2010, pp. 101–02). Permitted civilian uses of firearms typically include sport shooting, hunting, self-defence, and some types of professional work. Underpinning most national approaches to civilian firearm possession is an attempt to balance the prevention of social harm (crime, interpersonal violence, and suicide) with legitimate civilian use.

Although civilian firearm regulation has been debated in multilateral circles over the past two decades, it has largely eluded international control efforts. It is the prerogative of each country, based on its own mix of cultural, historical, and constitutional factors, to regulate civilian gun ownership as it sees fit. The resulting complexity and diversity of approaches make a comparative analysis of states' efforts to regulate civilian possession very difficult, and thus relatively few such studies have been undertaken.

This chapter seeks to fill this gap by analysing the legislation governing civilian access to and use of firearms in a sample of 42 jurisdictions (28 countries and 14 sub-national entities). The chapter aims to illustrate both the diversity of existing laws and their common features and foundations. The chapter does not, however, assess the efficacy or suitability of particular civilian possession laws, nor does it investigate the extent to which they have been implemented, enforced, or observed. Its principal conclusions regarding the jurisdictions under review include the following:

- A fundamental distinction can be made between jurisdictions that regard civilian firearm ownership as a basic right (two states) and those that treat it as a privilege (all others).
- Almost all states prohibit or restrict civilian access to weapons that they consider ill-suited to civilian use.
- The vast majority of states have a system of licensing in place to prevent certain types of civilians from owning firearms. In making their assessments, however, many use considerable discretion, rather than following specific criteria.
- Many states register firearms or maintain a record of firearms owned; these states tend to have centralized systems of registration.
- Most states require civilians to have a 'genuine reason' for owning a weapon.
- Some states permit the possession of firearms for self-defence, while others explicitly refuse licence applications for such purposes.

The first section of this chapter explains how the states in the sample were selected, provides a general overview of international and regional efforts to address civilian possession, and discusses legislative responses to mass shootings. The second section examines how states regulate the *firearm*,[2] providing an overview of what types of firearms civilians may be authorized to possess and which states keep centralized records of firearms in civilian hands. The third section explores how states regulate the civilian *user,* including the criteria used to determine eligibility to possess a firearm, how the licensing process works, and whether states permit private sales of firearms. The final section reviews how states regulate the *use* of civilian firearms, including the extent to which states require a reason to possess a firearm and the conditions that apply to firearm ownership. This section also considers the different approaches states take to the use of firearms in self-defence and their relationship to broader attitudes towards civilian gun access. The two online annexes to this chapter present details on owner licensing criteria and genuine reasons for owning a firearm (Annexe 9.1) and on conditions of firearm ownership (Annexe 9.2). Both annexes provide comprehensive referencing.

This study offers an overview of civilian possession laws in the selected countries using a series of tables to compare and summarize specific elements of national and sub-national controls. Although not agreed in any international negotiating forum, the elements of civilian gun control used to structure the tables have figured in various international reports and documents, such as the *United Nations International Study on Firearm Regulation* (UNCJIN, 1999), the UN Development Programme's *How to Guide: Small Arms and Light Weapons Legislation* (UNDP, 2008), and the draft International Small Arms Control Standard on national controls over the access of civilians to small arms and light weapons of the UN Coordinating Action on Small Arms (UNCASA, forthcoming).

SETTING THE STAGE

Countries and sub-national entities under review

To provide a balanced picture of national and sub-national laws covering civilian firearm possession, this chapter reviews countries from each region of the world; these states have varying rates of civilian gun ownership, distinct legal systems, and diverse attitudes towards firearm controls. National legislation (laws and regulations) has been identified through official government websites, where available, or through citations and references to national legislation contained in states' national reports on the implementation of the UN Programme of Action on Small Arms (PoA) (UNGA, 2001). Based on a review of available resources, the following 28 countries were included in the study:

- **Africa:** Egypt, Kenya, South Africa, Uganda;
- **Americas:** Belize, Brazil, Canada, Colombia, Dominican Republic, United States, Venezuela;
- **Asia:** India, Israel, Japan, Kazakhstan, Singapore, Turkey, Yemen;
- **Europe:** Croatia, Estonia, Finland, Lithuania, Russian Federation, Switzerland, United Kingdom;
- **Oceania:** Australia, New Zealand, Papua New Guinea.

In countries with federal systems, such as Australia and the United States, the primary regulators of civilian possession are typically the sub-national entities (states or territories), rather than the national (federal) government. In Australia, no single federal law covers all six states and two territories; in the United States, extensive state legislation supplements basic federal laws. Firearm legislation in all Australia's six states and two territories is included and analysed in this study. Six US states are included: three states with some of the most extensive controls (California, Massachusetts, and New Jersey) and three states with some of the least extensive controls (Arizona, Florida, and Texas).[3] All told, this chapter takes account of legislation in a total of 42 jurisdictions.[4]

Civilian possession in regional and international instruments

Over the past 20 years, UN member states have periodically highlighted the need for countries to review their national civilian possession laws. However, the focus of regional and international attention and efforts has generally been on combating the transnational illicit trade in small arms, with comparatively little consideration of regulating civilian possession at the national level. In fact, the issue of civilian possession was expressly removed from the discussion table during the 2001 PoA deliberations.

International level

International interest in and attention to the issue of civilian firearm regulation peaked in the mid-1990s with the adoption of a series of resolutions by the Economic and Social Council (ECOSOC) of the United Nations calling for the Secretary-General to initiate the exchange of data and other information on the regulation of firearms, including an international study of firearm regulation (ECOSOC, 1995, paras. IV.7–8). A July 1997 ECOSOC resolution emphasizes the importance of state responsibility for effective regulation of civilian possession of small arms, and encourages member states to consider regulatory approaches to the civilian use of firearms that include the following common elements:

> The focus of international attention has been on combating the illicit trade in small arms.

a) Regulations relating to firearm safety and storage;

b) Appropriate penalties and/or administrative sanctions for offences involving the misuse or unlawful possession of firearms;

c) Mitigation of, or exemption from, criminal responsibility, amnesty or similar programmes [. . .] to encourage citizens to surrender illegal, unsafe or unwanted firearms;

d) A licensing system […] to ensure that firearms are not distributed to persons convicted of serious crimes or other persons who are prohibited under the laws of the respective Member States from owning or possessing firearms;

e) A record-keeping system for firearms (ECOSOC, 1997, para. 5).

Based on these regulatory approaches, and in view of the four regional workshops on firearm regulation that were to be held in Brazil, India, Slovenia, and Tanzania in 1997, the resolution sought to include on the agenda the possible development of a UN declaration of principles regarding civilian firearm regulation (ECOSOC, 1997, para. 6).

That same year, the UN published the *United Nations International Study on Firearm Regulation,* based on a survey conducted at the request of ECOSOC on the recommendation of the UN Commission on Crime Prevention and Criminal Justice (UNCJIN, 1999). The study, which was updated in 1999, originated amid concerns over the high incidence of crime, accidents, and suicides involving firearms and over the lack of appropriate legal controls governing their possession and storage as well as training in their use. It was intended as a first step in a larger project involving the establishment of a database on firearm regulation, to be maintained by the Vienna-based Centre for International Crime Prevention, and the preparation of biennial reports on national firearm regulation.

The 1999 *Report of the UN Disarmament Commission* further encourages states to introduce appropriate national legislation, administrative regulations, and licensing requirements defining the conditions under which private citizens can acquire, use, and trade firearms. The report urges states to: 'consider the prohibition of unrestricted trade and private ownership of small arms and light weapons specifically designed for military purposes, such as automatic guns (e.g., assault rifles and machine-guns)' (UNGA, 1999, annex III, para. 36).

Although states continued to work together to address illicit trafficking and illicit manufacturing of small arms, with the eventual adoption of the Firearms Protocol in 2001, the discussion of civilian possession regulation effectively stalled. Plans to implement a second study were abandoned; the envisaged database and regular reports on firearm regulation never materialized; and the topic of 'measures to regulate firearms' ceased to appear on the agenda of the UN Commission on Crime Prevention and Criminal Justice after its seventh session in 1998. International attention shifted away from civilian possession towards illicit trafficking and international tracing of small arms, as well as the management of state-held stockpiles and surplus disposal.

Regional instruments

While civilian possession has faded from the global policy discussion, it has remained part of a number of regional discussions and agreements. Since 1991, at least eight regional agreements covering more than 110 countries have touched on elements of civilian possession—typically in the context of armed violence prevention or efforts to address illicit trafficking and manufacturing (see Table 9.1). These agreements are either intended to be incorporated into participating states' national law, to guide the adoption of legislation meeting minimum requirements, or to set broad standards and norms.[5] Four of the eight agreements are legally binding.[6]

Table 9.1 Civilian possession provisions of regional instruments

Instrument	Provisions
Africa	
Bamako Declaration (Bamako Declaration on an African Common Position on the Illicit Proliferation, Circulation and Trafficking of Small Arms and Light Weapons)[7]	Recommends that member states establish, among other things, illegal possession of small arms and light weapons, ammunition, and other related material as a criminal offence under national law (OAU, 2000, art. 3(A)(iii)).
SADC Firearms Protocol (Protocol on the Control of Firearms, Ammunition and Other Related Materials in the Southern African Development Community Region)[8]	Obliges states parties to incorporate the following elements into national law: prohibition of unrestricted possession by civilians; prohibition on possession and use of light weapons by civilians; registration of all civilian-owned firearms; provisions on safe storage and use, competency testing, and restrictions on owners' rights to relinquish control, use, and possession; the monitoring and auditing of firearm licences, and the restriction on the number of firearms that may be owned; prohibition on pawning and pledging of firearms, ammunition, and other related materials; and prohibition on misrepresentation or withholding of any information given during the application process (SADC, 2001, art. 5(3)). Also recommends a coordinated review of national procedures and criteria for issuing licences and establishing national electronic databases of licensed firearms, firearm owners, and dealers within their territories (art. 7).
Nairobi Protocol (Nairobi Protocol for the Prevention, Control and Reduction of Small Arms and Light Weapons in the Great Lakes Region and the Horn of Africa)[9]	Almost identical provisions to SADC Firearms Protocol but also requires registration and effective control of arms owned by private security companies and prohibits the civilian possession of semi-automatic and automatic rifles and machine guns and all light weapons (Nairobi Protocol, 2004, art. 5).
ECOWAS Convention (ECOWAS Convention on Small Arms and Light Weapons, Their Ammunition and Other Related Materials)[10]	Prohibits the possession, use, and sale of light weapons by civilians; encourages licensing systems, including the following criteria: minimum age; no criminal record or the subject of a morality investigation; proof of a legitimate reason to possess, carry, or use; proof of safety training and competency training; proof of safe storage and separate storage of ammunition. Also requires: limit on the number of weapons a licence may cover; waiting period of at least 21 days; expiration dates on licences and periodic reviews; seizure laws and revocation of licences for contraventions of possession laws; and adequate sanctions and penalties for illicit possession and use (ECOWAS, 2006, art. 14).

Americas[11]	
Andean Plan (Andean Plan to Prevent, Combat and Eradicate Illicit Trade in Small Arms and Light Weapons in All Its Aspects)[12]	Encourages improvement and reinforcement of civilian possession regulations (Andean Community, 2003, art. 4(e)); recommends the establishment of illicit possession, carrying, and use of small arms as a criminal offence under national law (art. 3).
Europe	
European Weapons Directive (Council Directive 91/477/EEC of 18 June 1991 on Control of the Acquisition and Possession of Weapons, as amended by Directive 2008/51/EC of the European Parliament and of the Council)[13]	Lays down minimum conditions for the acquisition and possession of firearms and ammunition including: prohibition on certain firearms (including automatic firearms) (Council of the European Union, 1991, art. 6); restrictions on certain categories of weapons (including semi-automatic firearms), for example, persons must have 'good cause'; must be at least 18 years old (except for hunting and target shooting); must not be a danger to themselves, to public order, or to public safety (art. 5); also requires member states to establish a computerized data filing system to record information on firearms, owners, and sellers (art. 4).
Middle East	
Arab Model Law on Weapons, Ammunitions, Explosives and Hazardous Material[14]	Contains provisions regarding the possession of weapons and ammunition including: requirement for a licence or permit to possess firearms (League of Arab States, 2002, art. 6); recommendations to restrict the amount of ammunition that can be possessed or carried (art. 8); limits on the number of weapons that an individual can be licensed to possess ('one rifle, one gun, two hunting weapons') (art. 10); licensing criteria including: age requirement (at least 21 years old); competency ('fully responsible/be of sound mind'); no criminal convictions; physical fitness (art. 23); as well as provisions regarding the cancellation of licences; the obligation to report lost or stolen weapons within three days; and associated criminal penalties.
Oceania	
Nadi Framework (in Legal Framework for a Common Approach to Weapons Control Measures)[15]	Establishes a requirement that civilians have a 'genuine reason' for possessing a gun and identifies acceptable reasons; lists firearms that should be prohibited for civilian possession; suggests a permit scheme and process for licensing civilians and keeping a register relating to possession and sale (SPCPC and OCO, 2000).

Approaches to regulating civilian possession

It is difficult to make generalizations about the legislative processes of countries with vastly different executive structures and political systems. However, it is clear that the process of passing laws is often a complex one, involving a range of interdependent factors—including public advocacy, private interests, social mobilization, prevailing national priorities, and even interpersonal relations between policy-makers, among many others. In the case of civilian possession laws, attitudes to and experiences of gun use and gun crime may be as important as all of these factors in shaping laws.

States take one of two general approaches to civilian possession of firearms: they treat it as a basic 'right' or consider it a 'privilege'. In some countries there is an automatic entitlement to have a gun unless certain factors apply (such as a serious criminal conviction), while in most others there is a presumption against civilians owning firearms unless certain conditions and requirements are met. Legislation reviewed for this chapter suggests that the vast majority of states fall into the latter category. This distinction informs the nature and, in some cases, the extent of the regulations states impose.

Box 9.1 Legislative and policy responses to mass shootings

Australia. Public mass shootings were instrumental in modifying Australia's civilian gun regulations in the late 1980s. The National Committee on Violence was established in Australia following two mass killings in Melbourne in 1987.[16] The Committee produced a report on violence reduction strategies, *Violence: Directions for Australia,* including recommendations for firearm controls. Between 1991 and 1995 the Australasian Police Ministers' Council (APMC)–responsible for coordinating gun control among Australia's six states and two territories–drew up a series of recommendations for harmonizing the different registration and licensing systems in these jurisdictions (Norberry, Woolner, and Magarey, 1996).

Concrete action was not taken until the Port Arthur massacre in Tasmania in April 1996, in which 35 people were shot and killed and many others were injured (Norberry, Woolner, and Magarey, 1996). During an emergency meeting of the APMC following the massacre, all state and territory governments committed to enacting uniform laws under what became known as the 'Nationwide Agreement on Firearms'. The agreement included the following significant changes to civilian gun laws: (a) a general ban on the use of semi-automatic rifles and shotguns and pump-action shotguns, except for specific purposes; (b) requirement of proof of a genuine reason to own, possess, or use a gun and special need for some licence categories; (c) basic licence requirements including a minimum age of 18 years, safety training as a prerequisite, and a 28-day waiting period on each purchase; (d) the introduction of nationwide registration of all firearms; and (e) strict storage requirements (APMC, 1996).

Canada. In the wake of Canada's worst mass shooting, in which 14 women were killed on the campus of the University of Montreal's Ecole Polytechnique in 1989, the government introduced a requirement that *all* firearms be registered. Prior to the incident, only restricted firearms (including handguns and assault weapons) had to be registered; non-restricted firearms (namely long guns used for hunting and sport shooting) did not (Canada, 1995, para. 83.1; Makarenko, 2010).

Germany. School shootings in Erfurt (2002) and Emsdetten (2006) prompted the government to ban TASERs and dummy guns, as well as several other weapons in 2008 (Harding, 2009).

Finland. Two school massacres stunned Finland in 2007 and 2008, with gunmen killing eight and ten people, respectively, before taking their own lives (Burridge, 2010). The events led to proposed revisions to the country's firearm laws, including provisions that prohibit persons under 20 from obtaining a handgun ('pistol, small-calibre pistol, revolver') (Harding, 2009; Finland, 2009b). Other amendments under consideration by the end of 2010 included a requirement that all licence applicants pass an aptitude test as a prerequisite and undergo a suitability test (similar to that used by the Finnish military) as well as increased scope for the police to obtain information on an applicant's health, drug habits, violent behaviour, and military service record and suitability (Finland, 2009a; 2009b).

New Zealand. The Aramoana massacre of 1990, in which 13 people were killed with a semi-automatic rifle, prompted a review of and amendments to the Arms Act of 1983. Some of the changes included: (a) the introduction of a new class of gun–military-style semi-automatic (MSSA) firearms– and additional restrictions imposed on them; (b) licences ceased to be

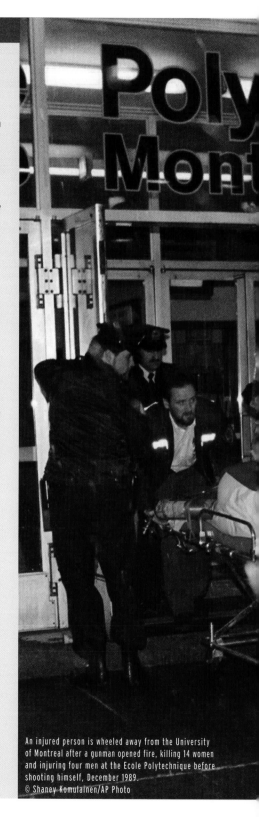

An injured person is wheeled away from the University of Montreal after a gunman opened fire, killing 14 women and injuring four men at the Ecole Polytechnique before shooting himself, December 1989.
© Shaney Komulainen/AP Photo

lifetime and were to be reviewed at ten-year intervals; (c) new restrictions on the mail-order purchase of firearms and ammunition; and (d) a prohibition on leaving firearms unattended in motor vehicles (New Zealand, 1992b; Newbold, 1998, pp. 116–17).

United Kingdom. Following the murder of 16 people in Hungerford, Berkshire, in August 1987, the Home Office introduced the Firearms (Amendment) Act of 1988, which expanded the class of prohibited weapons to include most semi-automatic rifles and smooth-bore shotguns, as well as self-loading or pump-action shotguns. Following the Dunblane (Scotland) massacre in March 1996, in which 16 primary school children and their teacher were murdered, a public inquiry on firearm control was undertaken that ultimately led to a ban on handguns (other than air guns, firearms for starting athletics races, and guns of historic interest) (UK, 1997, ss. 1(2), 2–8).

Civilian firearm regulations returned to the national agenda following a shooting in Cumbria in June 2010 in which 12 people were killed and 11 injured by a taxi driver in possession of a rifle and a shotgun. The Home Affairs Committee commenced a new inquiry to examine how firearm or shotgun certificates are issued, monitored, and reviewed, and whether revisions to these regulations were needed.[17]

United States. No country seems to have been more affected by public mass shootings than the United States—nor has any had such varied legal responses to them. The responses occur mainly at the state level, where most civilian gun regulations are enacted, and typically take two forms: increased criminal penalties for weapon-related offences conducted near schools and enhanced safe storage measures to prevent access of children and young adults to guns in the home. The latter reflects responses to mass shootings by young people at middle schools, high schools, and universities.[18] Many states have introduced specific penalties for adults who fail to prevent children from gaining access to loaded firearms (see Annexe 9.2[19]). In other states, the response has focused on the perpetrator rather than the means. In Arkansas, for example, legislative reforms following the Jonesboro school shooting in 1998 (carried out by two boys aged 11 and 13) allowed juveniles to be tried as adults (Scott and Steinberg, 2008, p. 18; NYT, 1999).

In August 2008, the board of the small rural Harrold Independent School District in Texas unanimously approved a plan to allow teachers to bring concealed handguns into classrooms (Reuters, 2008). The rationale behind the arming of teachers was deterrence and quick response time. Since then many state legislatures in the United States have debated this issue of 'campus carry', with proposals ranging from ones that allow anyone, including students, with a concealed handgun permit to carry guns on school premises to those that allow only full-time staff who have undergone specialized police training to possess guns on school premises.[20] The idea of allowing teachers and, potentially, students to carry firearms at school as a deterrent stands in stark contrast to the earlier policy response of banning firearms on school premises.[21]

Following the deadliest mass shooting in US history—the Virginia Tech massacre of 2007[22]—the federal government passed national legislation providing incentives for states to improve their criminal and mental health record reporting to the Federal Bureau of Investigation's National Instant Criminal Background Check System (NICS). Under federal law, states are encouraged but not required to report all relevant mental health information (LCAV, 2008, p. 115).

Regardless of which approach they take, all states recognize the need to adopt certain measures to promote the safe use of firearms and to prevent misuse and threats to public safety, as well as the safety of the person in possession of the firearm. Indeed, rising crime rates may spark legislative action; more frequently, news-grabbing public incidents of gun violence inspire changes in firearm law. This is most noticeable in countries where gun violence is perceived as relatively rare, where mass shootings tend to stir up strong feelings on all sides and push civilian gun regulations into the spotlight. Over the past 20 years, mass shootings have motivated changes to civilian possession laws in at least seven of the countries under review, many addressing specific factors that underpinned the shootings (see Box 9.1).[23]

The precise relationship between mass shooting incidents and particular legal responses is far from clear. In many cases, these incidents have acted as a driver to strengthen gun laws, but because legislative responses are ultimately political, and thus negotiated, outcomes vary widely.

The relationship between mass shootings and legal responses is far from clear.

REGULATING THE FIREARM

Prohibited and restricted firearms

There is no internationally agreed definition or classification of 'civilian' versus 'military' firearms. Nevertheless, most countries reviewed in this chapter prohibit or severely restrict civilian access to and ownership of weapons they deem inappropriate for civilians or better suited to military use. Bans and restrictions may also target specific types of firearms that are disproportionately involved in crime, or weapons deemed unsuitable for authorized civilian uses (such as hunting and self-defence).

The descriptions and classifications of firearms that are subject to restriction or prohibition vary enormously across jurisdictions; in the absence of an internationally agreed classification system, it is not possible to assert that all or most of the reviewed countries absolutely ban a particular type of weapon. Generally speaking, however, most countries reviewed (with the exception of **Yemen**) prohibit possession of automatic firearms and restrict possession of semi-automatic firearms and handguns.[24]

Handgun restrictions vary from limiting civilian possession of certain types of handguns to permitting the possession of any handguns for certain uses only (or both). For example, **Lithuania** prohibits the possession of 'short firearms'[25] whose calibre exceeds 9 mm for self-defence purposes (Lithuania, 2002, art. 7(4)), while **South Africa** grants licences for handguns (other than fully automatic ones) only for the purposes of hunting, sport shooting, and self-defence (South Africa, 2000, paras. 13, 16).

Brazil has two classifications: 'restricted-use weapons' and 'unrestricted-use weapons'. The use and possession of automatic firearms and certain handguns are restricted to the armed forces, law enforcement agencies, and, in certain cases, to sport shooters, hunters, and small arms collectors. Restricted-use weapons and ammunition cannot be sold in gun shops; they can only be purchased directly from the factory with a special authorization from the Brazilian Army. Other handguns can be sold to civilians (Dreyfus and Perez, 2007, p. 44). Similarly, in the **Dominican Republic**, certain firearms are considered 'war weapons' and can only be used by government forces, including .45 calibre pistols, rifles, machine guns, carbines, artillery pieces, and other heavy weapons (Inoa, 2010).

Table 9.2 identifies what categories of firearms are prohibited or restricted in each of the jurisdictions under review; it also lists the penalties associated with unlawful possession of such weapons. The table reveals that most countries ban the possession of automatic firearms and machine guns and that many also have restrictions on civilian access to handguns.

Table 9.2 Overview of prohibited and restricted firearms for civilian possession[26]

Country/state	Automatic firearms	Semi-automatic firearms[27]	Handguns	Penalty for possessing prohibited firearm	Registration
Australia	●[28]	○	○	2–20 years[29]	F, O, L, T
Australian Capital Territory	●[30]	○[31]	○[32]	1–2 prohibited firearms: 10 years; 3–9 prohibited firearms: 14 years; 10+ prohibited firearms: 20 years[33]	F, O, L, T[34]
New South Wales	●[35]	○[36]	○[37]	14 years[38]	F, O, L, T[39]
Northern Territory	●[40]	○[41]	○[42]	AUS 53,200 (USD 52,300) or 2 years[43]	F, O, L, T[44]
Queensland	●[45]	○[46]	○[47]	4–7 years[48]	F, O, L, T[49]
South Australia	●[50]	○[51]	○[52]	10 years or AUS 50,000 (USD 50,000)[53]	F, O, L, T[54]
Tasmania	●[55]	○[56]	○[57]	2 years and/or AUS 6,500 (USD 6,400)[58]	F, O, L, T[59]
Victoria	●[60]	○[61]	○[62]	2–7 years[63]	F, O, L, T[64]
Western Australia	●[65]	○[66]	○[67]	7 years[68]	F, O, L, T[69]
Belize	●[70]	● (semi-automatic handguns)[71]	● (handguns with calibre greater than 9 mm or 0.38 in. are prohibited)[72]	(summary) 3–7 years; (on conviction) 5–10 years[73]	L[74]
Brazil	●[75]		○[76]	3–6 years and a fine (not specified)[77]	F, L, T registered by SINARM;[78] restricted firearms to be registered with Army Command[79]
Canada	● (unless grand-fathered[80])	○[81]	● (some) ○ (some)[82]	5 years (10 years if knowingly)[83]	F, O, L, T, plus every application, import, export loss, theft, finding, destruction[84]
Colombia	●[85]	● (some)[86] ○ (some)[87]	○[88]	4–8 years for self-defence firearms; 5–15 years' for possession of military weaponry[89]	F[90]

▶▶

Country/state	Automatic firearms	Semi-automatic firearms	Handguns	Penalty for possessing prohibited firearm	Registration
Croatia	●[91]	● (some) ○ (some)[92]	○[93]	Up to 60 days or HRK 20,000-150,000 (USD 3,800-28,000)[94]	F, O, L[95]
Dominican Republic[96]	●		● (some)	3-10 years and USD 60-150 fine	L
Estonia	●[97]	n/a[98]	○[99]	1-5 years (5-15 if large quantity)[100]	F, O, L, T[101]
Finland	○[102]	[103]	[104]	4 months-4 years[105]	F, L, T[106]
India	●[107]	○[108]	○[109]	5+ years and fine (unspecified)[110]	F, L, plus dealers keep a record of transfers[111]
Israel[112]	●	●		7 years and fine (unspecified)	F, L, O
Japan	●[113]	●[114]	●[115]	Handgun: 1-10 years; firearms:[116] up to 3 years or fine up to JPY 500,000 (USD 600)[117]	No central register, but manufacturers and those who store handguns keep records, plus antique firearms are registered[118]
Kazakhstan[119]	●		●[120]	Unlawful possession generally: 1-5 monthly salaries (plus possible confiscation)	F
Kenya	●[121]	● (some)[122]		7-15 years (14+ if 'arms or munitions of war')[123]	F, plus dealers must keep a register of transactions[124]
Lithuania	●[125]	○[126]		Unlawful possession generally: up to 5 years[127]	F, O, plus import, export, and destruction[128]
New Zealand	○[129]	○[130]	○[131]	3 years and/or NZD 4,000 (USD 3,000)[132]	F (pistols, restricted firearms, and MSSAs; general registration of all firearms discontinued in 1983); plus dealers keep records of transactions[133]

Country/state	Automatic firearms	Semi-automatic firearms	Handguns	Penalty for possessing prohibited firearm	Registration
Papua New Guinea	●[134]		○[135]	6–12 months[136]	L[137]
Russian Federation[138]	●		●[139]	Up to 4 years and a fine up to RUB 80,000 (USD 2,800)[140]	F
Singapore	No specific prohibitions or restrictions (and automatic firearms and automatic pistols are included in the definition of 'arms'), but very strict provisions on who can possess a gun: only target shooters and, in very exceptional cases, for personal protection.[141]			Unlawful possession of any arm, including automatic arms: 5–10 years and 6+ strokes of the cane[142]	F, O, L, plus the nature of the arms licensed, any identification marks, and any licence conditions;[143] dealers keep records of sales[144]
South Africa	●[145]	○[146]	○[147]	25 years[148]	F, O, L, T, plus all other firearm records[149]
Switzerland	●[150]	○[151]		3 years or fine (unspecified)[152]	F, O, L, T[153]
Turkey[154]	●	●	○	Fully automatic: 5–8 years and fines of up to TRY 25,000 (USD 15,900); other: 1–3 years and fines of up to TRY 3,000–5,000 (USD 1,900–3,200)	L, T
Uganda	●[155]			5 years, UGX 20,000 (USD 8.50), or both[156]	O, L (firearm certificates)[157]
United Kingdom	●[158]	● (some)[159]	●[160]	5 years (with or without a fine, unspecified)[161]	Police keep a register of dealers; dealers and manufacturers must keep a register of all transactions[162]
United States	●[163]	[164]		(Determined by each state)	No federal register but the National Firearms Act (NFA) requires registration of certain firearms, including machine guns[165]

▶▶

Country/state	Automatic firearms	Semi-automatic firearms	Handguns	Penalty for possessing prohibited firearm	Registration
Arizona	● (unless federally registered under NFA)[166]			1–3.75 years for 1st offence (2.5 years is the presumptive sentence); increased (up to 15 years) for subsequent/repeat offences[167]	No[168] (other than federal requirements)
California	●[169]	○[170]		Assault weapons: up to 1 year (if 1st offence and no more than 2 weapons possessed: up to USD 500 fine if certain conditions apply);[171] machine gun: imprisonment (not specified) and/or up to USD 10,000 fine[172]	Registration of pre-ban assault weapons and .50 calibre rifles,[173] but registration of long guns is prohibited,[174] plus the licensing authority keeps records of handgun licences issued
Florida	● (except grand-fathered firearms)[175]		○ (3-day waiting period)[176]	Up to 15 years and up to USD 10,000[177]	No (other than federal requirements)
Massachusetts	● (only instructors and collectors)[178]	● (some)[179] ○ (some)[180]		Assault weapons: 1st offence: 1–10 years and/or USD 1,000–10,000; subsequent offences: 5–15 years and/or USD 5,000–15,000; machine gun: up to life imprisonment (if loaded, additional 2.5 years)[181]	No (other than federal requirements)
New Jersey	○[182]	○ (some)[183]	○[184]	5–10 years and fine up to USD 150,000[185]	Assault firearms owned before 1990 must be registered at police stations, plus dealers keep register of handguns sold and certification on rifle and shotgun sales[186]

Country/state	Automatic firearms	Semi-automatic firearms	Handguns	Penalty for possessing prohibited firearm	Registration
Texas	● (unless federally registered under NFA)[187]		○ (there are restrictions on the carrying of handguns)[188]		No (other than federal requirements)
Venezuela	Only the state may possess 'weapons of war', including: canon, rifles, mortars, machine guns, sub-machine guns, carbines, pistols, and revolvers, be they automatic or semi-automatic. Civilians are only authorized to hold .22 rifles and shotguns (repeating and hunting).[189]			5-8 years for firearms; 6-10 years for 'weapons of war'[190]	F[191]
Yemen	No types of firearm are restricted or prohibited for civilian use. The law divides 'arms' into two categories: 'personal' or civilian firearms, including 'rifles, machine guns, revolvers and hunting rifles' and 'military' weapons or weapons for state security institutions, which covers everything else.[192]				L, plus traders keep a record of transfers and report this to the licensing authority every 3 months; and the Ministry of Interior keeps a register of seized firearms[193]

Key

● = prohibited for civilian possession (with no or limited exceptions)

○ = special restrictions apply to some or all firearms in this category

F = details of the firearm must be recorded

O = details of the owner or licensed person must be recorded

L = details of the licence, permit, or authorization must be recorded

T = details of all transfers and other transactions must be recorded

Blank cell = no special restrictions apply

Registration

Registration involves keeping a record of certain information pertaining to firearms and sometimes their owners in an official register. Most of the countries in the sample have a system of registration in place, though the nature of the system and the type of information recorded vary considerably. While other countries may not have a formal, centralized register, they record information obtained through the owner licensing process. As indicated in Table 9.2, some states—such as **Australia**, **Brazil**, **Canada**, **Estonia**, **South Africa**, and **Switzerland**—require comprehensive records of *all* information relating to firearms, including details of licensed firearm owners, their firearms, every licence or authorization applied for, refused, and granted, and every firearm transaction or transfer, while other states only record limited information.

Mandatory registration is designed to facilitate the tracing of guns used in crime to their last known legal owner, the return of stolen guns to their owners, the seizing of firearms following criminal convictions, and the investigation of transfers that may be illegal. Some of the benefits that supporters associate with systems of registration include an increased ability to enforce firearm law, enhanced tracing of sources of illegally possessed or used firearms, and reduced availability of guns to criminals (Hahn et al., 2005, p. 53). Critics of registration focus on the financial cost of maintaining a registry and on the burden imposed on owners of non-restricted firearms (namely hunters and sport shooters). For example, there have been ongoing but unsuccessful efforts to abolish the long-gun register in **Canada**, established following the country's worst gun massacre in 1989 (see Box 9.1). The main criticisms concern the expense associated with the Canadian register[194] and the fact that non-restricted firearms (namely long guns— rifles and shotguns) must be registered even though it is claimed that they are relatively unimportant to the register's crime prevention goals.[195] The Royal Canadian Mounted Police dispute that the system is costly, noting that it is 'cost-effective in reducing firearms-related crime and promoting public safety' (RCMP, 2010, p. 59).[196]

With the adoption of its Arms Act in 1983, **New Zealand** moved away from a system of registration of all firearms towards improved licensing of individuals and the registration of certain types of firearms only. Prior to this, the country had a decentralized system of firearm registration (including for shotguns, rifles, pistols, and restricted weapons). However, efforts to assess the accuracy of the existing (pre-computer, card-based) registration system and an internal police report, entitled *Firearms Registration in New Zealand,* suggested that maintaining an accurate registry that would genuinely assist police investigations was an enormous and expensive task that depended on owners' attitudes and willingness to provide updated information. Ultimately, it was determined that the money could be spent to better advantage on other police work (New Zealand, 1997, s. 2.2). Accordingly, the system of registering long guns was abandoned under the 1983 law. Recommendations to reintroduce a system of registration by an independent inquiry commissioned by the minister of police in a 1997 report entitled *Review of Firearms Control in New Zealand* were not adopted, despite police support for the proposal.[197]

The **United States** does not have a national registry of firearms or their owners, although the National Firearms Act does require the registration of machine guns, short-barrelled shotguns, and rifles, as well as sound suppressors (silencers) (LCAV, 2008, p. 189; US, n.d.c, s. 5841). In fact, federal law specifically *precludes* the use of the National Instant Criminal Background Check System to create any system of registration of firearms or firearm owners by requiring the destruction of records pertaining to transfers that are permitted (US, n.d.b, s. 25.9(b)(3)), though some states have established registration systems for all types of firearms.

New Jersey requires assault weapons to be registered with the police; this US state also requires dealers and other firearm sellers to keep a register of handguns sold (including a description of the purchaser and the handgun) and to keep the certification provided by the buyer (including details of the purchaser's name, address, and purchaser identification number) on sales of shotguns and rifles (New Jersey, n.d.a, ss. 2C:58-2(a)(4), 58-2(b)).

Florida expressly *prohibits* the keeping of a register of legally owned firearms or firearm owners, asserting that such a list 'is not a law enforcement tool and can become an instrument for profiling, harassing, or abusing law-abiding citizens' and, further, that the list could 'fall into the wrong hands and become a shopping list for thieves' (Florida, n.d., s. 790.335(1)(a)(2)). Indeed, the legislation provides for redress for firearm owners whose names *have* been recorded in a list or registry (s. 790.335(1)(a)(4)). Yet the statute does permit information of firearms used in crime and persons involved in crime to be recorded; it also allows records of stolen firearms, though such lists are kept for a limited time.[198]

REGULATING THE USER

Who can possess a firearm?

In addition to regulating the types of weapons civilians can possess, states impose restrictions on who can lawfully possess firearms, primarily by establishing licensing systems and risk assessment criteria for determining eligibility. The criteria are designed to disqualify individuals who are considered more likely to misuse firearms and can involve: age restrictions; mental (and sometimes physical) health requirements; evidence of drug dependency; criminal records; protection orders; and general 'public interest' considerations. Many states are explicitly looking to evaluate whether an applicant is a 'fit and proper' person who can be trusted with the responsibility of owning a firearm or firearms.

Many countries have also established competency and pre-licensing training requirements to ensure that firearm owners are trained in firearm use and handling, in an effort to reduce the likelihood of accidents. Annexe 9.1 provides an overview of some of the eligibility criteria reviewed countries consider when evaluating a licence application or permitting firearm possession.

Age restrictions

Most countries prohibit the acquisition and ownership of guns by young people and minors, or at least restrict the types of firearm-related activities young people can engage in or the types of firearms they can possess. As illustrated in Annexe 9.1, most of the sample countries do not permit ownership of a firearm until a person has reached 18 years

An eight-year-old hunter-in-training aims her rifle with the help of her father in Minnesota, November 1998.
© Steve Liss/Time Life Pictures/Getty Images

of age (the age of majority or adulthood in many countries). Notable exceptions include **Croatia**, **Egypt**, **Estonia**, **India**, **Lithuania**, **South Africa**, and **Turkey**, which do not generally permit gun ownership until a person has reached the age of 21; **Brazil** and **Uganda**, which require a person to be 25 years of age; **Israel**, which requires a person to be 27 years of age (unless he or she has completed military service); and, at the other end of the spectrum, **New Zealand** and **Belize**, which allow gun ownership at the age of 16, and **Kenya**, which allows gun ownership at the age of 14.

Federal law in the **United States** prohibits the possession of handguns by those under the age of 18 but places no minimum age limit on the possession of long guns.[199] A number of US states do establish minimum age limits for the possession of long guns—typically 18 years of age, but sometimes 16 or 21—and raise the minimum age for handgun possession to 21 (LCAV, 2008, pp. 81–87).

Mental and physical health

Applicants with serious mental health issues may be ineligible to possess a firearm.

In some parts of the world, applicants with serious mental health issues may be ineligible for a firearm licence or to possess a firearm. In some cases, the regulatory provisions simply indicate the licensing authority must be satisfied that the applicant is not of 'unsound mind' (as in **India** and **Kenya**) or is of 'sound mind' (**Uganda**); elsewhere, applicants must show that they are not suffering from a mental disorder or that their mental health does not prevent them from handling a firearm in a responsible manner, as in the Australian Capital Territory (**ACT**), Australia. In **Belize**, though the legislation does not expressly state that mental health is a factor in the assessment of whether someone is a 'fit and proper' person to own a firearm, a firearm licence may be revoked if the licensee is found to be of 'unsound mind'. In other cases, an applicant may be asked to produce a medical certificate confirming he or she is capable of handling firearms and that he or she has no psychological—or physical—impediments (for example, in Australia's **ACT** and **Queensland** state as well as in **Croatia**, the **Dominican Republic**, **Kazakhstan**, the **Russian Federation**, **South Africa**, and **Turkey**); elsewhere, the licensing authority may investigate whether the applicant has been treated for mental illness (**Canada**).[200]

In the **United Kingdom**, applicants are required to indicate on their application form whether they have ever been treated for depression or other mental or nervous disorders, and they must give consent for the licensing authority to approach the relevant doctor to discuss the applicant's state of health (UK, 1998).[201] Indeed, in **Australia** and the **United Kingdom**, health practitioners who have reason to believe that patients may pose a threat to public safety (or their own safety) if in possession of a firearm may report their concerns to police, in spite of their confidentiality obligations (NSW, 1996, s. 79; Dodd, 2010). **Finland** is considering introducing similar legislation that would give doctors and other healthcare professionals a right under law to notify the police of a person who may, on the basis of his or her state of health, be considered unsuitable for possessing a firearm (Finland, 2009a, s. 114; 2009b).

In some cases, the focus of the mental health assessment is on the applicant's tendency towards violence, such as in **South Africa** (South Africa, 2000, s. 9(2)(d)), including a propensity to engage in self-destructive behaviour, such as whether the applicant has ever attempted suicide, as in **New South Wales** (NSW, 1996, s. 11(4)(b)) and **Northern Territory** (NT, n.d.b, s. 10(4)(b)), or is likely to use the firearm to harm himself, as in **Tasmania** (Tasmania, 1996, s. 29(2)(a)(ii)).

Some jurisdictions in the sample also consider whether the applicant is *physically* fit to possess a firearm (such as **Egypt**, **Lithuania**, **Switzerland**, and **Turkey**; Australia's **Queensland**, **South Australia**, **Tasmania**, and **Western Australia**; and the US state of **New Jersey**); these may require a medical certificate testifying to that effect.

In **Singapore**, applicants who are over a certain age (60 years for a target practice licence; 50 years for a self protection licence) must submit a medical report confirming they are fit to handle a firearm (Singapore, n.d.a, paras. A(3), B(2)). **Finland** simply indicates that an applicant's general state of health will be considered during the application process. Elsewhere, applicants suffering from specific conditions are not permitted to own firearms. For example, epileptics are prohibited from carrying a weapon in the **Dominican Republic** (Inoa, 2010); in **New Jersey**, when granting a handgun purchase permit or firearms purchaser identification card, authorities take into consideration any physical defect or disease that would make it unsafe for the applicant to handle a firearm (New Jersey, n.d.a, s. 2C:58-3(c)(3)).[202]

Drug dependency and intoxication

In some countries, licences or the purchase of a firearm will be refused if the applicant has a history of alcohol or substance abuse or drug or chemical dependency (as in **Croatia**, **South Africa**, **Turkey**, **the United States**, and **Yemen**).[203] In other countries, the fact that a person has a history of drug dependency or is an alcoholic is taken into consideration when determining whether he or she is 'fit and proper' to own a firearm, as in the **Dominican Republic** (Inoa, 2010), **New Zealand** (New Zealand, 2010, p. 40), and the **United Kingdom**. In the **United Kingdom**, applicants are required to indicate on their application forms whether they have a medical condition, including alcohol or drug dependency (UK, 1998).[204] In **Kazakhstan**, an applicant must certify that he or she is neither an alcoholic nor a drug addict (Karimova, 2010).

Most jurisdictions make it an offence to sell firearms to persons who are intoxicated or drunk at the time of sale; examples are the **United Kingdom** (UK, 1968, s. 25) and **Texas** (Texas, n.d.a, s. 46.06(a)(3)). Obviously, it is not possible to prevent people from getting drunk and using firearms in their possession once the licence has been granted and a gun has been acquired. But in some parts of the world, it is an offence to be in possession of a firearm while intoxicated, even if this does not result in harm or injury (or even use). That is the case in **Belize** (2003, s. 42(1)); **Lithuania** (2002, art. 30(2)(3)); **Kenya**, where offenders face a fine of KES 10,000 (USD 125) and/or up to one year's imprisonment (Kenya, 1954, s. 33); **New Zealand**, where the fine is NZD 3,000 (USD 2,300) or three months' imprisonment or both (New Zealand, 1983, s. 47); **Papua New Guinea**, where offenders face up to six months' imprisonment (PNG, 1978, s. 57); and **Uganda**, where the punishment is six months in jail and/or a fine of UGX 2,000 (USD 1) (Uganda, 1970, s. 31); as well as in Australia's **Victoria**, where the offender receives 120 penalty units or two years' imprisonment (Victoria, 1996, s. 132(1)), and **South Australia**, where the fine can reach AUD 10,000 (USD 10,000) or imprisonment for two years (South Australia, 1977, s. 29). In **Estonia**, a firearm licence (weapons permit) can be suspended for a year if the licensee is caught operating a car (or train or plane) under the influence of drugs or alcohol, even without being in possession of a firearm at the time (Estonia, 2002a, s. 43(1)(1)).

Criminal record

Most countries take account of an applicant's criminal record when evaluating a firearm licence application. Even in states where a firearms licence is not required, firearm sales to individuals who have a criminal record are prohibited, as in the United States (US, n.d.a, s. 922(d)(1)).

Approaches vary, with some states refusing a licence to anyone who has served a term of imprisonment for a certain period of time regardless of the nature of the offence; some considering whether the conviction occurred within a certain timeframe prior to the application, such as **Canada**—within the previous five years (Canada, 1995, s. 5(2)); and others refusing a licence to applicants previously convicted of certain classes of offences, such as firearm-related or drug-related offences, or felonies (more serious offences). Some countries employ a combination of these factors. For example, **Yemen** refuses a licence to anyone who has been convicted of a 'serious crime'; has been imprisoned for at least seven months for a crime involving assault, money, or honour; has been convicted twice of any of these crimes within the same year; or has been convicted of a crime involving the use of a firearm (Yemen, 1992, arts. 21(2)–(3)).

Authorities pay particular attention to whether an applicant has committed *violent* crimes.

In many jurisdictions, licensing authorities pay particular attention to whether the applicant has committed *violent* crimes, especially those involving sexual violence, domestic violence, family violence, or interpersonal violence; that is the case in Australia's **Victoria**, **Kenya**, **New Zealand**, **South Africa**, and **Switzerland**. For example, in **New Zealand** the licensing officer may reject a firearm licence if a protection order under the Domestic Violence Act 1995 is in force against the applicant (New Zealand, 1983, s. 27A), but he or she is not obliged to decide accordingly. Likewise, in the **United States**, federal law prohibits the purchase and possession of firearms and ammunition by those who have been convicted of a misdemeanour involving domestic violence or who are subject to certain kinds of domestic violence protective orders. Some states go further and strengthen these requirements, giving police the authority to remove firearms at the scene of a domestic violence incident, to remove firearms from the abuser when a protection order is granted, or to require the abuser to surrender his firearms (US, n.d.a, ss. 922(g)(8)–(9); LCAV, 2008, pp. 88–104).

In **South Africa**, in determining whether an applicant is a 'fit and proper' person to hold a licence, the registrar is encouraged to check whether there have been convictions involving violent behaviour, as well as whether in the past five years the applicant has been reported to the police or social services for alleged threatened or attempted violence, or whether in the past two years the applicant has experienced a divorce or separation from an intimate partner with whom the applicant resided and where there were written allegations of violence (South Africa, 2004, ss. 14(1)(e)–(f)). In **Canada**, applicants are required to give details of their conjugal status and information about their current and former conjugal partner(s) as part of the application form for a Possession and Acquisition Licence. In addition, the signature of the current *or former* spouse, common-law partner, or other conjugal partner is required on the application form, otherwise the chief firearms officer has a duty to notify them of the application (Canada, n.d.b, ss. E–F, boxes 18–19). In **New Zealand**, police will separately interview an applicant's spouse, partner, or next of kin as part of the application process to assess the applicant's suitability to hold a licence (New Zealand, 2010, p. 40).

Public interest

In many countries, in addition to the specific considerations described above that help determine whether an applicant is a 'fit and proper' person, licensing authorities are also directed to consider whether the applicant can possess a firearm without posing a danger to public peace or safety and whether the granting of a firearm licence to an applicant would be contrary to the 'public interest'. These are broad criteria that involve considerable discretion on the part of the licensing authority. Details of states that apply this criterion can be found in Annexe 9.1.

Competency

Many jurisdictions require prospective gun owners to undergo some kind of firearm training or competency testing prior to obtaining a firearm. The nature of the training or testing varies, but the general aim is to assess the applicant's practical and theoretical knowledge related to weapons use and safe handling and, in some cases, storage requirements and relevant laws governing acquisition, use, and the limits of legitimate self-defence (deadly force). For example, in **Brazil**, as part of the application, applicants must prove their technical ability in the handling of firearms, must be certified by an authorized firearm instructor, and must demonstrate their psychological aptitude for the handling of firearms, as certified by affidavit to be supplied by a Federal Police Department psychologist or a psychologist certified by the Federal Police (Brazil, 2004, art. 12). In **Estonia**, applicants must pass an examination regarding their knowledge of the firearm laws covering acquisition, storage, registration, carrying, transfer, and legal use of weapons as well as the provision of first aid to a victim with a shooting injury (there is also a practical test on firearm handling) (Estonia, 2002a, s. 35(5)). In **New Zealand**, all licensees must undergo training and pass theory tests on safe handling of firearms (New Zealand, 1992a, reg. 14).

In some instances, the nature of the training requirement depends on the type of firearm sought, or how the user intends to use or carry it. In the **United States**, for instance, there is no federal requirement for gun owners to have completed specific training to purchase firearms, although some states do have training requirements as part of licence or permit acquisition, including **California** (for handguns) and **Massachusetts** (LCAV, 2008, pp. 179, 210–11). Many US states have also introduced competency testing for licences to carry concealed handguns. For example, in **Texas**, anyone wishing to acquire a licence to carry a concealed handgun must complete a handgun proficiency course involving 10–15 hours of instruction on (1) the laws that relate to weapons and to the use of deadly force; (2) handgun use, proficiency, and safety; (3) non-violent dispute resolution; and (4) proper storage practices for handguns with an emphasis on storage practices that eliminate the possibility of accidental injury to a child (Texas, n.d.b, s. 411.188).

In other instances, the training requirement may depend on the activity or purpose for which the firearm is sought. For example, in **Croatia**, to obtain a permit to acquire a hunting firearm, the applicant must provide a certificate confirming having passed a hunting exam; for a sporting firearm, he or she must provide a certificate of active membership issued by a target shooting organization (Croatia, 1992, art. 19). In **Lithuania**, applicants are required to pass a competency test if the firearm licence is sought for self-defence purposes (Lithuania, 2002, art. 13(5)). Similarly, in **Singapore** applicants must pass a shooting proficiency test if they require a firearm for self-defence (Singapore, n.d.a, para. B(1)(d)).

In some jurisdictions, such as **Uganda**, the law indicates an applicant must be 'competent' to use a firearm, without specifying how competency is to be assessed (Uganda, 1970, s. 4).

Other criteria

In addition to the common criteria described above, states consider a range of other factors when determining whether an applicant is 'fit and proper' to own a firearm, including:

- *Lifestyle:* lifestyle or domestic circumstances (unspecified), such as in **Estonia** (Estonia, 2002a, s. 36(4)(3)) and Australia's **ACT** (ACT, 1996, s. 19(1)(a)), **New South Wales** (NSW, 1996, s. 11(4)(a)), and **Northern Territory** (NT, n.d.b, s. 10(4)(a)).
- *Associates:* reputation, honesty, integrity, and the nature of the applicant's close associates, for example in **South Australia** (South Australia, 1977, s. 5(13)), and whether the applicant's associates would be deemed unsuitable to obtain access to a firearm, as in **New Zealand** (New Zealand, 2010, p. 40); whether the applicant is living with someone with a criminal record, as in **Japan** (Japan, 1958, art. 5(6)(3)), or who is ineligible to own a firearm, as in **Lithuania** (Lithuania, 2002, art. 17(1)).
- *Loss of previous firearm:* whether the applicant has previously lost a firearm or had a firearm stolen through his or her negligence, as in **Belize** (Belize, 2003, s. 7(2)(g)), or has had a previous licence revoked due to the loss of a firearm, as in **Lithuania** (Lithuania, 2002, art. 17(1)(8)).

- *Military service record:* whether the applicant has evaded national service, as in **Estonia** (Estonia, 2002a, s. 36(4)), or has received a dishonourable discharge from the armed forces, as in the **United States** (US, n.d.a, s. 922(d)(6)).

- *Employment history:* whether the person has experienced forced job loss in the previous two years, as in **South Africa** (South Africa, n.d., question 16).

- *Outstanding warrants:* whether the applicant is a fugitive from justice, as in the **United States** (US, n.d.a, s. 922(d) (2)), has an outstanding arrest warrant against him or her, as in the US state of **Massachusetts** when granting a temporary licence to carry a firearm to a non-resident (Massachusetts, n.d., ch. 140, s. 131F(v)), or is in default of child support payments or taxes, as in the US state of **Texas** when granting a licence to carry a concealed handgun (Texas, n.d.b, ss. 411.172(10), (11)).

- *Number of firearms in the neighbourhood:* in **Papua New Guinea**, the registrar may refuse to grant a licence for any reason, including 'whether arising out of the number of firearms in the locality concerned or otherwise' (PNG, 1978, s. 9(2)(b)).

Owner licensing processes

Types of licences

Many countries—including **Croatia**, **Estonia**, **Finland**, **Kazakhstan**, **Lithuania, Papua New Guinea**, **Russian Federation**, **Singapore, Switzerland**, and **Turkey** (see Annexe 9.2)—have adopted a two-tier system of licensing, whereby a person is required to obtain a permit to acquire a firearm, and will then be granted a licence to possess or keep the firearm. In **Croatia**, for example, an applicant must obtain a permit to acquire a firearm, which involves scrutiny of the *applicant's eligibility* (Croatia, 1992, art. 17). He or she may then purchase a firearm, applying for registration of the firearm within eight days of acquisition, whereupon the licensing authority will grant a weapon licence for the firearm (art. 26). The registration and licence granting process involves the scrutiny of *the firearm:* Is the firearm marked? What are its origins? (art. 29). In countries where this system is in place, a licence will generally be valid for a specified length of time (usually in years), and the acquisition permit will be valid for a much shorter period of time (days or months rather than years).

In other countries, such as **Australia**, the reverse system is applied. A person must hold a firearm licence (having provided a 'genuine reason' for obtaining the licence) before applying for a permit to acquire a particular firearm, and can only acquire a particular type of firearm if he or she holds the corresponding category of licence. This is the case in **New South Wales**, for example (NSW, 1996, ss. 31(1)–(3)). In **Canada** a person must obtain a licence to possess a firearm (known as a 'Possession and Acquisition Licence'), which is issued with respect to the *person,* who must then acquire a registration certificate, which is issued with respect to a *firearm,* once the firearm is purchased (Canada, 1995, s. 13). Box 9.2 describes a number of different licences.

In other countries, a single licence or permit to possess is issued for each firearm, with no requirement to obtain a separate permit to acquire (as in **Belize** and the **United Kingdom**; see Annexe 9.2). Separate, specialized licences or permits may be required for certain types of weapons. For example, the **United Kingdom** issues two types of permits: firearm certificates and shotgun certificates, with slightly different procedures and conditions associated with each (UK, 1998, rr. 3(1), 5(1), schedules 1 (part 1), 2 (part 1)). Under **New Zealand's** licensing system, prospective firearm owners must obtain a firearm licence to possess a firearm. In order to possess a pistol, MSSA, or restricted firearm, however, a person must obtain special endorsement on his or her licence from the police and a permit to procure such a weapon (New Zealand, 1983, ss. 29, 30A).

Box 8.2 Licence terminology

Permit to acquire: a permit authorizing the holder to purchase or acquire a firearm. Usually valid for a short period of time (such as three months).

Licence or permit to possess: a licence or permit authorizing the licensee or permit holder to possess or own a weapon. In a few countries these are valid for the lifetime of the holder, but in most countries surveyed they are granted for a limited period of time (such as five years). Upon expiration, the licensee must apply to renew the licence, which generally involves going through the licence application process again.

Permit to carry: a permit that authorizes the holder to carry a firearm outside his place of residence. In some countries, a licence to possess also authorizes the holder to carry the firearm, and no separate permit to carry is required. In certain states, special permits are required for certain activities, such as a permit to carry a concealed weapon.

Parallel permit: a permit that entitles the holder to possess and use a specific firearm, even though he or she is not the licensed owner of the firearm. The consent of the firearm owner is generally sought as part of the application process.

In the **United States**, federal law does not require civilians to hold a licence to possess firearms as such (although in some US states they may be required to hold a licence to possess certain restricted firearms or to carry a concealed handgun). In many US states, however, civilians are required to obtain a firearm purchaser or owner identification card through their local police department, which undertakes a background check of the applicant on the National Instant Criminal Background Check System, an electronic database maintained by the Federal Bureau of Investigation.[205] This card is then presented to a licensed firearms dealer along with supporting identification when the holder goes to purchase a firearm. The licensed dealer places a call to the police to ensure the firearms identification card is valid, and the police will generally conduct another point-of-sale NICS check. In **Yemen**, where citizens also have the 'right' to possess weapons, civilians are not required to have a licence to hold or possess firearms, but they must have a licence to carry them (Yemen, 1992, arts. 9, 10; Sahouri, 2010b, p. 2).

In some jurisdictions, the licensing authority is required to give a reason for any licence refusal. For example, in **Lithuania** '[r]efusal to issue a permit must be grounded' (Lithuania, 2002, art. 12(5)). Similarly, in **Canada**, a chief firearms officer may refuse to issue an authorization to carry for 'any good and sufficient reason' (Canada, 1995, s. 68). Elsewhere, a reason is not required. For example, in **Uganda**, the chief licensing officer may, in his or her discretion, refuse to issue a firearms certificate 'without assigning any reason for the refusal' (Uganda, 1970, s. 4(4)).

Restrictions on quantities

In some jurisdictions in the sample, restrictions are imposed on the number of firearms a person can purchase at one time or can possess in total. For example, in the US state of **California**, an individual cannot apply to purchase more than one handgun within any 30-day period (California, n.d., s. 12072(a)(9)).

Some countries also limit the amount and type of ammunition a person can purchase or hold. For example, in **Lithuania**, a person can only acquire and keep 300 cartridges for each firearm held (and 1,000 cartridges if the person is a sport shooter) (Lithuania, 2002, art. 12(8)). Similarly, **Estonia** only permits civilians to store up to 100 cartridges for pistols, revolvers, gas weapons, and rifled barrel hunting guns; 300 cartridges for smoothbore-barrel hunting guns; up to 1,000 cartridges per sporting firearm; up to 1 kg of propellant per firearm, but not more than 5 kg in total; and up to 1,000 primers (Estonia, 2002a, s. 46). **Israel** allows individuals to hold a maximum of 50 bullets for a handgun.[206] **South Africa** restricts the amount of ammunition a person can have to a maximum of 200 cartridges, unless the person is a dedicated hunter or sport shooter (South Africa, 2000, s. 91).

In other jurisdictions, the legislation indicates that a person will only be entitled to possess a 'reasonable' amount of ammunition according to the needs of the owner and the type of the firearm, as is the case in **Tasmania** (Tasmania, 1996, s. 105(2)(b); 2006, reg. 14(a)(i)) and **South Australia** (South Australia, 2008, reg. 36). Many jurisdictions limit a licensee to purchasing and possessing ammunition of the type and calibre appropriate for the firearm for which he or she holds a licence. This calibre-specific licensing is usually policed by vendors who can only legally sell ammunition for a firearm type for which the purchaser displays a valid licence. Elsewhere, the legislation simply indicates that the firearm licence or authorization will indicate the quantity of ammunition that can be held by the licensee, as is the case in **Belize** (Belize, 2003, first schedule, form 1), **India** (India, 1962, r. 8(b)), **Kenya** (Kenya, n.d.a, second schedule, form 1), **Uganda** (Uganda, 1970, s. 4(8)(c)(iii)), and the **United Kingdom** (UK, 1968, s. 27(2)).

Waiting periods

A handful of the countries in the sample have mandatory waiting periods in place, meaning that applicants will not be granted a licence or permit to own a firearm until a certain period of time has elapsed or they must wait a certain number of days before they can collect a firearm once purchased. In **Australia**, for example, an acquisition permit can generally only be granted after 28 days from the date of the application (and if the applicant holds a firearm licence authorizing possession of the firearm sought).[207] Similarly, in **Canada** there is a 28-day waiting period between the application for a Possession and Acquisition Licence and the granting of the licence (Canada, 1998a, s. 5).

In the **United States**, several states impose a waiting period for the purchase of handguns—independently of a licensing requirement in some cases. For example, in **Florida** there is a three-day waiting period between purchase and delivery of handguns (although the county may shorten the waiting period for those who have a concealed weapon permit). Notably, in Florida applicants may be exempt from *any* waiting period if they or their families have been threatened with death or bodily injury and this threat has been reported to local law enforcement (Florida, n.d., s. 790.31(2)(d)(6)). In **New Jersey**, seven days must elapse after a permit to purchase a handgun has been issued before a purchaser can take possession (New Jersey, n.d.a, s. 2C:58-2.a(5)(a)). **California** imposes a waiting period of ten days for all firearms purchases (California, n.d., ss. 12071(b)(3)(A), 12072(c)(1)).

Penalties for unlawful possession

The penalties imposed for the unauthorized possession of a firearm vary from country to country (see Annexe 9.1). The penalties are generally higher if a person possesses a firearm with the intent to commit a crime, if the person is drunk or intoxicated while carrying the firearm, if the firearm is carried in a public place without authorization, or if the firearm is unlawfully discharged. In **Singapore**, the unlawful use of or attempt to use a firearm is punishable by death (Singapore, 1973, s. 4). Similarly, in **Egypt** the use of firearms against public order or the security of the state is punishable by death (Sahouri, 2010a).

Regulation of civilian transfer and retransfer

Civilians acquire firearms from a variety of sources. They may inherit them, purchase them privately from other civilians, or buy them from gun dealers, retailers, or, in some countries, at gun shows. Some states only allow firearms to be purchased from licensed dealers and retailers and do not permit transfers between private individuals, such as **Australia** (APMC, 1996, res. 9)[208] and **Singapore** (Singapore, 1913, s. 13(d)), or only permit private sales under restricted circumstances. In **South Africa**, for instance, a person may sell a firearm without the intervention of a dealer, but only on such conditions as the designated firearms officer may determine (South Africa, 2000, s. 31).

A Turkish boy looks at hunting guns during the Third Gun, Hunting & Nature Fair in Istanbul, September 2004.
© Mustafa Ozer/AFP Photo

Most countries do, however, allow private sales between civilians. Generally, where states permit civilians to buy and sell firearms privately, they stipulate that the purchaser must hold the relevant licence or acquisition permit (as is the case in **Canada**, **Estonia**, **Finland**, **Kenya**, **Papua New Guinea**, **Uganda**, and the **United Kingdom**) and the seller must notify the licensing authority, police department, or registrar, as the case may be, of the transfer. The latter is the case in **Belize**, where notification must occur within 14 days (Belize, 2003, s. 14(1)), **Canada** (Canada, 1995, s. 85(2)), **Finland**, where it must occur within 30 days (Finland, 1998, s. 89), **Kenya**, where it must occur within 48 hours (Kenya, n.d.a, form 2, s. 4(3)), **Uganda**, where it must occur within 48 hours (Uganda, 1970, s. 21(2)), and the **United Kingdom**, where it must occur within 7 days (UK, 1997, s. 33(2)).

In some instances, though there is no formal requirement to notify the authorities, a requirement that the seller cancel the registration of his or her firearm upon sale serves to alert the authorities that a transaction has taken place and the firearm has changed hands; in **Croatia**, for example, registration must be cancelled within eight days of sale or handover (Croatia, 1992, art. 25). In **Estonia**, the transfer of the firearm must take place in the presence of a police officer, and the transfer is formalized on the basis of a written application made by the owner of the firearm prior to the transfer (Estonia, 2002a, s. 64).

In the **United States**, federal laws govern sales by licensed dealers and retailers,[209] but the only restriction imposed on private sales by federal law is that a person (other than a licensed dealer, manufacturer, importer, or collector) cannot sell or transfer a firearm to someone who does not live in the same state as the seller (US, n.d.a, s. 922(a)(5)). It is also unlawful to sell or transfer a handgun or handgun ammunition to someone who is under 18 years of age (s. 922(x)(1)).

Nevertheless, some regulation of private sales does take place at the state level. A survey conducted by the US Department of Justice in 2005 indicates that the laws of 17 states regulate at least some private sales by requiring that purchasers obtain a permit or undergo a background check before receiving a firearm (US, 2005, p. 9). Among US states selected for this chapter, state law does not require background checks on prospective purchasers in private sales in **Arizona**, **Florida**, or **Texas**; nor are permits required to purchase firearms in these states (see Annexe 9.2). **Californian** law requires private sales to be processed through a licensed dealer, who must conduct a background check (LCAV, 2008, pp. 164–65). In **Massachusetts**, a purchaser must have a permit to obtain a firearm from a private seller (LCAV, 2008, pp. 164–65). In **New Jersey**, a purchaser must have a permit to buy a handgun or an identification card to receive a long gun from a private seller, and he or she will have been screened as part of the process of obtaining those documents (New Jersey, n.d.a, s. 2C:58-3).

REGULATING THE USE

'Genuine reason'

Often applicants must demonstrate that they have a *reason* for acquiring a weapon.

Most countries require a person to have a 'genuine reason' for acquiring a firearm or they permit civilians to possess firearms only for certain purposes, such as hunting, sport shooting, and, in some instances, self-defence. Many countries require applicants to demonstrate that they have a reason for acquiring a weapon, though the level of 'evidence' required to prove the need for a firearm varies from case to case. Annexe 9.1 indicates which of the surveyed countries require applicants to demonstrate that they have a 'genuine reason' for possessing a firearm and identifies the reasons accepted in each case.

The *reason* given for acquiring a weapon will often be taken into consideration by the official who determines what type of weapon a person can acquire. For example, in **Finland**, before granting an acquisition permit, the authorities will consider whether the firearm sought is not unnecessarily powerful or efficient and is suitable for the purpose for which the permit is granted (Finland, 1998, s. 44). In **Western Australia**, the commissioner who considers a licence application must be satisfied with the reason for the acquisition but also 'that the kind of firearm or ammunition can be reasonably justified' (Western Australia, 1973, s. 11A(3)).

Some countries that require applicants to have a good or genuine reason to possess a firearm specify what constitute good or genuine reasons in their national legislation. In **Croatia**, a person must have a 'justifiable reason' for acquiring a firearm and the legislation stipulates the following reasons for acquisition: hunting, target shooting, and self-defence (see Annexe 9.1). Other countries that require applicants to have a good or genuine reason for possessing a firearm leave it to the discretion of the licensing authority to decide on a case-by-case basis, as in **Kenya**, **Papua New Guinea**, and the **United Kingdom**. In the **United Kingdom**, applicants will not be granted a firearm certificate or a shotgun certificate unless they have a 'good reason' for possessing the firearm. The legislation indicates that if a shotgun is to be used for sporting or competition purposes or for pest control, this constitutes a 'good reason' for having such a gun (UK, 1968, s. 28(1B)); however, it does not specify what constitutes a 'good reason' for possessing other types of firearms, for which licensing officers have discretion in awarding or rejecting a licence. Similarly, the law in **Yemen** indicates that a person must state 'sufficient reasons and justifications' for granting a licence as part of their application for a licence to carry; it states that a person must have 'a political and social position or his work requires carrying a personal weapon' but otherwise gives the licensing authority 'the right to estimate the worthiness' of the applicant (Yemen, 1994, art. 6).

In some countries, the legislation does not expressly state that a person must have a 'good' or 'genuine' reason for owning a firearm, but in practice a reason is required. For example, **South African** legislation does not state that a person must have a reason for possessing a firearm, but firearm licences are only granted for specific activities, such as 'licence to possess firearm for occasional hunting and sports-shooting' (South Africa, 2000, s. 15).

In other cases, certain uses of firearms are automatically approved, while others are subject to government approval. For example, the law in **Switzerland** provides that 'any person who requests a license to acquire a firearm for a purpose other than sport, hunting or collection must give reasons for their request' (Switzerland, 1997, art. 8). In **Canada**, an applicant does not need a reason to possess a firearm; however, a person who holds a licence to possess a restricted firearm (including certain handguns and certain semi-automatic firearms) and who wishes to take it to a place other than where it is authorized to be possessed must prove he or she *needs* the restricted firearm to protect his or another person's life, or in connection with his or her lawful occupation (Canada, 1995, s. 20).

Similarly, in **New Jersey** the legislation indicates applicants must have a 'justifiable need' before they will be granted a permit to carry a handgun (New Jersey, n.d.a, s. 2C:58-4(c)). Anyone wishing to obtain a licence to purchase, possess, and carry a machine gun or assault firearm must submit a written application 'setting forth in detail his reasons for desiring such a license' (s. 2C:58-5(a)). These provisions do not specify what reasons are acceptable in either case, but others indicate ownership of certain assault rifles may be authorized for the purposes of competitive shooting matches, provided the owner is a member of a rifle or pistol club (s. 2C:58-12). The legislation also indicates that 'no license shall be issued unless the court finds that the public safety and welfare so require' (s. 2C:58-5(b)).

The reasons specified by countries as constituting valid or lawful justifications for possessing a firearm generally fall into the following categories: hunting; sport shooting and competition; employment or profession (such as farming and rural purposes or pest control); performance or art, including theatre and film; collection or exhibition, such as in a museum; and protection of person or property.

Hunting, target practice, and sport shooting

Hunting, target practice, and sport shooting are the primary reasons most countries surveyed permit civilian ownership of firearms; **Singapore** is a notable exception as it does *not* permit firearm possession for hunting purposes, other than a speargun for fishing purposes (Singapore, n.d.d). In some jurisdictions an applicant must demonstrate that he or she is a dedicated hunter or sport shooter, for example by providing evidence that he or she is a member of a sport shooting club or has a hunting licence. In **Uganda**, a person must obtain a bird or game licence within three months of receiving a firearm certificate or else the certificate ceases to be valid (Uganda, 1970, s. 4(5)). In **Singapore**, a person must be a member of a registered gun club to obtain a firearm for target shooting and must attend at least 12 shooting practices per year to have a licence renewed (Singapore, n.d.a, para. A(1); n.d.c, para. 2).

Profession

In many jurisdictions, firearm possession is authorized for certain employment-related or professional purposes, including pest control and farming or rural occupations. In some countries, there is simply a general acknowledgement that a firearm may be required for a professional or work purpose; examples include **Estonia** (Estonia, 2002a, s. 28(1)(4)) and **Finland**, whose law refers to 'work where a weapon is necessary' (Finland, 1998, s. 43(3)). In others, licences will only be granted for specific work-related activities such as nature research and conservation, as in **Croatia** (Croatia, 1992, art. 16), or where the work is of an especially dangerous or risky nature, such as remote wilderness work or the cash-transport business, as in **Canada** (Canada, 1995, s. 20(b); 1998c, s. 3).

Performance or art

Most countries have provisions authorizing the use of firearms for performance purposes, such as theatrical or cinematic productions. Often this is subject to certain conditions, such as that the firearm must be deactivated, as in **Croatia** (Croatia, 1992, art. 33), or that it may only be *borrowed* from a licensed dealer under a special permit, as in **Victoria** (Victoria, 1996, s. 92A(1)). Elsewhere, a person may seek permission to use a firearm in a film or theatrical production without having to obtain a formal licence or certificate, as in **Uganda** (Uganda, 1970, s. 7(1)(l)) and the **United Kingdom** (UK, 1968, s. 12).

Collection or museum

Many countries permit museums and individuals to possess firearms, including restricted and prohibited firearms, for the purposes of display and collection. In some cases, as in **Croatia** and **Lithuania**, such firearms must be deactivated or rendered inoperable (see Annexe 9.1). In others, collectors will be required to prove they are genuine collectors of firearms, and collection will be the only purpose endorsed on their firearm licence, as in **South Australia** (1977, ss. 12(7b), 13(2a)).

Self-defence

As indicated in Annexe 9.1, at least 16 of the reviewed jurisdictions explicitly contemplate self-defence or personal protection as a 'genuine reason' for possessing a firearm;[210] others explicitly prohibit acquisition of a firearm for self-defence, such as **Australia**. Many countries that permit the possession of firearms for self-defence restrict the type of weapon that can be held under such a licence and require some level of proof that the applicant needs a self-defence weapon. In **South Africa**, for example, the legislation stipulates that a person can hold only one licence to possess a firearm for self-defence, and that such a licence can only be granted for shotguns that are not fully or semi-automatic or for handguns that are not fully automatic. Furthermore, the applicant must demonstrate that he or she needs a firearm for self-defence and cannot reasonably satisfy that need by means other than the possession of a firearm (South Africa, 2000, s. 13). A person may be granted a licence to possess a *restricted* firearm for self-defence (including a semi-automatic rifle or shotgun) if he or she can demonstrate that a non-restricted firearm would not provide sufficient protection (s. 14). Similarly, in **Singapore**, a person must be able to show proof that there is a serious threat to his or her life and that there is no other way of overcoming the threat (Singapore, n.d.a, para. B(1)(c)).

Firearms and self-defence

The right to defend oneself against a physical threat is a universally recognized principle in all legal systems, provided (generally speaking) that the threat is 'immediate' and the response is 'necessary' and 'proportionate'.[211] There is no universally recognized right to *possess* a firearm to defend oneself, however. The 2006 report prepared by the UN Special Rapporteur on Human Rights discusses the principle of self-defence under human rights law; it appraises claims that the principle provides legal support for a 'right' to possess small arms, which, in turn, would negate or substantially minimize the duty of states to regulate possession. Yet the report concludes that, though the principle has an important place in international human rights law, 'it does not provide an independent, legal supervening right to small arms possession, nor does it ameliorate the duty of States to use due diligence in regulating civilian possession' (UNHRC, 2006, para. 19).

Although some countries do not accept self-defence as a genuine reason for possessing a firearm (as illustrated in Annexe 9.1), they do not necessarily preclude the use of a firearm to defend oneself, provided such use is consistent with criminal law provisions governing the use of force. Conversely, just because a state *does* accept self-defence as a genuine reason for possessing a firearm does not mean the use of that firearm in self-defence will always be justified.

This section explores the different policy approaches to the use of firearms for self-defence in two countries that explicitly permit the acquisition of firearms for self-defence (**Lithuania** and the **United States**), one country that permits the acquisition of firearms for self-defence in certain, limited circumstances (**Canada**), and two countries that do *not* permit the acquisition of firearms for self-defence (**Australia** and **New Zealand**).

> There is no universally recognized right to possess a firearm for self-defence.

Lithuania

The relevant legislation in Lithuania states that:

> *A person may use any arm in self-defence or in defence of another person, property, inviolability of one's home, other rights, interests of society or the state interests from an imminent or direct threat,* regardless of whether he has the possibility of avoiding the attempt or calling for assistance from another person or authority (Lithuania, 2002, art. 34(2), emphasis added).

The wording of the legislation indicates the right to use a firearm in self-defence or in defence of property is not dependent on *necessity*. A person is entitled to use a firearm in self-defence against a threat that is 'imminent or direct', even if he or she can avoid the threat or call for assistance from authority (presumably including the police).

United States

The Second Amendment to the US Constitution protects and preserves the right of US citizens to 'keep and bear' arms, although it does not elaborate on the legitimate purpose or use to which those arms can be put. Federal legislation indirectly confirms the right to use firearms for self-defence by virtue of the notes to the section of the United States Code addressing crimes and criminal procedure associated with firearms, which state that one of the purposes of the section is 'to avoid hindering industry from supplying firearms to law abiding citizens for all lawful purposes, including hunting, *self-defense,* collecting, and competitive or recreational shooting' (US, n.d.a, note on 'Purposes', emphasis added).

Two landmark US Supreme Court cases in 2008 and 2010 put it beyond doubt that the Second Amendment protects the individual right to keep and bear arms for the purpose of self-defence. In *District of Columbia* v. *Heller,* the Court ruled that a law introduced by the District of Columbia banning the possession of handguns in the home and requiring any lawful firearm in the home to be disassembled or rendered inoperable by a trigger lock was unconstitutional. With respect to the handgun ban, the Court noted:

the American people have considered the handgun to be the quintessential self-defense weapon. There are many reasons that a citizen may prefer a handgun for home defense [. . .]. Whatever the reason, handguns are the most popular weapon chosen by Americans for self-defense in the home, and a complete prohibition of their use is invalid (District of Columbia v. Heller, 2008, pp. 56, 57).

In *McDonald* v. *Chicago*—another case involving a challenge to the constitutionality of a handgun ban—the Court determined that the Second Amendment limits state and local governmental authority to the same extent that it limits federal authority (a question that the *Heller* case had left unanswered since it only considered the effect of the Second Amendment on federal law) (*McDonald* v. *Chicago,* pp. 1–2).

It is not yet clear what the impact of these decisions will be on firearm regulation in the United States. In *Heller* the Court did acknowledge that gun rights are not unlimited, and it was careful to point out that its ruling was not to be taken as an indication that all firearm laws are unconstitutional (*District of Columbia* v. *Heller,* pp. 22, 54–56). In both cases, however, the question of what kinds of gun laws *can* be reconciled with Second Amendment protection was left open.[212]

At the US state level, the majority of jurisdictions have adopted the so-called 'Castle doctrine', also known as 'Stand your ground' laws. This doctrine (the title of which is derived from the adage: 'an Englishman's home is his castle') permits a person to use deadly force to defend him- or herself and anyone inside his or her home from an attack by an intruder if he or she reasonably believes that force is immediately necessary, and provides that a person in lawful residence does not have a duty to retreat before using deadly force.

Gun rights supporters hold up banners outside the Supreme Court in Washington, DC, after the court ruled that US citizens have a constitutional right to keep guns in their homes for self-defence, June 2008. © Jose Luis Magana/AP Photo

All of the US states under consideration have adopted the doctrine into their penal codes, though there are some subtle differences in the legal provisions. For example, in **Texas** a person is justified in using deadly force to prevent a *robbery,* whereas in most other states the use of deadly force is only justified to protect against the unlawful use of force by the intruder. The Texas provisions stipulate that the 'finder of fact' (judge or jury) in court proceedings may not consider whether the person retreated when deciding whether he or she was justified in shooting the intruder (Texas, n.d.a, s. 9.32(d)).

Canada

The only specific reference made to the use of firearms for self-defence in Canada's Firearms Act 1995 is in relation to restricted and prohibited firearms. Section 20 provides that individuals who hold licences for restricted or prohibited firearms may only carry such firearms '(a) to protect the life of that individual or of other individuals; or (b) for use in connection with his or her lawful profession or occupation', such as cash handling, working in a remote wilderness area, or working as a licensed trapper (Canada, 1995, s. 20; 1998c, reg. 3). Similarly, section 28 provides that a chief firearms officer may approve the transfer to an individual of a restricted firearm or certain handguns only if the buyer needs the restricted firearm or handgun for the reasons identified above.

The regulations stipulate that, in order to possess a restricted firearm or prohibited handgun, it must be needed to protect the life of the applicant or of other individuals when, inter alia, the person's life is in imminent danger and 'police protection is not sufficient in the circumstances' (Canada, 1998c, reg. 2).

Otherwise, the question of whether and when firearms can be used for self-defence in Canada is determined on a case-by-case basis in accordance with the principles governing the right to self-defence contained in the Criminal Code. The Code stipulates that a person may use force to repel an attack 'if the force he uses is not intended to cause death or grievous bodily harm and is no more than is necessary to enable him to defend himself'. If, in repelling the attack, the person causes the death of the attacker, the act is justified if the person is under the 'reasonable apprehension' of his or her own death or grievous bodily harm and 'cannot otherwise preserve himself' from such death or grievous bodily harm (Canada, n.d.a, ss. 34(1), 35).

There is no explicit reference to the use of a firearm in the Criminal Code; nor does the Firearms Act 1995 expressly acknowledge a right to possess or use firearms for purposes of self-defence. The storage requirements for firearms in Canada are such that firearms are not readily usable for self-defence. As noted in Annexe 9.2, even non-restricted firearms are to be stored unloaded; they must be 'rendered inoperable' by means of a secure locking device or by the removal of the bolt or bolt-carrier. Alternatively, they are to be stored in a container of some kind that is securely locked, with associated ammunition not readily accessible (Canada, 1998b, reg. 5).[213]

Australia

As noted in Annexe 9.2, self-defence is not recognized as a 'good reason' for possessing a gun in any state or territory in Australia. Nevertheless, as discussed above, this does not categorically preclude a person from using a firearm in self-defence if justified in the circumstances.

One aspect of the criminal law that is particularly relevant in this regard is the partial defence of excessive force (also known as 'excessive defence'), applicable in cases of murder and recognized in **New South Wales** and **South Australia**.[214] This defence can be invoked when a person uses force that is excessive or not reasonable in the circumstances, but where the requisite intent for murder may not be present. In a sense it is an acknowledgement that, in the heat of the moment, it may not be possible to measure precisely what would constitute 'reasonable' force to repel an attacker, allowing such persons to be charged with manslaughter (or culpable homicide), which carries a lesser sentence than murder.

One notable difference between the two states with respect to this plea is that the New South Wales legislation only permits offenders to rely on it where their actions resulted in the death of another person if they were acting to protect themselves or another person, not property (NSW, 1900, s. 421). The South Australian legislation allows partial defence if the offender kills another person while protecting property or preventing trespass, even if the conduct is excessive (not 'reasonably proportionate'), provided the offender genuinely believes the conduct was reasonable and necessary *and* did not intend to kill the person (South Australia, 1935, s. 15A(2)).

Courts have considered a number of factors when determining whether fatal shootings were conducted in self-defence, including: whether the firearm was brought to the scene by the defendant or belonged to the deceased and was seized by the defendant opportunistically; whether the deceased was armed; whether the defendant was carrying a gun, which involves a risk in and of itself and adds to the culpability of the crime; and whether the shooting took place in public. In other words, the fact that a person was carrying a gun or was in possession of a gun may reduce his or her ability to rely on the partial defence of 'excessive defence', since the presence and use of a firearm will inevitably attract a risk that it will potentially be used for deadly consequences, thus making it difficult to argue the defendant did not intend to kill the deceased.

New Zealand

The New Zealand Arms Code includes the following self-explanatory 'important note' regarding firearms for self-defence:

> *Self-defence is not a valid reason to possess firearms. The law does not permit the possession of firearms 'in anticipation' that a firearm may need to be used in self-defence.*
>
> *Citizens are justified in using force in self defence in certain situations. The force that is justified will depend on the circumstances of the particular case. Every person is criminally responsible for any excessive use of force against another person.*
>
> *A firearm is a lethal weapon. To justify the discharge of a firearm at another person the user must hold a[n] honest belief that they or someone else is at imminent threat of death or grievous bodily harm.*
>
> *Discharge of a firearm at another person will result in a Police investigation and what ever the consequences of the incident you may face serious criminal charges* (New Zealand, 2010, pp. 41–42).

Conditions of firearm possession

Countries impose a range of conditions on licensees. Annexe 9.2 provides an overview of the common conditions imposed on licensees and firearm owners and indicates which jurisdictions impose them.

Reporting requirements

In many jurisdictions, firearm owners have an obligation to report the theft or loss of any firearm in their possession to the police or other authority within a short period of time. Owners may have an obligation to report the loss or theft of their licence, but not the firearm itself, as in **Papua New Guinea** (PNG, 1978, s. 67). Reporting requirements are thought to serve several public safety functions: timely reporting enables police to trace guns more effectively and increases the chances of recovering the weapon; reporting requirements make gun owners more accountable for their weapons and protect them from accusations in the event their stolen or lost firearm is used in a crime; and timely reporting helps enforcement authorities identify incidents of trafficking and 'straw purchasing'.[215]

Most reviewed states require firearm owners to report lost or stolen firearms, with some specifying the timeframe for reporting (such as 'within 24 hours') and others simply indicating if reporting should take place 'immediately' or 'forthwith' (see Annexe 9.2). Most states with reporting obligations impose a fine on owners who fail to report theft or loss; moreover, in several cases, owners may face a prison sentence. **South Africa**, for example, carries the highest possible penalty—ten years' imprisonment (South Africa, 2000, ss. 120(11), 121).

In **New Jersey**, if the owner of a registered assault weapon (firearm) fails to report the theft within 24 hours of his or her knowledge of the theft, not only is he or she liable to pay a fine, but if the firearm is used in the commission of a crime, the owner is civilly liable for any damages resulting from that crime (New Jersey, n.d.a, s. 2C:58-12(g)).

Safe storage

Safe storage requirements, such as storing the firearm unloaded, storing ammunition separately, and ensuring the firearm is in a locked receptacle, purportedly help reduce the risk that firearms will be stolen or misused, intentionally or accidentally. The emphasis in most jurisdictions is on preventing access to the firearm by unauthorized persons—including thieves or young children. What constitutes 'safe storage' varies from country to country, as does the extent of the obligation. In some instances, there is no specific standard to be applied, and licensees are simply required to take 'reasonable precautions', as in **Papua New Guinea** (PNG, 1978, s. 51), or to store their firearms 'with caution', as in **Switzerland**

A gun owner puts a firearm in a gun safe at his residence in Colorado, March 2001.
© Michael Smith/Newsmakers/Getty Images

(Switzerland, 1997, art. 34(1)(e)), or 'securely and in safe custody', as in **Kenya** (Kenya, 1954, s. 18(3)), to prevent them from being accessible to unauthorized persons.

In the **United Kingdom**, one of the conditions printed on the firearm certificate is that the firearm must, at all times, be stored securely 'so as to prevent, so far as is reasonably practicable, access to the firearms or ammunition by an unauthorised person' (UK, 1998, schedule 1, part II, condition 4(a)). The legislation does not impose specific requirements in this regard, although the Home Office has produced guidelines listing options and recommended practices, but emphasizing that it is the responsibility of certificate holders to ensure their compliance with the condition (UK, 2005, para. 1.4). For example, the guidelines recommend storing ammunition and other removable parts separate from the firearms (para. 2.2), adding that, although it is not a requirement of the Firearms Act, it is 'sensible' that they be locked away for security and safety reasons, especially if there are children in the house (para. 2.47).

Separate storage of ammunition is a common feature wherever safe storage requirements exist. The regulations in **Canada** require that firearms be stored in such a way that they are not within reach of ammunition, unless the ammunition is stored, together with or separately from the firearm, in a locked container (Canada, 1998b, s. 5(1)(c)). In **New Zealand**, owners of pistols, MSSA firearms, or restricted firearms must store their ammunition separately (New Zealand, 1992a, reg. 28). For other firearms, the ammunition must be stored so as to be inaccessible to someone who obtains access to the firearm, or, if the firearm and ammunition are stored together, the firearm must be rendered incapable of firing (for example, by removing the bolt and magazine, if possible, and locking them away separately from the firearm; making sure both the chamber and the magazine are empty; or using trigger-locking devices for firearms that cannot be taken apart) (New Zealand, 1992a, reg. 19; 2010, p. 12).

Another common feature is that firearms must be *unloaded* or must be incapable of discharging when they are stored. An exception to this requirement is often permitted in countries where firearms may be possessed for self-defence. Under **South African** law, firearms must be unloaded when stored unless they are held under a licence to possess a firearm for self-defence (South Africa, 2004, s. 86(11)).

Attempts to introduce safe storage requirements have had interesting repercussions in the **United States** because of their implications for self-defence. As discussed above, the US Supreme Court struck down as unconstitutional a Washington, DC, law that required firearms in the home to be rendered inoperable (*District of Columbia* v. *Heller*, 2008, p. 58). Some US states, including **Florida** and **Texas**, require firearm owners to take precautions to prevent children from gaining access to loaded firearms, including keeping the firearm in a locked container or securing it with a trigger lock (or other means to render it inoperable) (Florida, n.d., s. 790.174(1); Texas, n.d.a, s. 46.13(b)).

In **New Jersey**, keeping firearms securely stored is not a condition of ownership or possession; however, it is a criminal offence for an adult to leave a loaded firearm within easy reach of a minor (persons under 18 years of age) (New Jersey, n.d.b, s. 9:17B-3), unless the person stores the firearm in a securely locked box or container, secures the firearm with a trigger lock, or stores the firearm in a location which a reasonable person would believe to be secure (s. 2C:58-15(a)). Firearm owners are eligible for a USD 5 instant rebate when they purchase a trigger-locking device along with their firearm; retailers must display a sign announcing the rebate and the following warning:

> *Remember—the use of a trigger lock is only one aspect of responsible firearm storage. Firearms should be stored, unloaded and locked in a location that is both separate from their ammunition and inaccessible to children* (New Jersey, n.d.b, s. 2C:58-17(b)).

Separate storage of ammunition is a common feature wherever safe storage requirements exist.

Though some states recommend that certain safety measures be adopted, few countries prescribe the exact storage conditions. Exceptions include **Kazakhstan**, which requires owners to store their firearms in a safe deposit or metallic cabinet or any other storage that excludes the possibility of access to it by third persons (Karimova, 2010). **Estonia** requires owners who hold two or more firearms to store them in a specially adapted steel cabinet, permanently attached to the floor, a wall, or another structural element. Individuals who possess more than eight firearms must establish a specially adapted storage room (Estonia, 2002a, s. 46). In **Croatia**, weapons and ammunition 'shall be kept in such a manner so as to be inaccessible to persons not authorised to own them, locked and isolated, unless they are kept in a metal cabinet, safe box, or similar storing place that may not be opened by a tool in common use' (Croatia, 1992, art. 36).

In **New Zealand**, owners of pistols, restricted firearms, and MSSA firearms must store them in a steel and concrete strongroom, steel safe, steel box, or steel cabinet bolted to the building; the storage unit must be approved, either generally or specifically, by a police officer (New Zealand, 1992a, reg. 28(1)(a)). The regulations go on to specify in extensive detail the nature of the fixtures and fittings of such strongrooms, including the condition of the locks, bolts, and hinges, and the security of any windows or skylights. With respect to non-restricted firearms, owners must not store them 'in such a place that a young child has ready access to it' and must store ammunition separately or must ensure the firearm cannot be discharged (New Zealand, 1992a, regs. 19(1)(a)–(b)).

In some jurisdictions, the licensing authority must be satisfied that the applicant is capable of storing the firearm safely *before* granting a licence. For example, in **Kenya**, **New Zealand**, **South Australia**, and **Uganda**, the licensing authority must be satisfied that the applicant will take the necessary steps to keep the firearm secure and in safe custody, to prevent theft, loss, or access by an unauthorized person.[216] In **Belize**, applicants are asked to describe the security arrangements they have in place for the storage of their weapon (though the relevant act does not specify what storage arrangements are required) (Belize, n.d., question 7).

Singapore has some of the most restrictive storage provisions, requiring anyone who possesses a firearm for target practice to store the firearm at the gun club where he or she is a member. If a person owns a firearm for self-protection, he or she must keep it 'in a safe with combination and key locks' when not in use, and the safe must weigh no less than 70 kg (Singapore, n.d.b, para. 2). If the owner is planning to leave the country for more than one month, he or she must deposit the firearm at a police station for safekeeping (or with another authorized person) (Singapore, 1913, s. 19(3)). Similarly, in **Croatia**, if firearm owners intend to leave their place of residence for more than six months without taking the weapon with them, they must hand it over to a person who is entitled to hold such a weapon and to notify the authorities (Croatia, 1992, art. 37).

Carrying in public

Some jurisdictions have imposed an absolute prohibition on the carrying of firearms in public places by civilians, such as **Brazil** (Dreyfus and Perez, 2007) and the **Russian Federation** (Pyadushkin, 2008), though exceptions are made for private security guards, hunters, and sport shooters. Others permit the carrying of weapons, provided the carrier has reasonable grounds for carrying the weapon and the weapon is unloaded, as in **Finland** (Finland, 1998, s. 106)), or has obtained a permit or special (one-off) permission to carry, as in **Croatia** (Croatia, 1992, art. 27), **Kazakhstan** (Karimova, 2010), **Lithuania** (Lithuania, 2002, art. 12(10)), **New Zealand**—for pistols and restricted firearms (New Zealand, 1983, s. 36(1))—and **Switzerland** (Switzerland, 1997, art. 27). Other jurisdictions prohibit the carrying of firearms in certain locations, such as government premises, in and around schools and

A gun owner wears a .45 calibre semi-automatic pistol as he barbecues at a gun-rights rally in Portland, Maine, April 2010.
© Joel Page/AP Photo

Box 9.3 Concealed carry laws in the United States

The **United States** has four standards of law governing the carrying of firearms (typically handguns) in public. These are known as 'no issue', 'may issue', 'shall issue', and unrestricted laws. Two states currently have 'no issue' laws, which do not allow civilians to carry concealed weapons at all. Twelve states and the District of Columbia have 'may issue' laws, which require a permit to be obtained to carry a concealed handgun and give the licensing authority some level of discretion with respect to the granting of such a permit. In other words, the licensing authority *may* issue a permit to carry a concealed handgun if certain criteria are met. Finally, 34 states with 'shall issue' laws require that a permit be obtained to carry a concealed handgun. The licensing authority has no discretion in granting the permit, however; the permit *shall* be issued if certain objective criteria are met (typically, the applicant must be at least 21 years old; must pass a fingerprint-based background check; and may have to complete a safety class).

In three states (Alaska, Arizona, and Vermont) no permit is required to carry a concealed handgun or carrying is lawful without a permit when the carrier is engaged in certain activities (such as hunting, or when working as a security guard). **Arizona** deregulated the carrying of concealed weapons in April 2010, passing a law that removed the requirement to obtain a permit to carry a concealed handgun (Davenport and Cooper, 2010). Such unrestricted approaches to concealed carry are sometimes referred to as 'constitutional carry', as some perceive it as more closely aligned to the constitutional right to bear arms.

Different states also have varying laws concerning 'open carry' in the United States, that is, the act of carrying a firearm in a public place in plain view (as opposed to concealed from view). Some states allow open carry without a permit; some allow open carry provided the person has a permit or licence; and others may prohibit open carry altogether or restrict it to certain circumstances, such as when a person is hunting or on their own property. Annexe 9.2 indicates the provisions regarding concealed and open carry with respect to the US states reviewed for this chapter. In general, the trend over the last ten years has been towards less restrictive concealed carry state laws.

Source: LCAV (2008, pp. 203-11)

Authorities remove a man from the Texas Capitol after he fired off a few rounds into the air in January 2010. A few months later, legislators ordered magnetometers installed at all four public entrances to the capitol. © Vida Walker Burtis/AP Photo

churches, or during specific holidays and events, such as elections. For example, the **United States** has restrictions on carrying guns in or near federal buildings and school premises (see Box 9.3); and the cities of Bogotá and Cali in **Colombia** have both experimented with bans on the carrying of handguns on holidays, which reportedly led to a lower incidence of homicide during periods when the firearm-carrying ban was in effect (Villaveces et al., 2000, cited in HDC, 2005, p. 20). **South Africa's** law permits any premises or categories of premises to be declared 'firearm-free zones', if doing so is determined to be in the public interest (South Africa, 2000, s. 140).

CONCLUSION

This chapter has reviewed national controls over the civilian possession of firearms in 42 jurisdictions with a view to identifying some of the differences and similarities among them. Its first observation is that national approaches to civilian firearm regulation turn on the question of whether civilian ownership is seen as a basic right or a privilege. In two of the states the chapter reviews (the **United States** and **Yemen**) civilians have a basic right to own firearms and, accordingly, regulation tends to be more permissive; legislation limiting the right of possession is narrowly phrased. In all other jurisdictions, firearm possession is regarded as a privilege, and states place greater restrictions on ownership.

The chapter's second observation is that, despite the lack of international standards in this area, and irrespective of whether countries see civilian firearm ownership as a right or as a privilege, the reviewed jurisdictions share many elements of civilian firearm control. These include licensing systems that regulate access, gun registration or record-keeping, and restrictions and prohibitions on the possession of certain weapons. More fundamentally, national controls on civilian firearm access are generally three-pronged, simultaneously regulating the *type* of firearm civilians can possess, the *user*, and the permitted *use* of firearms.

By and large, states share the same underlying objectives—to prevent gun misuse and improve public security. In some instances, they pursue these goals through strong firearm controls; in others they favour more permissive gun laws. Only a few states worldwide prohibit civilians, as a group, from owning firearms, and none permit unrestricted civilian possession and use. In fact, nearly all seek a balance—one that is shaped by the unique history and culture of each country, and by its legal (constitutional) system. Simply put, there is no one-size-fits-all approach to the issue: authorities must consider many context-specific factors when designing national controls on civilian possession.

Yet civilian firearm controls are not only a reflection of geography. As with other types of social regulation, civilian gun laws change over time. Revisions are sparked not only by high-profile mass shootings, such as those reviewed in this chapter, but also by broader shifts in public attitudes towards armed violence—and towards regulation itself. As a result, the chapter can only offer a snapshot of civilian firearm legislation at a particular time (late 2010). States will undoubtedly continue to fine-tune their civilian gun laws as they seek to balance permission and restriction. ◾

LIST OF ABBREVIATIONS

.22 L.R.	.22 calibre long rifle
ACT	Australian Capital Territory (Australia)
APMC	Australasian Police Ministers' Council
ECOSOC	United Nations Economic and Social Council

ECOWAS	Economic Community of West African States
EU	European Union
MSSA	Military-style semi-automatic firearm (New Zealand)
NICS	National Instant Criminal Background Check System (US)
NFA	National Firearms Act (United States)
OAU	Organization of African Unity
PoA	Programme of Action to Prevent, Combat and Eradicate the Illicit Trade in Small Arms and Light Weapons in All Its Aspects
SADC	Southern African Development Community

ANNEXES

Online annexes at <http://www.smallarmssurvey.org/de/publications/by-type/yearbook/small-arms-survey-2011.html>

Annexe 9.1 Overview of owner licensing criteria and genuine reasons for owning a firearm

Annexe 9.1 identifies the eligibility criteria for owning a firearm in each of the jurisdictions under review (such as age, mental health, drug dependency, criminal record, public interest, and competency); it also identifies the 'genuine reasons' considered as part of the licensing process (such as hunting, sport shooting, professional requirements, performance, collection, and self-defence).

Annexe 9.2 Overview of conditions of firearms possession

Annexe 9.2 lists the duration of licences and permits to acquire firearms in the jurisdictions under review; it also identifies the conditions associated with licences and firearms ownership generally (such as the obligation to report theft or loss, safe storage requirements, and rules governing carrying weapons in public), as well as the penalties associated with a failure to fulfil these conditions.

ENDNOTES

1 Brunei Darussalam, Cambodia, and Taiwan (Republic of China) have exceptionally stringent civilian gun laws, prohibiting civilian gun possession in all but a few cases. Author correspondence with the Government of Brunei Darussalam; Cambodia (2005, art. 3); *China Post* (2009). In addition, Eritrea and Liberia report that they prohibit civilian ownership of firearms in their national reports on the implementation of the UN Programme of Action on Small Arms (Eritrea, 2010; Liberia, 2010); and the Solomon Islands has reported that only members of the Regional Assistance Mission are allowed to own and carry firearms (Solomon Islands, 2004).

2 The categorization of national arms control into regulation of the *firearm*, regulation of the *user* of the firearm, and regulation of the *use* of the firearm is based on Zimring (1991).

3 These six US states were selected on the basis of information and comparative analysis obtained from the websites of the Bureau of Alcohol, Tobacco, Firearms and Explosives (ATF), the Legal Community against Violence (LCAV), and the National Rifle Association (NRA).

4 While the sample is diverse and balanced, it may not be representative of the systems in place in countries outside the sample.

5 This study has not specified the extent to which the regional agreements touching on civilian possession have been incorporated into the national laws of participating states.

6 The four legally binding agreements are: the European Weapons Directive (Council of the European Union, 1991), the Firearms Protocol of the Southern African Development Community (SADC, 2001), the Nairobi Protocol (2004), and the Convention of the Economic Community of West African States (ECOWAS, 2006).

7 The Bamako Declaration was agreed among the member states of the Organization of African Unity (OAU) in 2000. In 2002, the African Union was formed as a successor to the OAU. The member states of the African Union are listed on the organization's website (AU, n.d.).

8   The following countries are member states of the Southern African Development Community (SADC): Angola, Botswana, the Democratic Republic of the Congo, Lesotho, Madagascar, Malawi, Mauritius, Mozambique, Namibia, Seychelles, South Africa, Swaziland, Tanzania, Zambia, and Zimbabwe. The Standard Operating Procedures for the Implementation of the SADC Protocol on the Control of Firearms, Ammunition and Other Related Materials further elaborate on the implementation of the provisions regarding civilian possession (SADC, 2008).

9   The following countries are states parties to the Nairobi Protocol: Burundi, the Democratic Republic of the Congo, Djibouti, Eritrea, Ethiopia, Kenya, Rwanda, Seychelles, Sudan, Tanzania, and Uganda.

10 The following countries are members of the Economic Community of West African States (ECOWAS): Benin, Burkina Faso, Cape Verde, Gambia, Ghana, Guinea-Bissau, Liberia, Mali, Nigeria, Senegal, Sierra Leone, and Togo. Côte d'Ivoire, Guinea, and Niger have all been suspended from the regional group since the adoption of the ECOWAS Convention in 2006.

11 The main regional agreement in the Americas—the Inter-American Convention against the Illicit Manufacturing of and Trafficking in Firearms, Ammunition, Explosives, and Other Related Materials, or CIFTA—does not include provisions regarding civilian possession. It explicitly notes in the preamble that: 'this Convention does not commit States Parties to enact legislation or regulations pertaining to firearms ownership, possession, or trade of a wholly domestic character' (OAS, 1997).

12 The following countries are members of the Andean Community: Bolivia, Colombia, Ecuador, and Peru.

13 The Directive binds all European Union (EU) member states. At the time of writing, EU member states included: Austria, Belgium, Bulgaria, Cyprus, Czech Republic, Denmark, Estonia, Finland, France, Germany, Greece, Hungary, Ireland, Italy, Latvia, Lithuania, Luxembourg, Malta, Netherlands, Poland, Portugal, Romania, Slovakia, Slovenia, Spain, Sweden, and the United Kingdom.

14 The following are members of the League of Arab States: Algeria, Bahrain, Comoros, Djibouti, Egypt, Iraq, Jordan, Kuwait, Lebanon, Libya, Mauritania, Morocco, Oman, Palestine (which is a Permanent Observer to the United Nations), Qatar, Saudi Arabia, Somalia, Sudan, Syria, Tunisia, United Arab Emirates, and Yemen.

15 The Nadi Framework was signed by representatives from the following countries: Australia, Cook Islands, Fiji, New Zealand, Samoa, Tonga, and Vanuatu.

16 In August 1987 seven people were shot in Melbourne in the Hoddle Street killings; in December 1987, eight people were killed in Melbourne in the Queen Street massacre.

17 For more information, see UK (2010).

18 Some of the deadliest US school shootings in recent years occurred at the Westlake Middle School (1998), Columbine High School (1999), Red Lake Senior High School (2005), and Virginia Tech (2007).

19 In Annexe 9.2, see Texas and New Jersey (columns on 'safe storage' and 'penalties for unsafe storage'). Child access prevention laws are discussed further in the section on 'safe storage'.

20 See Kopel (2009, pp. 517–18).

21 The US Gun-Free Schools Act (1994) required the expulsion for at least one year of any student who brought a weapon to school (US, 1994a, para. (b)(1)). Earlier legislation that sought to regulate guns in school zones—the Gun-Free School Zones Act (1990)—was struck down by the Supreme Court as unconstitutional in the case of *United States* v. *Lopez*, and an amended version of the legislation was introduced as the Gun-Free School Zones Act of 1995, as an amendment to title 18 of the United States Code (US, n.d.a, s. 922(q)).

22 The massacre took place on 16 April 2007 on the campus of the Virginia Polytechnic Institute and State University in Blacksburg, Virginia. Thirty-two people, and the shooter himself, were killed during the incident.

23 This chapter uses the term 'mass shooting' to refer to events in which more than five firearm-related homicides are committed by one or two perpetrators in proximate events in a civilian setting, not counting any perpetrators killed by their own hand or otherwise. This definition is taken from Chapman et al. (2006, table 1, p. 367).

24 The term 'handgun' is generally used to define a firearm that is designed to be held and discharged with one hand, such as pistols and revolvers. Many states do not use the term 'handgun' in their legislation, but rather prohibit or restrict firearms under a certain barrel length or overall length.

25 Short firearms are defined as 'a firearm with a barrel not exceeding 30 cm or whose overall length does not exceed 60 cm' (Lithuania, 2002, art. 2).

26 As of March 2011, the Egyptian government had not responded to requests for copies of the annexes to Law No. 394/1945, thus precluding Egypt's inclusion in Table 9.2. The Egyptian Ministry of Interior presumably keeps record of licences granted.

27 States were included in this category if they explicitly used the term 'semi-automatic' or 'self-loading'.

28 All of Australia's states and territories prohibit the possession of fully automatic weapons. Certain exceptions are made for the purposes of collection or historical re-enactments, but in many cases the firearm must be deactivated. For example, in Victoria automatic long guns may be possessed for historical re-enactments or public ceremonial events, but they must have been modified to be incapable of firing cartridge ammunition first (Victoria, 2008, reg. 24(3)).

29 Australia: penalties vary depending on the state or territory (see below).

30 ACT (1996, schedule 1, item 1, s. 42).

31 ACT: some semi-automatic firearms are prohibited, including self-loading rimfire rifles, self-loading centre-fire rifles, self-loading or pump-action shotguns, self-loading centre-fire rifles if designed or adapted for military purposes, and self-loading shotguns if designed or adapted for military purposes (ACT, 1996, schedule 1, items 2–6). Others are restricted and the 'genuine reasons' that apply to this category are more limited than those concerning other long guns; for example, recreational hunting does not constitute a 'genuine reason' for owning such firearms (ss. 64, 65).

32 ACT: different restrictions apply to handgun licensing and the 'genuine reasons' that apply to this category are more limited than those concerning long guns (other than semi-automatic long guns); for example, recreational hunting does not constitute a 'genuine reason' for owning handguns (ACT, 1996, s. 66).

33 ACT (1996, s. 42).

34 ACT (1996, ss. 156, 157, 193).

35 NSW (1996, schedule 1).

36 New South Wales: self-loading pistols with a barrel length of less than 120 mm are prohibited, but licences may be issued for sport/target shooting purposes (NSW, 1996, ss. 4C(1)(b), 16B(1)(a)); self-loading rimfire rifles with a magazine capacity of no more than ten rounds and self-loading shotguns with a magazine capacity of no more than five rounds are prohibited (except for limited purposes, including collection, primary production, and target shooting) (ss. 8(1), 14(a), 17A, 20); self-loading centre-fire rifles, self-loading rimfire rifles with a magazine capacity of more than ten rounds, and self-loading shotguns with a magazine capacity of more than five rounds are prohibited except for official purposes (s. 8(1)).

37 New South Wales: self-loading pistols with a barrel length of less than 120 mm are prohibited, but licences may be issued for sport/target shooting purposes (NSW, 1996, ss. 4C(1)(b), 16B(1)(a)); self-loading rimfire rifles with a magazine capacity of no more than ten rounds and self-loading shotguns with a magazine capacity of no more than five rounds are prohibited (except for limited purposes, including collection, primary production, and target shooting) (ss. 8(1), 14(a), 17A, 20); self-loading centre-fire rifles, self-loading rimfire rifles with a magazine capacity of more than ten rounds, and self-loading shotguns with a magazine capacity of more than five rounds are prohibited except for official purposes (s. 8(1)).

38 NSW (1996, s. 7(1))

39 NSW (1996, ss. 33, 45; 2006, cl. 36).

40 NT (n.d.b, schedule 1, item 1).

41 Northern Territory: some semi-automatic firearms are prohibited, including self-loading rimfire rifles, self-loading centre-fire rifles, self-loading or pump-action shotguns, self-loading centre-fire rifles if designed or adapted for military purposes, and self-loading shotguns if designed or adapted for military purposes (NT, n.d.b, schedule 1, items 2–6). Others are restricted and the 'genuine reasons' that apply to this category are more limited than those concerning other long guns; for example, recreational hunting does not constitute a 'genuine reason' for owning such firearms (NT, n.d.a, s. 12(1)(a)).

42 Northern Territory: different restrictions apply to handgun licensing and the 'genuine reasons' that apply to this category are more limited than those concerning long guns (other than semi-automatic long guns); for example, recreational hunting does not constitute a 'genuine reason' for owning handguns (NT, n.d.a, s. 14(a)).

43 NT (n.d.b, s. 58(6); 2010, s. 3).

44 NT (n.d.b, ss. 7, 18).

45 Queensland (1990, s. 4(a); 1997, s. 8(a)).

46 Queensland: for example, in order to obtain a licence with an endorsement for a semi-automatic shotgun (with a magazine capacity no greater than 5 rounds) for clay target shooting, an applicant must be a member of an approved shooting club that takes part in, or is affiliated with a body that takes part in, national and international clay target shooting competitions, and must, because of a lack of strength or dexterity, have a physical need for such a shotgun to enable the applicant to take part in clay target shooting, which must be supported by a doctor's statement (Queensland, 1996, s. 20; 1997, ss. 4, 5).

47 Queensland: certain handguns are 'prohibited handguns', including firearms with a calibre of more than .38 inches, semi-automatic weapons with a barrel length of less than 120 mm (unless they have an overall length of at least 250 mm measured parallel to the barrel), weapons that are not semi-automatic and have a barrel length of less than 100 mm (unless they have an overall length of at least 250 mm measured parallel to the barrel), weapons with a magazine with a maximum capacity of more than ten rounds, and weapons designed to be used without a magazine that have a maximum capacity of more than 10 rounds (Queensland, 1990, ss. 132, 174).

48 Queensland (1990, s. 50(1)(c)(i)).

49 Queensland (1990, s. 49).

50 South Australia (2008, reg. 4(1)(a)).

51 South Australia: semi-automatic firearms can only be possessed for farming purposes and clay target shooting (South Australia, 1977, s. 15A(3)).

52 South Australia: handguns are generally only permitted for sport shooting and collection purposes (South Australia, 1977, ss. 15A(4a)–(4e)).

53 South Australia (1977, s. 11(7)(a)).

54 South Australia (1977, s. 6A; 2008, reg. 17).

55 Tasmania (1996, schedule 1).

56 Tasmania: certain semi-automatic weapons may only be possessed for animal population control and collection purposes (Tasmania, 1996, s. 32(1)(a)) and others are prohibited entirely (schedule 1).

57 Tasmania: different restrictions apply according to the type of handgun sought, and whether the intended use is sport or target shooting, security guarding, or other purposes (Tasmania, 1996, s. 18); certain types of handguns are prohibited, including pistols with a calibre exceeding .38 inches (Tasmania, 1996, schedule 1; 2006, reg. 5).

58 Tasmania (1987, s. 4A; 1996, s. 9(1)).

59   Tasmania (1996, ss. 83, 89).

60   Victoria (2008, reg. 24(3)).

61   Victoria: semi-automatic long guns can only be possessed for certain activities, such as primary production, professional hunting, and clay target shooting (Victoria, 1996, s. 11(1)(a)), and certain types of semi-automatic weapons (such as semi-automatic centre-fire rifles) can only be possessed for professional hunting (s. 12 (1)(a)(i)).

62   Victoria: automatic handguns can only be possessed by collectors, and specific storage requirements and penalties apply. Other types of hand-guns (including semi-automatic firearms with a barrel length of less than 120 mm) can be possessed for historical re-enactments, for public ceremonial events, for starting or finishing sporting events, or for the purposes of the training and trialling of dogs (Victoria, 1996, ss. 3 (definitions), 15(1); 2008, reg. 24(4)).

63   Victoria (1996, ss. 6, 7).

64   Victoria (1996, ss. 87, 113).

65   Western Australia (1974, reg. 26, table).

66   Western Australia: certain semi-automatic long guns can only be possessed for training for international sport shooting contests, others can only be possessed for farming purposes (Western Australia, 1974, schedule 3, items 3, 6).

67   Western Australia: there are restrictions on the reasons for which a handgun can be possessed. For example, hunting, recreational shooting, and pest control are not considered 'genuine reasons' for obtaining a licence for a handgun (Western Australia, 1974, schedule 3, item 11(2)).

68   Western Australia (1973, s. 19(1ac)(b)).

69   Western Australia (1973, s. 31).

70   Belize (2003, s. 35(1), second schedule), as amended by Belize (2008, item 13).

71   Belize: schedule 2 of the act also indicates 'all assault rifles' are prohibited. The term 'assault rifles' is not defined in the act but presumably includes semi-automatic rifles (Belize, 2003, s. 35(1), second schedule), as amended by Belize (2008, item 13).

72   Belize (2003, s. 35(1), second schedule), as amended by Belize (2008, item 9).

73   Belize (2003, s. 35(3)).

74   Belize (2003, s. 10 (1)).

75   Brazil (2000, art. 16(V)); Dreyfus and Perez (2007).

76   Brazil (2000, art. 16(III)); Dreyfus and Perez (2007).

77   Brazil (2000, art. 16); Dreyfus and Perez (2007).

78   SINARM is Brazil's National Firearms System (Sistema nacional de armas).

79   Brazil (2000, art. 17(2)); Dreyfus and Perez (2007).

80   Canada (1995, s. 12). Grandfathered status allows an individual to possess a prohibited firearm. Generally, it applies in situations where a person legally owned a machine gun or other prohibited firearm before such firearms were banned or prohibited. Additional restrictions on the transfer of such firearms, restrictions on where such firearms may be possessed, and registration requirements often apply in jurisdictions that allow grandfathered firearms.

81   Canada: certain semi-automatic firearms are restricted, that is, firearms that have a barrel length of less than 470 mm and are capable of discharg-ing centre-fire ammunition in a semi-automatic manner (Canada, n.d.a, s. 84(1), interpretation).

82   In Canada certain handguns are prohibited (handguns with a barrel length of 105 mm or less and handguns designed or adapted to discharge .25 or .32 calibre ammunition)—other than handguns used in international sporting competitions—and all other handguns are restricted (Canada, n.d.a, s. 84(1)).

83   Canada (n.d.a, ss. 91(3), 92(3)).

84   Canada (1995, s. 83).

85   Nowak (2010a).

86   Colombia: automatic and semi-automatic long rifles with a calibre of more than .22 (.22 L.R.) are prohibited for civilian use (Nowak, 2010a).

87   Colombia: semi-automatic weapons (rifles and handguns) other than .22 L.R. are allowed if they are defined as firearms for personal defence (though there are limitations to this; for example, the maximum calibre is 9.652 mm and the maximum length of canon is 15.24 cm) (Nowak, 2010a; Colombia, 1993, arts. 8, 9).

88   Colombia: handguns may be acquired for personal defence, sporting, and collection purposes (Colombia, 1993, arts. 10–13).

89   Colombia: the penalty doubles if the firearm possession is aggravated, such as if i) the firearm was used in a motorized vehicle; ii) the firearm was used in a previous felony; iii) the owner resisted authorities; and iv) the owner used firearms while hiding his or her face or identity (such as by wearing a mask) (Nowak, 2010a).

90   Nowak (2010a).

91   Croatia (1992, art. 11).

92   Croatia: semi-automatic firearms are generally prohibited (Croatia, 1992, art.11), but certain semi-automatic firearms can be possessed and used for hunting and sport shooting purposes, such as long semi-automatic firearms whose magazine cannot be loaded, or is blocked so it is impos-sible to load, with more than three rounds can be used for hunting (art. 12(1)).

93   Croatia: semi-automatic firearms, including handguns, are generally prohibited. But semi-automatic 'short firearms' (those with a barrel length of less than 30 cm and an overall length of less than 60 cm (Croatia, 1992, art. 4) may be permitted for target shooting (art. 12(2)).

94   Croatia (1992, art. 74), as amended by Croatia (1999, art. 6).

95   Croatia (1992, arts. 12(4), 26, 71).

96   Inoa (2010).

97   Estonia (2002a, s. 20(5)).

98   Estonia: the Weapons Act lists all the weapons and ammunition prohibited for civilian purposes and includes 'particularly powerful firearms which are generally used as military weapons' (Estonia, 2002a, s. 20(7)). The provision goes on to state that the list of models of particularly powerful firearms generally used as military weapons and prohibited for civilian purposes shall be established by a regulation of the minister of Internal Affairs. At the time of writing, the existence and content of such regulations could not be confirmed, and it is possible that Estonia prohibits or restricts all or some semi-automatic weapons as 'military weapons'.

99   Estonia bans civilian possession of 'smoothbore guns with an overall length of less than 840 mm or where the length of each barrel is less than 450 mm' (Estonia, 2002a, s. 20(1)(2)), but does allow 'guns with a smoothbore barrel; guns with a rifled barrel; pistols; revolvers' owned or possessed for the provision of security services or for internal guarding (s. 31(3)).

100   Estonia (2002b, s. 418(superscript 1)(1)).

101   Estonia (2002a, ss. 24(1); 33(1); 33(5)–(7)).

102   Finland: a permit for an automatic firearm will only be granted for performance purposes, or for museums or collections. If there is a 'special reason', it may also be granted for work in which a weapon is necessary (Finland, 1998, s. 44).

103   Finland: semi-automatic firearms are not categorically prohibited; however, an acquisition permit will only be granted for a firearm that 'on the basis of the number of cartridges in the magazine, the calibre or other properties, and with regard to the purpose of use notified by the applicant, is not unnecessarily powerful or efficient, and which is suitable for the purpose of use notified by the applicant' (Finland, 1998, s. 44).

104   See endnote 103. An inquiry commission created following shootings at two Finnish schools in 2007 and 2008 had recommended the banning of semi-automatic handguns, but the proposal was rejected (*Helsinki Times,* 2010).

105   Finland (1998, s. 102).

106   Finland (1998, s. 113).

107   India (1959, s. 7(a)).

108   India: semi-automatic firearms (other than semi-automatic revolvers and pistols) as well as smoothbore guns with a barrel of less than 20 inches in length are classified as Category I firearms, to which slightly different regulations apply (India, 1962, schedule I). For example, a licence to acquire or possess ammunition for such firearms will only be granted if the arms are for sporting purposes and the amount of ammunition the licensee may possess is entered on the licence (India, 1962, r. 8(b)).

109   India (1962, schedule I).

110   India (1959, s. 25(1A)).

111   India (2010).

112   Author interview with a representative of the Israeli government, Geneva, 11 March 2011; Israel (2008).

113   Japan (1958, art. 3); author correspondence with the Japanese Ministry of Foreign Affairs.

114   Japan: possession of automatic or semi-automatic firearms is generally prohibited, together with other firearms (Japan, 1958, art. 3). Although a licence for possessing semi-automatic firearms for civil use can theoretically be obtained for hunting purposes, obtaining such a licence is extremely difficult in Japan for security reasons (author correspondence with the Japanese Ministry of Foreign Affairs).

115   Japan: there is a total ban on handguns, except for legitimate antique gun collectors and licensed shooting teams (Japan, 1958, art. 3-2).

116   Japan: the term 'firearms' is defined to include machine guns; the penalties that apply to unlawful possession of firearms other than machine guns thus also apply to machine guns (Japan, 1958, art. 2).

117   Japan (1958, arts. 31-3, 31-16).

118   Author correspondence with the Japanese Ministry of Foreign Affairs; Japan (1958, art. 14).

119   Karimova (2010).

120   Kazakhstan: the regulations prohibit civilian ownership of firearms with barrels shorter than 500 mm and with an overall length of less than 800 mm. This excludes handguns (Karimova, 2010).

121   Kenya (1954, ss. 2, definition of 'prohibited weapon', subsection (b); 26(1)).

122   Kenya: the possession of semi-automatic self-loading military assault rifle of 7.62 mm or 5.56 mm calibre is prohibited (Kenya, 1954, ss. 2, definition of 'prohibited weapon', subsection (b); 26(1)).

123   Kenya (1954, ss. 4(3)(a), 26(2)(a)–(aa)).

124   Kenya (n.d.b, para. 2.4(b)(1)(ii); 1954, s. 17). In its 2010 national report on PoA implementation, Kenya notes that the Central Firearms Bureau maintains a register of all small arms in licensed civilian possession and that this will be upgraded under the new national policy to include information on transfers (Kenya, 2010, p. 19).

125   Lithuania (2002, art. 7(2)(1)); however, employees of the Bank of Lithuania may carry automatic weapons acquired and possessed by the Bank of Lithuania, provided they are at least 21 years of age, have passed an examination, and have obtained a permit (Lithuania, 2002, art. 13(7)).

126   Lithuania: there are 'restrictions' on semi-automatic weapons in the sense that certain firearms do not require a permit or registration to be owned, while semi-automatic firearms do (Lithuania, 2002, arts. 12(2), 13).

127   Lithuania (2000, art. 253; 2010, p. 7).

128 Lithuania (2002, art. 8).

129 New Zealand (1984, schedule, item 4; 1983, s. 30). Automatic firearms are classified as 'restricted weapons' and can only be possessed for specific purposes and under certain conditions (see Annexes 9.1 and 9.2).

130 New Zealand: MSSA firearms are subject to certain restrictions and a person must have a special endorsement on his or her firearm licence to possess an MSSA (New Zealand, 1983, ss. 29, 30A, 32). The term MSSA does not include pistols (handguns) or semi-automatic firearms that are maintained at all times in a sporting configuration (New Zealand, 1983, s. 2).

131 New Zealand: pistols are defined as 'any firearm that is designed or adapted to be held and fired with 1 hand; and includes any firearm that is less than 762 millimetres in length' (New Zealand, 1983, s. 2); they are subject to similar restrictions that apply to 'restricted weapons' and a person must have a special endorsement on his or her firearm licence to possess a pistol (s. 29).

132 New Zealand (1983, s. 50(1)).

133 New Zealand (1992a, reg. 26; 1983, s. 12).

134 PNG (1978, s. 62).

135 Papua New Guinea: a special pistol licence is required to own a pistol in Papua New Guinea (PNG, 1978, part VII).

136 PNG (1978, s. 62).

137 PNG (1978, s. 8).

138 Pyadushkin (2008).

139 Russian Federation: the only exception is honorary handguns that can be owned and carried (in a holster, with general limits for carrying a firearm), but cannot be sold, gifted, or inherited, as stipulated in section 20.1 of the Arms Law of 1996 (Pyadushkin, 2008).

140 Russian Federation (1996, art. 222). This penalty can be increased to up to eight years if committed by an organized group.

141 Singapore (n.d.a, paras. A, B).

142 Singapore (1973, s. 3(1)).

143 Singapore (2006, r. 6).

144 Singapore (2006, r. 15).

145 South Africa: automatic firearms may only be possessed by collectors or for theatrical purposes (South Africa, 2000, ss. 4(1)(a), 17(1), 20(1)(b), 20(2)(c)).

146 South Africa: 'restricted firearms' (semi-automatic rifles and shotguns) cannot be legally owned under a licence to possess a firearm for self-defence, unless the applicant proves he or she needs such a weapon (South Africa, 2000, ss. 13, 14). In addition, restricted firearms may not be possessed by occasional hunters or sport shooters, though semi-automatic shotguns manufactured to fire no more than five shots in succession without having to be reloaded may be possessed by dedicated hunters and sport shooters under certain circumstances (ss. 15, 16).

147 South Africa: handguns that are not fully automatic may be possessed for self-defence, by occasional hunters, and by sport shooters, as well as by dedicated hunters and sport shooters (South Africa, 2000, ss. 13(1)(b), 15(1)(a), 16(1)(a)).

148 South Africa (2000, s. 4, schedule 4).

149 South Africa: other information recorded in the Central Firearms Registry includes: competency certificates, renewals, cancellations, refusals, fingerprints of applicants, imports, exports, loss and theft or destruction, seizures, all original documentation, a record of all licensed dealers, manufacturers, gunsmiths, importers, exporters, and a record of all firearms held by official institutions (South Africa, 2000, s. 125).

150 Switzerland (1997, art. 5.2a). Exceptional authorizations permitting the possession of prohibited firearms (including automatic weapons) can be granted for 'good reasons', including requirements inherent to the profession; use for industrial purposes; offsetting of a physical handicap; and for collection purposes (art. 28b).

151 Switzerland: the acquisition of automatic firearms that have been transformed into semi-automatic firearms is prohibited (Switzerland, 1997, art. 5.1a); however, Swiss military automatic firearms that have been transformed into semi-automatic firearms do not fall into this category (art. 5.6). Conversely, certain manual repetition rifles (such as semi-automatic firearms) can be acquired without a licence to acquire, including '(a) military repetition rifles; b) sports rifles functioning with military calibre ammunition generally used in Switzerland or with sport-calibre ammunitions, such as standard rifles with repetition breech system; c) hunting arms that are accepted for hunting as defined under federal legislation on hunting; d) sports rifles that are accepted at national and international sport hunting shooting competitions' (Switzerland, 2008, art. 19).

152 Switzerland (1997, art. 33.1a).

153 Switzerland (1997, arts. 11.3, 31b, 31c, 32a, 32b).

154 Pehlevan (2010).

155 Uganda (1970, s. 25(1)).

156 Uganda (1970, s. 25(2)).

157 Uganda (1970, s. 4(2); n.d., form FA 3).

158 UK: a person may be authorized to possess an automatic firearm in certain circumstances, such as for the purpose of keeping or exhibiting it as part of a collection (UK, 1968, s. 5A(1)) or for use in a film or theatrical production (s. 12).

159 In the UK, the following semi-automatic firearms are prohibited: any self-loading or pump-action rifled gun other than one which is chambered for .22 rimfire cartridges; any self-loading or pump-action smoothbore gun that is not an air weapon or chambered for .22 rimfire cartridges and either has a barrel of less than 24 inches in length or is less than 40 inches in length overall (UK, 1968, ss. 5(1)(ab)–(ac)).

160 UK (1968, s. 5(1)(aba)).

161   UK (1968, s. 51A).

162   UK (1968, ss. 33(1), 40).

163   US (n.d.a, s. 922(o)(1)). Short-barrelled shotguns and short-barrelled rifles are also prohibited; however, there are exceptions: machine guns and other prohibited weapons may be specifically authorized by the attorney general in a manner consistent with public safety and necessity (US, n.d.a, s. 922(b)(4)). In addition, machine guns possessed before the introduction of the ban may be kept (so-called 'grandfathered' weapons) (US, n.d.a, s. 922(o)(2)(B)). See LCAV (2008, pp. 20–23) for more information on, and examples of the types of restrictions that apply to, grandfathered firearms in the United States.

164   United States: the Violent Crime Control and Law Enforcement Act of 1994 introduced a federal prohibition on the manufacture for civilian use, transfer, and possession of certain semi-automatic firearms—so-called 'assault weapons' (known as the 'Federal Assault Weapons Ban') (US, 1994b, subtitle A). The ban was put in place for ten years on 13 September 1994 and expired on 13 September 2004, as part of the law's sunset provision (s. 110105). So far, attempts to renew the ban have been unsuccessful. Accordingly, there are no federal restrictions on the possession of semi-automatic firearms in the United States.

165   US (n.d.c, s. 5841).

166   Arizona (n.d., s. 13-3101(8)(a)(iii)).

167   Arizona (n.d., ss. 13-3102(A)(3), 13-3102(K), 13-702, 13-703).

168   Arizona: in fact, the Criminal Code prohibits any political subdivision of the state from requiring the licensing or registration of firearms or ammunition (Arizona, n.d., s. 13-3108(B)).

169   California: the California Penal Code specifies that the Department of Justice may issue a permit for the possession of a machine gun upon a satisfactory showing that 'good cause' exists and provided the applicant is at least 18 years old. In practice, however, such permits are rarely issued and generally only law enforcement personnel can possess machine guns (California, n.d., s. 12230).

170   California: certain semi-automatic firearms designated as 'assault weapons' are banned, unless they are grandfathered. A detailed list of assault weapons banned in California is available in the state Penal Code (California, n.d., s. 12276).

171   California: the conditions include that the person lawfully possessed firearm(s) prior to them being classified as 'assault weapons' and that he or she has no prior convictions for unlawful possession (California, n.d., s. 12280(b)).

172   California (n.d., s.12220).

173   California (n.d., ss. 12285(a), 12053). As noted in the 'semi-automatic firearms' column, California has banned assault weapons but allows continued possession of such weapons if they were owned before a certain date and are registered. In addition, California bans .50 calibre rifles but allows anyone who owned one before 1 January 2005 to retain the weapon provided it was registered by 30 April 2006 (s. 12285(a)).

174   California: the Penal Code prohibits the retention or compilation of information by the attorney general on firearms that are not handguns, unless the records are required for criminal prosecution or investigations (California, n.d., s. 11106(b)).

175   Florida (n.d., s. 790.221(1)).

176   Florida (n.d., s. 790.0655(1)(a)).

177   Florida (n.d., ss. 790.221(2), 775.082(3)(c), 775.083(1)(b)).

178   Massachusetts (n.d., ch. 140, s. 131(o)).

179   Massachusetts: only law enforcement officers or retired officers can legally possess assault weapons, defined to include certain semi-automatic firearms, unless they have been grandfathered (Massachusetts, n.d., ch. 140, ss. 121; 131M).

180   Massachusetts: large-capacity weapons, including other semi-automatic weapons, are restricted; exceptions for grandfathered weapons apply (Massachusetts, n.d., ch. 140, ss. 121; 131M, 131(a)).

181   Massachusetts (n.d., ch. 140, s. 131M; ch. 269, ss. 10(c), 10(n)).

182   New Jersey (n.d.a, s. 2C:58-5).

183   New Jersey: certain semi-automatic firearms designated as 'assault weapons' are restricted, unless grandfathered (New Jersey, n.d.a, ss. 2C:39-1(w), (x); 2C:58-5).

184   New Jersey (n.d.a, s. 2C:58-3(a)).

185   New Jersey (n.d.a, ss. 2C:39-5, 2C:43-6(a)(2), 2C:43-3(a)(2)).

186   New Jersey (n.d.a, ss. 2C:58-12, 2C:58-2(a)(4), 2C:58-2(b)).

187   Texas (n.d.a, s. 46.05).

188   Texas (n.d.a, ss. 46.02, 46.035).

189   Venezuela (1939, art. 3).

190   Nowak (2010b).

191   Nowak (2010b).

192   Yemen (1992, art. 2(5)).

193   Yemen, (1992, arts. 14, 28, 29, 51).

194   In 1994, the Canadian Department of Justice estimated that the net cost of the new programme would be about CAD 2 million (USD 1.5 million) on the basis that expenditures of about CAD 119 million (USD 89.6 million) were expected to be offset by licensing and registration fees of approximately CAD 117 million (USD 88.1 million) (Canada, 2002, para. 10.27). However, implementation is estimated to have cost more than CAD 1 billion (USD 750 million) (para. 10.3).

195   That said, the website of the Royal Canadian Mounted Police indicates: 'The Government has announced its intention to simplify licence requirements for firearm owners and to remove the requirement to register non-restricted firearms' (RCMP, 2011). In addition, following the defeat of
the private member's bill to abolish the long-gun registry in September 2010 (CBC News, 2010), the prime minister indicated that the government remained committed to its repeal, although, at the time of writing, there was no bill before Parliament and no indication that one would
be introduced soon.

196   In Canada, '[l]ong guns [rifles and shotguns] had been used in 72% of the firearm deaths in 2001. This decreased to 69% of deaths by 2004.
Handguns by comparison were used in 25% of the deaths in 2001. This increased to 26% in 2004' (RCMP, 2010, p. 21).

197   Retired judge Sir Thomas Thorp was appointed to conduct the independent inquiry (hence the report is often referred to as the 'Thorp Report').
The report's recommendation 14 concerns reintroducing registration. In addition, the report notes that: 'Police support the view that all firearms
of every type should be registered in a central database for tracking and control of use and ownership' (New Zealand, 1997, p. 50).

198   In Florida, records can only be kept for up to ten days after firearms are recovered and official documents recording the theft of a recovered
weapon can only be maintained for the remainder of the year they are recovered, plus two more years (Florida, n.d., ss. 790.335(3)(a)–(c)).

199   Nevertheless, US federal law does impose restrictions on the *sales* of long guns—as well as handguns—to minors. For example, federally
licensed firearm dealers may not sell shotguns or rifles to those under the age of 18, or handguns or handgun ammunition to those under the
age of 21 (US, n.d.a, s. 922(b)(1)). Private sales of handguns can be made to persons who are under the age of 21, provided they are over 18
years of age (ss. 922(x)(1)(a), 922(x)(5)).

200   See Annexe 9.1 for details and sources.

201   See, in particular, UK (1998, schedules 1–2, forms 101, 103, part A).

202   These requirements have not prevented authorities from permitting a legally blind man from owning and collecting firearms; indeed, the firing
of these weapons is permitted in the presence of an adult trained in the use of firearms (despite evidence suggesting the blind man also had
alcoholic tendencies) (Horowitz, 2010).

203   Croatia (Croatia, 1992, art. 17(4)); South Africa (South Africa, 2000, s. 9(2)(e)); Turkey (Pehlevan, 2010); United States (US, n.d.a, s. 922(d)(3));
and Yemen (Yemen, 1992, art. 21(4)). In Yemen, such a refusal would relate only to a licence to carry. Yemen does not place any restrictions
on buying a weapon; there are restrictions on obtaining a licence to carry a weapon, however.

204   See, in particular, UK (1998, schedules 1–2, forms 101, 103, part A, question 15(a)).

205   The NICS was introduced under the Brady Handgun Violence Prevention Act of 1993 (US, 1993). For more information on the NICS, see FBI (n.d.).

206   Author interview with a representative of the Israeli government, Geneva, 11 March 2011.

207   See, for example, NT (n.d.b, ss. 35(3)(a), 35(4)); Queensland (1990, ss. 39(2), 42(1); 1996, s. 56(2)); Victoria (1996, ss. 103, 107(1)).

208   The specific provisions within each Australian state and territory are: NSW (1996, s. 51A); NT (n.d.b, s. 63); ACT (1996, s. 226); Queensland (1990,
s. 36); South Australia (1977, s. 15B); Tasmania (1996, ss. 24, 25); Victoria (1996, ss. 95, 96); Western Australia (1973, ss. 16(1)(d), 19(1)).

209   For example, licensed dealers and collectors may not sell shotguns or rifles or associated ammunition to anyone under 18 years of age, and
may not sell other firearms (including handguns) or associated ammunition to anyone under 21 years of age (US, n.d.a, s. 922(b)(1)).

210   The 16 states are Belize, Canada, Colombia, Croatia, Dominican Republic, Estonia, India, Israel, Kenya, Lithuania, Papua New Guinea, Singapore,
South Africa, Switzerland, Venezuela, and Yemen.

211   See, for example, European Convention on Human Rights (Council of Europe, 1950, art. 2), and subsequent case law.

212   At the time of writing, there were two bills before Congress of relevance to civilian possession for self-defence purposes. In April 2010 Senator
John McCain introduced a bill that seeks to prohibit the District of Columbia from enacting legislation to curtail civilian gun ownership and use
(US, 2010). The second bill, entitled the Citizens' Self-Defense Act of 2009, aims to enshrine the use of firearms for self-defence in national law
(US, 2009).

213   Exceptions apply if, for example, the firearm is needed to fend off predators and other animals (Canada, 1998b, reg. 5(2)).

214   For more information on the doctrine of excessive defence, see the Law Commission (2003, p. 52).

215   A 'straw purchaser' is a person who buys firearms on behalf of a criminal, a minor, or other purchaser not eligible to buy a gun.

216   Kenya (1954, ss. 5(2), 18(3)); New Zealand (2010, p. 40); South Australia (2008, reg. 21(1)(d)); Uganda (1970, ss. 4, 30).

BIBLIOGRAPHY

ACT (Australian Capital Territory). 1996. Firearms Act 1996. A1996-74. Effective 28 May 2010.
    <http://www.austlii.edu.au/au/legis/act/consol_act/fa1996102/>

Andean Community. 2003. Andean Plan to Prevent, Combat and Eradicate Illicit Trade in Small Arms and Light Weapons in All Its Aspects.
    <http://www.comunidadandina.org/ingles/normativa/D552e.htm>

APMC (Australasian Police Ministers' Council). 1996. Resolutions of the Special Firearms Meeting, 10 May 1996. Canberra: Australasian Legal Information
    Institute. <http://www.austlii.edu.au/au/other/apmc/>

Arizona. n.d. Criminal Code, title 13, ch. 31. <http://www.azleg.gov/ArizonaRevisedStatutes.asp?Title=13>

AU (African Union). n.d. 'Member States.' <http://www.au.int/en/member_states/countryprofiles>

Belize. 2003. Firearms Act, ch. 143. Revised edn. <http://www.belizelaw.org/lawadmin/PDF%20files/cap143.pdf>

—. 2008. Firearms (Amendment of Schedules) Order 2008. S.I. No. 70 of 2008.

—. n.d. 'Application for Firearm License.' <http://www.belize.gov.bz/public/Attachment/81223153771.pdf>

Brazil. 2000. Decreto Nº 3665 de 20 de Novembro de 2000, Da nova redação ao Regulamento para a Fiscalização de Produtos Controlados (R-105).

—. 2004. Decreto No. 5123, de 1 de Julho de 2004. Regulamenta a Lei no 10.826, de 22 de dezembro de 2003, que dispõe sobre registro, posse e comercialização de armas de fogo e munição, sobre o Sistema Nacional de Armas (SINARM) e define crimes.

Burridge, Tom. 2010. 'Finland Reviews Its Gun Laws after Mass Shootings.' BBC News. 16 August.
    <http://news.bbc.co.uk/2/hi/programmes/world_news_america/8913432.stm>

California. n.d. Penal Code, part 4, title 2, chs. 1–2. <http://www.leginfo.ca.gov/cgi-bin/calawquery?codesection=pen&codebody=&hits=20>

Cambodia. 2005. Law on the Management of Weapons, Explosives and Ammunition.
    <http://www.interior.gov.kh/uploads/files/Law_on_Arm_control.pdf>

Canada. 1995. Firearms Act 1995. Current to 14 December 2010. <http://laws-lois.justice.gc.ca/PDF/Statute/F/F-11.6.pdf>

—. 1998a. Firearms Licenses Regulations SOR/98-199. Current to 3 February 2011. <http://laws-lois.justice.gc.ca/PDF/Regulation/S/SOR-98-199.pdf>

—. 1998b. Storage, Display, Transportation and Handling of Firearms by Individuals Regulations, SOR/98-209. Current to 25 February 2011.
    <http://laws-lois.justice.gc.ca/PDF/Regulation/S/SOR-98-209.pdf>

—. 1998c. Authorizations to Carry Restricted Firearms and Certain Handguns Regulations, SOR/98-207. Current to 31 December 2010.
    <http://laws-lois.justice.gc.ca/PDF/Regulation/S/SOR-98-207.pdf>

—. 2002. *Report of the Auditor General of Canada to the House of Commons.* Ottawa: Office of the Auditor General of Canada, ch. 10.
    <http://www.oag-bvg.gc.ca/internet/English/parl_oag_200212_10_e_12404.html>

—. n.d.a. Criminal Code, ch. C-46. Current to 16 January 2011. <http://laws-lois.justice.gc.ca/PDF/Statute/C/C-46.pdf>

—. n.d.b. 'Information Sheet: Application for a Possession and Acquisition Licence under the Firearms Act.' Ottawa: Royal Canadian Mounted Police.
    <http://www.rcmp-grc.gc.ca/cfp-pcaf/form-formulaire/pdfs/921-eng.pdf>

CBC News. 2010. 'Gun Registry Survives Commons Vote.' 22 September. <http://www.cbc.ca/canada/story/2010/09/22/gun-registry-vote-results.html>

Chapman, Simon, et al. 2006. 'Australia's 1996 Gun Law Reforms: Faster Falls in Firearm Deaths, Firearm Suicides, and a Decade without Mass Shootings.'
    Injury Prevention, Vol. 12, Iss. 6, pp. 365–72.

China Post. 2009. 'Strict Gun Control to Stay.' 23 October.
    <http://www.chinapost.com.tw/taiwan/national/national-news/2009/10/23/229755/Strict-gun.htm>

Colombia. 1993. Decreto 2535 de 1993. Published 17 December 1993 in *Gaceta Oficial,* No. 41.142.

Council of Europe. 1950. Convention for the Protection of Human Rights and Fundamental Freedoms ('European Convention on Human Rights').
    Rome, 4 November. <http://conventions.coe.int/treaty/Commun/QueVoulezVous.asp?NT=005&CL=ENG>

Council of the European Union. 1991. Council Directive of 18 June 1991 on Control of the Acquisition and Possession of Weapons ('European Weapons
    Directive'). 91/477/EEC of 18 June. Brussels: Council of the European Communities.
    <http://eur-lex.europa.eu/LexUriServ/LexUriServ.do?uri=CELEX:31991L0477:EN:HTML>

Croatia. 1992. Weapons Law. <http://www.seesac.org/uploads/armslaws/croatia.pdf>

—. 1999. Law on Amendments and Supplements to the Weapons Law. <http://www.seesac.org/uploads/armslaws/croatia.pdf>

Davenport, Paul and Jonathan Cooper. 2010. 'Arizona Gun Law: Concealed Weapons Allowed without Permit under New Law.' 16 April.
    <http://www.huffingtonpost.com/2010/04/17/arizona-gun-law-concealed_n_541445.html>

District of Columbia v. *Heller.* 2008. 554 US 570. <http://www.lcav.org/pdf/DCv.Heller.Opinion.07-290.pdf>

Dodd, Vikram. 2010. 'GPs Agree to Waive Privacy of Mentally Ill Gun Owners.' *Guardian.* 14 June.
    <http://www.guardian.co.uk/world/2010/jun/14/gps-mental-health-gun-owners>

Dreyfus, Pablo and Rebecca Perez. 2007. *Firearms Legislation Project: Questionnaire (Brazil).* Unpublished background paper. Geneva: Small Arms Survey.

ECOSOC (United Nations Economic and Social Council). 1995. Implementation of the Resolutions and Recommendations of the Ninth United Nations
    Congress on the Prevention of Crime and the Treatment of Offenders. Resolution 1995/27 of 24 July.
    <http://www.un.org/documents/ecosoc/res/1995/eres1995-27.htm>

—. 1997. Firearm Regulation for Purposes of Crime Prevention and Public Health and Safety. Resolution 1997/28 of 21 July.
    <http://www.un.org/documents/ecosoc/res/1997/eres1997-28.htm>

ECOWAS (Economic Community of West African States). 2006. ECOWAS Convention on Small Arms and Light Weapons, Their Ammunition and Other
    Related Materials. Abuja, Nigeria, 14 June. <http://www.oecd.org/dataoecd/56/26/38873866.pdf>

Eritrea. 2010. *National Report on the Implementation of the United Nations Programme of Action to Prevent, Combat and Eradicate the Illicit Trade in
    Small Arms and Light Weapons.* <http://www.un-casa.org/CASACountryProfile/PoANationalReports/2010@62@PoA-Eritrea-2010.pdf>

Estonia. 2002a. Weapons Act. As amended up to June 2004. <http://www.legaltext.ee/en/andmebaas/ava.asp?m=022>

—. 2002b. Penal Code. As amended up to April 2008. <http://www.legaltext.ee/en/andmebaas/ava.asp?m=022>

FBI (Federal Bureau of Investigation). n.d. 'National Instant Criminal Background Check System: Fact Sheet.' Accessed 16 February 2011.
    <http://www.fbi.gov/about-us/cjis/nics/general-information/fact-sheet>

Finland. 1998. Firearms Act 1998. As of January 1998; amendments up to 804/2003 included.

—. 2009a. Draft Act on Amending the Firearms Act.

—. 2009b. 'Proposals for Amending the Firearms Act Are Being Finalised: Significant Restrictions Proposed on the Availability of Handguns.' Helsinki: Ministry of the Interior. 11 March. <http://www.intermin.fi/intermin/bulletin.nsf/PublicbyIdentifierCode/20090311013?opendocument&3>

Florida. n.d. Florida Statute, title XLVI, ch. 790.
    <http://www.leg.state.fl.us/Statutes/index.cfm?App_mode=Display_Statute&URL=0700-0799/0790/0790.html>

Hahn, Robert, et al. 2005. 'Firearms Laws and the Reduction of Violence: A Systematic Review.' *American Journal of Preventive Medicine*, Vol. 28, Iss. 2, Suppl. 1. February, pp. 40–71.

Harding, Luke. 2009. 'Strict German Gun Laws Fail to Prevent School Shooting.' *Guardian*. 11 March.
    <http://www.guardian.co.uk/world/2009/mar/11/germany-school-shooting-laws>

HDC (Centre for Humanitarian Dialogue). 2005. *Missing Pieces: Directions for Reducing Gun Violence through the UN Process on Small Arms Control.* Geneva: Centre for Humanitarian Dialogue. <http://www.hdcentre.org/publications/missing-pieces>

Helsinki Times. 2010. 'Finland Not to Ban Semiautomatic Firearms–YLE.' 21 June.
    <http://www.helsinkitimes.fi/htimes/domestic-news/politics/11546-finland-not-to-ban-semiautomatic-firearms-yle-.html>

Horowitz, Ben. 2010. 'Rockaway Township Blind Man Who Shot Himself with Gun May Keep Collection, Judge Rules.' *Star-Ledger.* 19 August.
    <http://www.nj.com/news/index.ssf/2010/08/rockaway_township_blind_man_wh.html>

India. 1959. The Arms Act 1959 (Act No. 54 of 1959). <http://www.cyberabadpolice.gov.in/Acts%20and%20Laws/Arms%20Act.pdf>

—. 1962. The Arms Rules 1962. <http://www.lawsindia.com/Advocate%20Library/C18.htm>

—. 2010. *Integrated National Report on the Implementation of International Instrument to Enable States to Identify and Trace, In a Timely and Reliable Manner, Illicit Small Arms and Light Weapons and the United Nations Programme of Action to Prevent, Combat and Eradicate the Illicit Trade in Small Arms and Light Weapons in All Its Aspects.*
    <http://www.un-casa.org/CASACountryProfile/PoANationalReports/2010@90@India's%20Integrated%20National%20Report.pdf>

Inoa, Orlidy. 2010. *Firearms Legislation Project: Questionnaire (Dominican Republic).* Unpublished background paper. Geneva: Small Arms Survey.

Israel. 2008. *National Report on the Implementation of the United Nations Programme of Action to Prevent, Combat and Eradicate the Illicit Trade in Small Arms and Light Weapons in All Its Aspects and the Implementation of the International Instrument to Enable States to Identify and Trace, in a Timely and Reliable Manner, Illicit Small Arms and Light Weapons.*
    <http://www.un-casa.org/CASACountryProfile/PoANationalReports/2008@95@Israel(E).pdf>

Japan. 1958. Firearms and Swords Control Law. Law No. 6 of 10 March.

Karimova, Takhmina. 2010. *Firearms Legislation Project: Questionnaire (Kazakhstan).* Unpublished background paper. Geneva: Small Arms Survey.

Kenya. 1954. Firearms Act, ch. 114. <http://www.kenyalaw.org/kenyalaw/klr_app/frames.php>

—. 2010. *Country Report to the Fourth UN Biennial Meeting of States on the Status of Implementation of the UN Program of Action on Illicit Small Arms and Light Weapons and the Implementation of International Tracing Instrument.*
    <http://www.un-casa.org/CASACountryProfile/PoANationalReports/2010@101@COUNTRY%20REPORT%20-%20Kenya.pdf>

—. n.d.a. Firearms Rules. <http://www.kenyalaw.org/kenyalaw/klr_app/frames.php>

—. n.d.b. National Policy on Small Arms and Light Weapons. Draft. <http://www.kecosce.org/downloads/Final_SALW_policy_draft.pdf>

Kopel, David. 2009. 'Pretend "Gun-Free" School Zones: A Deadly Legal Fiction.' *Connecticut Law Review*, Vol. 42, No. 2. December, pp. 515–84.
    <http://connecticutlawreview.org/documents/Kopel.pdf>

Law Commission. 2003. *Partial Defences to Murder: Overseas Studies.* Consultation Paper No. 173. London: Law Commission, appendices.
    <http://www.lawcom.gov.uk/docs/cp173apps.pdf>

LCAV (Legal Community Against Violence). 2008. *Regulating Guns in America: An Evaluation and Comparative Analysis of Federal, State and Selected Local Gun Laws.* San Francisco: LCAV. February. <http://www.lcav.org/publications-briefs/reports_analyses/RegGuns.entire.report.pdf>

League of Arab States. 2002. Arab Model Law on Weapons, Ammunitions, Explosives and Hazardous Material. Tunisia: Interior Ministers Council, League of Arab States. <http://www.poa-iss.org/RegionalOrganizations/LeagueArab/Arab%20Model%20Law.doc>

Liberia. 2010. *Report on the Implementation of the United Nations Programme of Action to Prevent, Combat and Eradicate Illicit Trade Small Arms and Light Weapons in All Its Aspects (UNPoA).*
    <http://www.un-casa.org/CASACountryProfile/PoANationalReports/2010@111@2010-National-Report-Liberia(en).pdf>

Lithuania. 2000. Penal Code. No. VIII-1968. 26 September. <http://www3.lrs.lt/pls/inter3/dokpaieska.showdoc_l?p_id=360699>

—. 2002. Law on the Control of Arms and Ammunition. No. IX-705. 15 January.
    <http://www.opbw.org/nat_imp/leg_reg/lithuania/Arms_Ammunition.pdf>

—. 2010. *Report on Implementation of the United Nations Programme of Action to Prevent, Combat and Eradicate the Illicit Trade in Small Arms and Light Weapons in All Its Aspects.*
    <http://www.un-casa.org/CASACountryProfile/PoANationalReports/2010@113@100201_SALW_PoA_report_Lithuania_2009.pdf>

Makarenko, Jay. 2010. 'The Long-Gun Registry in Canada: History, Operation and Debates.' Mapleleafweb.com.
    <http://www.mapleleafweb.com/features/long-gun-registry-canada-history-operation-and-debates>

Massachusetts. n.d. Massachusetts General Law. <http://www.malegislature.gov/Laws/Search>

McDonald v. *Chicago.* 2010. 561 US ___, 130 S.Ct. 3020. <http://www.law.cornell.edu/supct/pdf/08-1521P.ZO>

Nairobi Protocol. 2004. Nairobi Protocol for the Prevention, Control and Reduction of Small Arms and Light Weapons in the Great Lakes Region and the Horn of Africa ('Nairobi Protocol'). Nairobi, 21 April. <http://www.recsasec.org/pdf/Nairobi%20Protocol.pdf>

Newbold, Greg. 1998. 'The 1997 Review of Firearms Control: An Appraisal.' Social Policy Journal, Iss. 11, pp. 115–30. <http://www.msd.govt.nz/documents/about-msd-and-our-work/publications-resources/journals-and-magazines/social-policy-journal/spj11/spj11-1997-review.doc>

New Jersey. n.d.a. New Jersey Statutes. Code of Criminal Justice, title 2C, chs. 39, 43, 58. <http://law.onecle.com/new-jersey/>

—. n.d.b. New Jersey Statutes. Children-Juvenile and Domestic Relations Courts, title 9. <http://law.onecle.com/new-jersey/>

New Zealand. 1983. Arms Act 1983. No. 44.
<http://www.legislation.govt.nz/act/public/1983/0044/latest/DLM72622.html?search=ts_act_arms_resel&p=1&sr=1>

—. 1984. Arms (Restricted Weapons and Specially Dangerous Airguns) Order 1984. SR 1984/122.
<http://www.legislation.govt.nz/regulation/public/1984/0122/latest/DLM95640.html?search=ts_regulation_arms_resel&p=1&sr=1>

—. 1992a. Arms Regulations 1992. SR 1992/346.
<http://www.legislation.govt.nz/regulation/public/1992/0346/latest/DLM168889.html?search=ts_regulation_arms_resel&p=1&sr=1>

—. 1992b. Arms Amendment Act. No. 95. <http://www.legislation.govt.nz/act/public/1992/0095/latest/whole.html#dlm278351>

—. 1997. Review of Firearms Control in New Zealand. Wellington: Minister of Police. June.
<http://www.police.govt.nz/resources/1997/review-of-firearms-control/review-of-firearms-control-in-new-zealand.pdf>

—. 2010. Arms Code: Firearms Safety Manual. Wellington: New Zealand Police.
<http://www.police.govt.nz/sites/default/files/services/firearms/NZP-Arms-Code-R3.pdf>

Norberry, Jennifer, Derek Woolner, and Kirsty Magarey. 1996. After Port Arthur: Issues of Gun Control in Australia. Current Issues Brief 16. Canberra: Parliamentary Library, Parliament of Australia. 7 May. <http://www.aph.gov.au/library/pubs/cib/1995-96/96cib16.htm>

Nowak, Matthias. 2010a. Firearms Legislation Project: Questionnaire (Colombia). Unpublished background paper. Geneva: Small Arms Survey.

—. 2010b. Firearms Legislation Project: Questionnaire (Venezuela). Unpublished background paper. Geneva: Small Arms Survey.

NSW (New South Wales). 1900. Crimes Act 1900. No. 40. <http://www.legislation.nsw.gov.au/viewtop/inforce/act+40+1900+FIRST+0+N/>

—. 1996. Firearms Act 1996. No. 46. <http://www.legislation.nsw.gov.au/fullhtml/inforce/act+46+1996+pt.6-sec.54+0+N>

—. 2006. Firearms Regulation 2006. <http://www.legislation.nsw.gov.au/fullhtml/inforce/subordleg+512+2006+FIRST+0+N>

NT (Northern Territory). 2010. Penalty Units Regulations No. 7 of 2010. <http://www.austlii.com/au/legis/nt/num_reg//pur7o2010340/>

—. n.d.a. Firearms Regulations. In force as of 29 November 2007. <http://www.austlii.edu.au/au/legis/nt/consol_reg/fr211/>

—. n.d.b. Firearms Act. As in force as from the 1 July 2010. <http://www.austlii.edu.au/au/legis/nt/consol_act/fa102/>

NYT (The New York Times). 1999. 'National News Briefs: Arkansas Law to Permit Life Sentence for Youths.' 8 April.
<http://www.nytimes.com/1999/04/08/us/national-news-briefs-arkansas-law-to-permit-life-sentence-for-youths.html?ref=mitchelljohnson>

OAS (Organization of American States). 1997. Inter-American Convention against the Illicit Manufacturing of and Trafficking in Firearms, Ammunition, Explosives, and Other Related Materials ('CIFTA'). Washington, DC, 14 November. <http://www.oas.org/juridico/english/treaties/a-63.html>

OAU (Organization of African Unity). 2000. Bamako Declaration on an African Common Position on the Illicit Proliferation, Circulation and Trafficking of Small Arms and Light Weapons ('Bamako Declaration'). Bamako, 1 December.
<http://www.armsnetafrica.org/content/bamako-declaration-african-common-position-illicit-proliferation-circulation-and-trafficking>

Pehlevan, Berna Capcioglu. 2010. Firearms Legislation Project: Questionnaire (Turkey). Unpublished background paper. Geneva: Small Arms Survey.

PNG (Papua New Guinea). 1978. Firearms Act 1978, ch. 310. <http://www.paclii.org/pg/legis/consol_act/fa1978102/>

Pyadushkin, Maxim. 2008. Firearms Legislation Project: Questionnaire (Russian Federation). Unpublished background paper. Geneva: Small Arms Survey.

Queensland. 1990. Weapons Act 1990. Reprinted as in force on 1 July 2010. <http://www.austlii.edu.au/au/legis/qld/consol_act/wa1990107/>

—. 1996. Weapons Regulation 1996. <http://www.austlii.edu.au/au/legis/qld/consol_reg/wr1996198/>

—. 1997. Weapons Categories Regulation 1997. <http://www.austlii.edu.au/au/legis/qld/consol_reg/wcr1997290/>

RCMP (Royal Canadian Mounted Police). 2010. RCMP Canadian Firearms Program: Program Evaluation. Ottawa: National Program Evaluation Services, RCMP. 1 February. <http://www.rcmp-grc.gc.ca/pubs/fire-feu-eval/eval-eng.pdf>

—. 2011. 'Registration of Firearms (Individuals).' Accessed 4 March. <http://www.rcmp-grc.gc.ca/cfp-pcaf/online_en-ligne/reg_enr-eng.htm>

Reuters. 2008. 'Texas School District to Let Teachers Carry Guns.' 15 August. <http://www.reuters.com/article/idUSN1538661720080815>

Russian Federation. 1996. The Criminal Code of the Russian Federation. No. 63-FZ of 13 June.
<http://www.legislationline.org/documents/section/criminal-codes/country/7>

SADC (Southern African Development Community). 2001. Protocol on the Control of Firearms, Ammunition and Other Related Materials in the Southern African Development Community (SADC) Region ('SADC Firearms Protocol'). 14 August.
<http://www.sadc.int/index/browse/page/125>

—. 2008. Standard Operating Procedures for the Implementation of the SADC Protocol on the Control of Firearms, Ammunition and Other Related Materials. Windhoek, Namibia, August.
<http://www.armsnetafrica.org/content/standard-operating-procedures-implementation-sadc-protocol-control-firearms-ammunition-and-o>

Sahouri, Nadine. 2010a. Firearms Legislation Project: Questionnaire (Egypt). Unpublished background paper. Geneva: Small Arms Survey.

—. 2010b. Firearms Legislation Project: Questionnaire (Yemen). Unpublished background paper. Geneva: Small Arms Survey.

Scott, Elizabeth and Laurence Steinberg. 2008. 'Adolescent Development and the Regulation of Youth Crime.' Juvenile Justice, Vol. 18, No. 2. Fall 2008, pp. 15–33. <http://futureofchildren.org/futureofchildren/publications/docs/18_02_02.pdf>

Singapore. 1913. Arms and Explosives Act, ch. 13. Last updated 2009.
    <http://statutes.agc.gov.sg/non_version/cgi-bin/cgi_retrieve.pl?&actno=Reved-13&date=latest&method=part>
—. 1973. Arms Offences Act, ch. 14. Last updated 1993.
    <http://statutes.agc.gov.sg/non_version/cgi-bin/cgi_retrieve.pl?&actno=Reved-14&date=latest&method=part>
—. 2006. Arms and Explosives (Arms) Rules, ch. 13, s. 46.
—. n.d.a. 'Guidelines on Application for a Licence to Possess Arm.' <http://www.spf.gov.sg/licence/AE/Guidelines/Possess_Arm.html>
—. n.d.b. 'Licensing Conditions for Licence to Possess Arm for Self-Protection.' <http://www.spf.gov.sg/licence/AE/Conditions/Selfprotection.html>
—. n.d.c. 'Licensing Conditions for Licence to Possess Arm for Target Practice.' <http://www.spf.gov.sg/licence/AE/Conditions/Target_Practice.html>
—. n.d.d. 'Licensing Conditions for Licence to Possess Speargun.' <http://www.spf.gov.sg/licence/AE/Conditions/Speargun.html>
Small Arms Survey. 2007. *Small Arms Survey 2007: Guns and the City*. Cambridge: Cambridge University Press.
—. 2010. *Small Arms Survey 2010: Gangs, Groups, and Guns*. Cambridge: Cambridge University Press.
Solomon Islands. 2004. *National Report on the Implementation of the United Nations Program of Action to Prevent, Combat and Eradicate the Illicit
    Trade in Small Arms and Light Weapons in All Its Aspects.*
    <http://www.un-casa.org/CASACountryProfile/PoANationalReports/2004@175@solomonislands.pdf>
South Africa. 2000. Firearms Control Act 2000. <http://www.acts.co.za/firearms/whnjs.htm>
—. 2004. Firearms Control Regulations, 2004. <http://www.armsnetafrica.org/content/south-africa-firearms-control-regulations-2004>
—. n.d. 'Application for a Competency Certificate.' Johannesburg: South African Police Service.
    <http://www.services.gov.za/services/WebDav/Documents/SAPS/Competency%20certificate.pdf>
South Australia. 1935. Criminal Law Consolidation Act.
    <http://www.legislation.sa.gov.au/LZ/C/A/CRIMINAL%20LAW%20CONSOLIDATION%20ACT%201935/CURRENT/1935.2252.UN.PDF>
—. 1977. Firearms Act 1977. Version 27.11.2008. <http://www.legislation.sa.gov.au/LZ/C/A/FIREARMS%20ACT%201977/CURRENT/1977.26.UN.PDF>
—. 2008. Firearms Regulations 2008. Under the Firearm Act 1977. Version 1.7.2010. <http://www.austlii.edu.au/au/legis/sa/consol_reg/fr2008211/>
SPCPC and OCO (South Pacific Chiefs of Police Conference and Oceania Customs Organisation). 2000. Legal Framework for a Common Approach to
    Weapons Control Measures: The 'Nadi Framework.' Nadi, Fiji: Pacific Islands Forum Secretariat. 10 March.
Switzerland. 1997. Federal Law on Arms, Arms Accessories, and Ammunition. Status as of 1 December 2010.
    <http://www.admin.ch/ch/f/rs/c514_54.html>
—. 2008. Ordinance on Arms, Arms Accessories and Ammunition. Status as of 28 July 2010. <http://www.admin.ch/ch/f/rs/c514_541.html>
Tasmania. 1987. Penalty Units and Other Penalties Act 1987. Act 13 of 1987. <http://www.austlii.edu.au/au/legis/tas/consol_act/puaopa1987354/>
—. 1996. Firearms Act 1996. Act 23 of 1996. <http://www.austlii.edu.au/au/legis/tas/consol_act/fa1996102/>
—. 2006. Firearms Regulations 2006. S.R. 2006, No. 109. <http://www.austlii.edu.au/au/legis/tas/consol_reg/fr2006211/>
Texas. n.d.a. Texas Penal Code, ch. 46. <http://www.statutes.legis.state.tx.us/Docs/PE/htm/PE.46.htm#46.06>
—. n.d.b. Texas Government Code, ch. 411. <http://www.statutes.legis.state.tx.us/Docs/GV/htm/GV.411.htm>
Uganda. 1970. The Firearms Act, ch. 299. <http://www.ulii.org/ug/legis/consol_act/fa1970102/>
—. n.d. The Firearms (Fees and Forms) Regulations.
UK (United Kingdom). 1968. Firearms Act 1968, ch. 27. <http://www.legislation.gov.uk/ukpga/1968/27/contents>
—. 1997. Firearms (Amendment) Act 1997, ch. 5. <http://www.legislation.gov.uk/ukpga/1997/5/data.pdf>
—. 1998. The Firearms Rules 1998. No. 1941. <http://www.legislation.gov.uk/uksi/1998/1941/contents/made>
—. 2005. Firearms Security Handbook 2005.
    <http://www.homeoffice.gov.uk/publications/police/operational-policing/firearms-handbook-2005/firearms-security-handbook?view=Binary>
—. 2010. 'Firearms Control.' London: Commons Select Committee, UK Parliament.
    <http://www.parliament.uk/business/committees/committees-a-z/commons-select/home-affairs-committee/inquiries/firearms-control/>
UNCASA (United Nations Coordinating Action on Small Arms). Forthcoming. 'National Controls over the Access of Civilians to Small Arms and Light
    Weapons.' CASA Project on International Small Arms Control Standards (ISACS). Unpublished draft. <http://www.un-casa-isacs.org>
UNCJIN (United Nations Crime and Justice Information Network). 1999. *United Nations International Study on Firearm Regulation.*
    <http://www.uncjin.org/Statistics/firearms/index.htm>
UNDP (United Nations Development Programme). 2008. *How to Guide: Small Arms and Light Weapons Legislation*. Geneva: Bureau for Crisis Prevention
    and Recovery, UNDP. <http://www.undp.org/cpr/documents/sa_control/SALWGuide_Legislation.pdf>
UNGA (United Nations General Assembly). 1999. *Report of the Disarmament Commission*. Supplement No. 42. A/54/42 of 6 May.
    <http://www.nti.org/h_learnmore/nwfztutorial/pdfs/un_res/Disarmament%20Cmsn%20Report,%201999.pdf>
—. 2001. Programme of Action to Prevent, Combat and Eradicate the Illicit Trade in Small Arms and Light Weapons in All Its Aspects ('UN Programme
    of Action'). A/CONF.192/15 of 20 July. <http://www.poa-iss.org/PoA/PoA.aspx>
UNHRC (United Nations Human Rights Council). 2006. *Prevention of Human Rights Violations Committed with Small Arms and Light Weapons: Final Report
    Submitted by Barbara Frey, Special Rapporteur, in Accordance with Sub- Commission Resolution 2002/25*. A/HRC/Sub.1/58/27 of 27 July 2006.
United States v. *Lopez*. 1995. 514 US 549. <http://www.supremecourt.gov/opinions/boundvolumes/514bv.pdf>
US (United States). 1993. Brady Handgun Violence Prevention Act. Codified in United States Code, title 18, part 1, ch. 44, sec. 922.
    <http://uscode.house.gov/download/pls/18C44.txt>

—. 1994a. Gun-Free Schools Act. Codified in United States Code, title 20, ch. 70, subch. IV, part A, subpart 3, s. 7151.
<http://www.gpo.gov:80/fdsys/pkg/USCODE-2009-title20/pdf/USCODE-2009-title20-chap70-subchapIV-partA-subpart3-sec7151.pdf>

—. 1994b. Violent Crime Control and Law Enforcement Act of 1994. Title XI, subtitle A. Expired.
<http://thomas.loc.gov/cgi-bin/query/F?c103:1:./temp/~c103ZcrJnF:e644150:>

—. 2005. *Survey of State Procedures Related to Firearm Sales, 2005.* Washington, DC: Bureau of Justice Statistics, Office of Justice Programs, US Department of Justice. <http://bjs.ojp.usdoj.gov/content/pub/pdf/ssprfs05.pdf>

—. 2009. Citizens' Self-Defense Act of 2009. H.R. 17. Washington, DC: House of Representatives, US Congress. Introduced 6 January.
<http://www.gpo.gov:80/fdsys/pkg/BILLS-111hr17ih/pdf/BILLS-111hr17ih.pdf>

—. 2010. Second Amendment Enforcement Act. H.R. 5150. Washington, DC: House of Representatives, US Congress. Introduced 27 April 2010.
<http://www.gpo.gov:80/fdsys/pkg/BILLS-111hr5150ih/pdf/BILLS-111hr5150ih.pdf>

—. n.d.a. United States Code, title 18, part I, ch. 44. As of 1 February 2011.
<http://www.gpo.gov:80/fdsys/pkg/USCODE-2009-title18/pdf/USCODE-2009-title18-partI-chap44.pdf>

—. n.d.b. Code of Federal Regulations, title 28, part 25. As of 3 March 2011.
<http://ecfr.gpoaccess.gov/t/text/text-idx?c=ecfr&rgn=div6&view=text&node=28:1.0.1.1.26.1&idno=28>

—. n.d.c. United States Code, title 26, subtitle E, ch. 53. As of 1 February 2011.
<http://www.gpo.gov:80/fdsys/pkg/USCODE-2009-title26/pdf/USCODE-2009-title26-subtitleE.pdf>

Venezuela. 1939. Ley sobre Armas y Explosivos de 1939. Published 19 June 1939 in *Gaceta Oficial* 19.900.

Victoria. 1996. Firearms Act 1996. Version No. 058, No. 66. Incorporating amendments as of 1 November 2010.
<http://www.austlii.edu.au/au/legis/vic/consol_act/fa1996102/>

—. 2008. Firearms Regulations 2008. S.R. No. 22/2008. Version as of 20 April. <http://www.austlii.edu.au/au/legis/vic/consol_reg/fr2008211/>

Villaveces, Andrés, et al. 2000. 'Effect of a Ban on Carrying Firearms on Homicide Rates in 2 Colombian Cities.' *Journal of the American Medical Association*, Vol. 283, No. 9, pp. 1205–09.

Western Australia. 1973. Firearms Act 1973. As of 13 December 2009. <http://www.austlii.edu.au/au/legis/wa/consol_act/fa1973102/>

—. 1974. Firearms Regulations 1974. As of 1 July 2010. <http://www.austlii.edu.au/au/legis/wa/consol_reg/fr1974211/>

Yemen. 1992. Law No. 40 of 1992: On Regulating Carrying Firearms and Ammunitions and Their Trade.

—. 1994. Republican Decree No (1) of 1994 on the Executive Regulation of Law No. (40) of 1992 on Regulating Carrying Firearms and Ammunitions and Their Trade.

Zimring, Franklin. 1991. 'Firearms, Violence, and Public Policy.' *Scientific American*. November, pp. 48–54.

ACKNOWLEDGEMENTS

Principal author

Sarah Parker

Contributors

Pablo Dreyfus, Orlidy Inoa, Takhmina Karimova, Paavo Kotiaho, Elkana Laist, Matthias Nowak, Halim Ozatan, Berna Çapçıoğlu Pehlevan, Rebecca Perez, Maxim Pyadushkin, and Nadine Sahouri

INDEX